Sources and Studies
in the History of Mathematics and
Physical Sciences

Springer
New York
Berlin
Heidelberg
Barcelona
Hong Kong
London
Milan
Paris
Singapore
Tokyo

Sources and Studies in the History of Mathematics and Physical Sciences

K. Andersen
Brook Taylor's Work on Linear Perspective

Henk J.M. Bos
Redefining Geometrical Exactness: Descartes' Transformation of the Early Modern Concept of Construction

J. Cannon/S. Dostrovsky
The Evolution of Dynamics: Vibration Theory from 1687 to 1742

B. Chandler/W. Magnus
The History of Combinatorial Group Theory

A.I. Dale
A History of Inverse Probability: From Thomas Bayes to Karl Pearson, Second Edition

A.I. Dale
Pierre-Simon Laplace, Philosophical Essay on Probabilities, Translated from the fifth French edition of 1825, with Notes by the Translator

P.J. Federico
Descartes on Polyhedra: A Study of the *De Solidorum Elementis*

B.R. Goldstein
The Astronomy of Levi ben Gerson (1288–1344)

H.H. Goldstine
A History of Numerical Analysis from the 16th through the 19th Century

H.H. Goldstine
A History of the Calculus of Variations from the 17th through the 19th Century

G. Graßhoff
The History of Ptolemy's Star Catalogue

T. Hawkins
The Emergence of the Theory of Lie Groups: An Essay in the History of Mathematics 1869–1926

A. Hermann, K. von Meyenn, V.F. Weisskopf (Eds.)
Wolfgang Pauli: Scientific Correspondence I: 1919–1929

C.C. Heyde/E. Seneta
I.J. Bienaymé: Statistical Theory Anticipated

J.P. Hogendijk
Ibn Al-Haytham's *Completion of the Conics*

A. Jones (Ed.)
Pappus of Alexandria, Book 7 of the *Collection*

Continued after Index

Henk J.M. Bos

Redefining Geometrical Exactness

Descartes' Transformation of the Early Modern Concept of Construction

With 95 Illustrations

Springer

Henk J.M. Bos
Faculty of Mathematics and Computer Science
University of Utrecht
PO Box 80.010
3508 TA, Utrecht
The Netherlands

Sources and Studies Editor:
Gerald J. Toomer
2800 South Ocean Boulevard, 21F
Boca Raton, FL 33432
USA

Library of Congress Cataloging-in-Publication Data
Bos, H.J.M.
Redefining geometrical exactness : Descartes' transformation of the early modern concept of construction / Henk J.M. Bos.
p. cm. — (Studies and sources in the history of mathematics and physical sciences)
Includes bibliographical references and index.
ISBN 978-1-4612-6521-4 ISBN 978-1-4613-0087-8 (eBook)
DOI 10.1007/978-1-4613-0087-8
1. Geometrical constructions—History—17th century. 2. Geometry—History—17th century. 3. Descartes, René, 1596–1650. I. Title. II. Series.
QA464 .B69 2001
516—dc21 00-056269

Printed on acid-free paper.

Production managed by Allan Abrams; manufacturing supervised by Jacqui Ashri.
Photocomposed copy prepared from the author's TEX files.

9 8 7 6 5 4 3 2 1

ISBN 978-1-4612-6521-4 SPIN 10746064

Preface

In his first letter to Johann (I) Bernoulli, Leibniz wrote that it was better to reduce quadratures to rectifications than *vice versa*, "because surely the dimension of a line is simpler than the dimension of a surface."[1] I first read this passage in the late 1960s and it struck me as strange. Apparently Leibniz considered the arclength integral $\int \sqrt{1 + (\frac{dy}{dx})^2} dx$ as more basic than the area integral $\int y dx$. Why? And what support for his opinion did he glean from the dimension argument?

Looking back I think that the encounter with this argument was the beginning of my interest in the concept of construction in early modern mathematics. Its study proved extremely useful and revealing for understanding the conceptual developments in mathematics during the crucial period between the Renaissance and the Enlightenment. At first I concentrated on the treatment of curves (especially transcendental curves) in early analytic geometry and calculus, and I emphasized the representational aspects of geometrical construction. It soon became clear that Descartes' ideas were decisive in the matter. For a long time after c. 1650 virtually all statements on mathematical construction could be understood as critical or concurring responses to a canon of geometrical construction laid down in Descartes' *Geometry* of 1637.

Descartes' canon itself was also a response, namely to classical and previous early modern ideas about the proper solution of geometrical problems. Indeed the half-century before the publication of the *Geometry* witnessed an ongoing debate about this question. The debate centred around the concept of construction, because geometrical problems required a solution in the form of a construction.

In studying Descartes' ideas and the previous discussions on construction I found that the representational aspects, on which I had concentrated at first,[2] did not provide a satisfactory entry into the matter. Rather than as a means of representing objects, such as curves, construction here functioned as the key concept in determining proper procedure in geometry and in demarcating between

[1] Letter of 21-03-1694 [Leibniz 1961–1962], vol. 3 pp. 135–137, passage on p. 137: "... praestat reducere Quadraturas ad Rectificationes Curvarum, quam contra, ut vulgo fieri solet; eaque de re dudum cum successu cogitavi: nam simplicior utique est dimensio lineae, quam dimensio superficiei." For the context of the statement cf. [Bos 1974] pp. 7–8 (p. 105 of ed. in [Bos 1993c]) and [Bos 1987] pp. 1638–1640 (pp. 34–35 of ed. in [Bos 1993c]).

[2] Notably so in [Bos 1981].

genuine knowledge and inadequate understanding in geometry and mathematics generally.

No convincing answers have been found to these questions of propriety and demarcation, and personally I don't think that any will be found in the conceivable future. Thus the historical study of the reactions to these and similar questions cannot be studied in terms of right or wrong answers. Nonetheless, the matter presents a challenge to the historian because many mathematicians have found the questions important, or at least unavoidable, and opinions on the matter have deeply influenced the evolution of the field. I have found that such an historical study is best undertaken as the investigation of the reactions of professional mathematicians to ultimately unanswerable, and yet unavoidable, extra-mathematical questions. Thus I have studied the early modern conception and practice of geometrical construction primarily as the actualized answers to the questions about proper procedure and the demarcation of geometry. Behind these answers I have tried to detect the various attitudes or strategies that mathematicians have adopted in responding to these questions, and the role and effectiveness of these strategies in the development of mathematics. I use the term *exactness* to denote the complex of qualities that were (and are) invoked with respect to propriety and demarcation in mathematics. Accordingly, I call the activity of mathematicians engaged in defining boundaries of proper mathematical procedure and of legitimately mathematical fields the *interpretation of exactness.* The present book, then, is a study of the concept of construction and the interpretation of geometrical exactness as developed and employed by Descartes and by early modern mathematicians active before him.

In the introductory chapter I further explain the historiographical scheme that I have adopted in elaborating my research interest; the above outline of its earlier development and of the main concerns of the investigations exposed here may explain some of the present book's features, which I like to mention at the outset in order to avoid disappointing my prospective readers.

It is not a book about the geometry of Descartes and the early modern mathematicians before him. There is no attempt at such generality; even the area of geometry that formed the setting of most of the developments described, the *early modern tradition of geometrical problem solving*, is treated only in the period mentioned and primarily as context for the mathematical developments related to construction and exactness. Thus, although the *Geometry* is analyzed in great detail, some themes of Descartes' book are disregarded, because they are outside the focus of my enquiry. This applies in particular for Descartes' determination of tangents (which he himself once mentioned as his most valuable contribution), his treatment of ovals and his remarks about curves on non-plane surfaces. Otherwise important geometrical activities in the period, notably the reception of Archimedes' techniques, occupy a much smaller place in the present book than in general histories of mathematics. The same applies to some mathematicians of the period, and in particular to Fermat, whose importance in early modern mathematics is beyond doubt. Indeed Fermat's name is so often paired with Descartes' in connection with analytic geometry that a book with a strong interest in Descartes may surprise in bringing so little (a short chapter and a few remarks in the *Epilogue*) about Fermat. But the latter's interest in mathematical exactness (in the sense explained above) and geometrical construction was minimal, and he wrote about the theme primarily in reaction to Descartes' *Geometry*, whereby his opinions fall outside the period studied here.

Although I have found that the conception and the practices of construction provide a revealing entrance into the world of early modern mathematics, I do not view mathematics essentially in constructivist terms, nor do I consider the conceptual challenges of construction as a principal driving force in the development of mathematics. As will become clear in the Introduction, I link the principal dynamics in the developments studied here to analytical rather than to construction-related interests.

The book is essentially restricted to the period from c. 1590 to c. 1650, the boundaries being marked by the publication of Pappus' *Collection* in 1588 and by Descartes' death in 1650. The subject matter is confined to the early modern tradition of geometrical problem solving from c. 1590 to c. 1635, and to the mathematical activities of Descartes from c. 1620 onward, culminating in the publication of the *Geometry* in 1637. The book's themes would be better presented in a cadre extending t0 c. 1750, including the developments around construction and exactness within the investigation of curves by means of the new infinitesimal analysis. Reasons of time and of the size of the material have made me decide to postpone the treatment of the additional period.

⋆ ⋆ ⋆

I started writing this book in 1977. During the years that followed Kirsti Andersen was a singular and invaluable companion in this scholarly enterprise — as in so many other things! She discussed with me the ideas, the plans and the changes of plan, she read and critically commented all the drafts (and there were many), and she supported me in the times the undertaking seemed hopelessly misconceived. I am deeply thankful to her.

⋆ ⋆ ⋆

On receiving the manuscript of the book in early 1999, Springer Verlag invited Joella Yoder to read and comment on it. She did so, devoting to the task an amount of time and energy in no proportion to the remuneration which is customary in academia for such activities. Her wise and detailed comments were very helpful and led to changes that I'm sure are for the better.

Many audiences have lent me their attention during lectures on subjects treated in the present book. I am thankful for the stimulus these lectures and the ensuing discussions provided for shaping, changing, and refining my ideas.

The book would never have been finished without a number of temporary fellowships at illustrious research institutes, which offered facilities for extended periods of concentration on thinking and writing. I am thankful to the Institutes of Mathematics and of History of Science of the University of Aarhus for repeatedly providing such a supportive environment. I also express my gratitude for their hospitality and support to the *Institut des Hautes Études Scientifiques* in Bures-sur-Yvette (1979), the *Institute for Advanced Study* (School of Mathematics) in Princeton (1988), and the *Dibner Institute* in Cambridge (MA.) (1997).

The periods of concentrated work on the book elsewhere were generously made possible by my own academic base, the Department of Mathematics of Utrecht University, which granted me sabbaticals and other variants of leave of absence. And although, naturally, uninterrupted concentration is a rare commodity at one's own base, the department offered many other benefits for my research: good facilities, interested and stimulating students and colleagues, electronic media of constant reliability despite continuous change, and above all an environment of mathematical professionalism and collegiality in which the value of historical awareness for the well-being of mathematics was a matter of course. I am pleased to be able here to express my appreciation.

Contents

List of Figures

List of Tables

Part I

The Early Modern Tradition of Geometrical Problem Solving

Chapter 1

General introduction

1.1 Construction and representation

Exactness

Mathematics is an exact science — but what does exactness mean? Throughout history mathematicians have repeatedly raised this question and reshaped their science in order to meet more appropriate and higher standards of exactness. Several such endeavors have been highly successful. Exploring the intricacies of incommensurability and the infinite, Greek mathematicians created theories whose rigor is still impressive to the modern mind. The rigorization of analysis in the nineteenth century provided new standards of proof, which led to a deeper understanding of numbers and functions, as well as to powerful new analytic methods.

Exactness of constructions

The early modern period[1] witnessed another endeavor to clarify and institute exactness, but a less successful one. This endeavor occupied the minds of many mathematicians from the sixteenth to the eighteenth centuries, and it concerned, in a primarily geometrical context, the question what it meant for a mathematical entity to be "known" or "given," and what it meant for a problem to be "solved," its solution to be "found." Classical Greek geometry provided answers to these questions: geometrical figures were "known" or "given" if they could be *constructed* starting from elements that were considered given at the outset; similarly a problem was considered solved if the required configuration was geometrically *constructed*.

During the Renaissance revival of Greek geometry mathematicians accepted the main lines of this classical answer, but they also found themselves confronted with questions about construction for which the available ancient mathematical texts did not provide sufficient guidance. Such a confrontation occurred in particular with respect to the acceptability of means of construction: Which means of construction can be accepted as sufficiently exact for figures constructed by

[1] I use the term "early modern" to denote the period between the Renaissance and the Enlightenment, i.e., roughly from c. 1550 to c. 1750.

them to be considered genuinely known? Ruler and compass, or rather straight lines and circles,[2] were certainly acceptable means of construction. But the classical geometers already knew from experience that many problems remained unsolvable if geometry were restricted to these means only. They had suggested various additional means of construction beyond straight lines and circles, but their texts did not clearly explain why these new means should be acceptable.

Algebra Yet classical geometry had flourished despite an apparent lack of clarity about the acceptability of means of construction. Why did the issue become acute at the beginning of the early modern period? One reason was that mathematicians began using algebra as an analytical tool for the solution of geometrical problems. It was a powerful tool; more problems could be solved, old problems were solved more easily, and at the same time the new method suggested many new problems. Thus the field of geometrical problem solving grew considerably, and this growth necessitated a rethinking of the question what it meant to solve a problem.

Curves Another reason for considering the concept of construction in a new light was connected with special geometrical objects, namely curves. Curves occurred in geometry in three main roles: as objects of study, as means of construction, and as solutions of problems. In the first case they were usually supposed to be given, known, or constructed beforehand, and the mathematicians investigated their particular properties. In the two other cases, however, they were not necessarily known beforehand. Because of the growth of the field of geometrical problem solving, the question of when a curve was sufficiently known, or how it could acceptably be constructed, acquired a new urgency.

Representation of curves At the beginning of the early modern period this new urgency mainly concerned the use of curves as means of construction. The problems that were studied usually required the construction of one special point or straight line in a given configuration. If such problems could not be solved by circles and straight lines, other curves had to be used, and this procedure obviously raised the question of how these curves themselves were to be constructed.

After c. 1640 problems whose solutions were curves moved to the foreground. Such problems even more necessitated a rethinking of the question how curves themselves were to be constructed. These problems were of two types: the locus problems, in which a curve was to be determined from a given property shared by all its points, and the so-called "inverse tangent problems" in which a curve had to be determined from a given property of its tangents. Locus problems were known from classical sources. Inverse tangent problems (equivalent to first-order differential equations) were new; they arose especially in connection with the study of non-uniform and curvilinear motion in mechanics, and their

[2] I use the terminology "by straight lines and circles" rather than "by ruler and compass" for constructions according to the Euclidean postulates, because Euclid did not mention instruments to perform these constructions.

treatment became possible through the various new methods of infinitesimals and indivisibles that were explored during the seventeenth century.

Sometimes it turned out that the solution of a problem was a curve with which mathematicians were already familiar. In that case only the parameters of the curve and its position in the plane had to be determined. But, in particular for inverse tangent problems, it could also happen that the solution curve was hitherto unknown. In such cases the question arose of how to make unknown curves known. I use the expression *representation of curves* as a technical term to denote descriptions of curves that were considered to be sufficiently informative to make the curves *known*. For representing curves mathematicians resorted to the means which geometry offered for making objects known: the conceptual apparatus of *construction*.

Conceptual and operative aspects

The questions which confronted mathematicians when compelled to reconsider the meaning of geometrical construction were primarily conceptual: What does "exactness" mean? When is an object "known"? But because of the confrontation with new types of problems and with new objects, these questions acquired an additional operative aspect. They arose in connection with mathematical problems that had to be solved. So the answers to the questions of exactness and known-ness could not remain abstract, they had to offer guidance in a very concrete process: the solution of problems. In other words, the answers had to provide a viable canon of acceptable procedures within the practice of problem solving.

A forgotten issue

The story of construction in early modern mathematics, then, is one about mathematicians confronted with methodological problems. They did not find satisfactory answers. There is no danger here for a "Whig interpretation" of history.[3] The endeavors of early modern mathematicians to clarify the questions about geometrical construction cannot be seen as the gradual realization and growth to maturity of established modern mathematical insights. On the contrary, the issue hardly occurs in present-day mathematics. Indeed the questions were not resolved; they were dissolved by later disregard and oblivion.

A research project

The conceptual and technical developments concerning geometrical construction and representation in the early modern period invite a historical research project: to analyze the developments and to explore their relations to other aspects of early modern mathematics, notably the adoption of algebraic methods of analysis in geometry at the beginning of the period and the later emancipation of analysis from its geometrical context. The project should also deal with the questions why no satisfactory solution to the problem of geometrical construction was achieved, and what was, nevertheless, the function and the in-

[3]The term taken in the now usual, derogatory sense of a historiography of science, which finds interest in past events only if they can be argued to have led directly to valuable achievements of modern science, cf. [Bynum 1984] s.v. "Whig history," pp. 445–446.

fluence of the debates on construction and representation within early modern mathematics.

In the present study I pursue this research project with respect to the period c. 1590 – c. 1650. I hope to deal with the later period (c. 1650 – c. 1750) in a sequel to the present study.[4]

1.2 The interpretation of exactness

Solving problems, knowing objects

The present study, then, is about the opinions and arguments of mathematicians concerning the acceptability of geometrical procedures, in particular procedures of construction. These procedures served to *solve* problems and to make geometrical objects *known*, and so the issue of their acceptability hinged on the questions: When is a problem solved? When is an object known? Which procedures are acceptable in mathematics to solve problems and represent objects?

Early modern mathematicians agreed that approximative procedures, although very useful in practice, were not acceptable in genuine geometry. I use the terms *exact* and *exactness* for describing the ideal situation presupposed in the questions above about solving problems and knowing objects. "Exact" is indeed one of the terms which early modern mathematicians used in this connection.[5] Other contemporary phraseology with this connotation included terms like "pure geometry," "true knowledge," "geometrical," "certain," "perfect" and expressions invoking virtuous or lawful behaviour.[6] Thus "exactness," in my present study, stands for a quality of mathematical procedures that, in the opinion of some mathematicians, makes them acceptable as leading to genuine (as opposed to flawed) and precise (as opposed to approximate) mathematical knowledge.

Mathematicians confronted with the question of which procedures are acceptable had to explain, to themselves or to others, what requirements would make mathematical procedures exact in the above sense. Thus they had to *interpret* what it meant to proceed exactly in mathematics. I call this activity the *interpretation of exactness*.

I should note that I avoid the terms "rigor" and "rigorous" in this respect, because they are commonly used in connection with proofs rather than with constructions, whereas my study concerns constructions primarily. However, I do think that many of the processes I discuss in connection with the interpretation of the exactness of constructions have their counterparts in the historical

[4] See [Bos 1987] for a sketch of a research program concerning the second period.

[5] For instance Descartes, [Descartes 1637], p. 316: ". . . il est, ce me semble, tres clair, que prenant comme on fait pour Geometrique ce qui est precis et exact, et pour Mechanique ce qui ne l'est pas . . .."

[6] Compare the quotations at Chapter 12 Note 11 ("pure geometry"); Chapter 10 Note 23 ("true knowledge"); Chapter 9 Note 15 ("geometrical"); Chapter 2 Note 29 ("certain"); Chapter 7 Note 19 ("perfect"); Chapter 3 Note 27 ("a considerable sin among geometers"); Chapter 10 Note 22 ("limits set by postulates").

episodes in which mathematicians were forced to interpret or reinterpret the rigor of proofs.

Existence

There is a tradition in the historiography of mathematics that interprets the ancient Greek interest in geometrical construction as related to existence proofs. In this view, first advocated by Zeuthen,[7] the postulates and constructions of Euclid's *Elements* served to prove the existence of the geometrical objects about which the subsequent theorems made assertions. This interpretation has been — convincingly, in my opinion — refuted by recent historians of mathematics, such as Mueller[8] and Knorr.[9] But, independently of the status of constructions in classical Greek geometry, the interpretation of constructions as existence proofs is inapplicable for the early modern period. I have found no evidence that early modern mathematicians doubted the existence of the solutions of the problems whose solution, by construction, they pursued with such intensity. Nor does it appear that they understood the classical interest in construction in terms of existence.[10] Arguments, based on continuity, were known and accepted, for instance, to prove the existence of a square equal in area to a given circle,[11] but there was common agreement that such arguments did not count as solution of the problem to find, determine, or indeed construct such a square. Similarly, the existence of two mean proportionals[12] between two given line segments was not doubted in the early modern period, but their existence left unanswered the question how, in pure, exact geometry, these proportionals should be constructed.

An extra-mathematical question

It should be stressed that an answer to this question cannot be derived from axioms within an accepted corpus of mathematical knowledge. In classical geometry as codified in the Euclidean *Elements*, figures are considered known if they can be constructed by straight lines and circles from other figures assumed known. This interpretation of exactness is codified in postulates; it cannot itself be derived from other axioms or postulates. Similarly, in the 1690s Christiaan Huygens advocated the acceptance of a particular curve, the "tractrix", in the construction of other logarithmic curves. There is no way of proving mathematically that the tractrix is acceptable for that purpose; Huygens used extra-mathematical arguments in support of his suggestion.[13]

Thus any answer to the question of acceptable means of construction necessarily has the nature of a chosen postulate; the reasons for its choice lie outside the realm of proven argument. The question whether these reasons are correct or valid has, strictly speaking, no meaning. Mathematicians are free to accept

[7] [Zeuthen 1896].

[8] Cf. [Mueller 1981] pp. 15–16, 27–29.

[9] Cf. [Knorr 1986], Ch. 8; in particular p. 374, note 77.

[10] Cf. Section 11.3 for an explicit early modern statement concerning the difference between existence and constructibility.

[11] Cf. Section 2.3.

[12] Cf. Section 4.4.

[13] Cf. [Bos 1988].

or reject any proposed decision on the question of which means of construction are legitimate in geometry and which are not.

The fact that choices of postulates and axioms in mathematics are based on extra-mathematical arguments does not necessarily involve great arbitrariness or ambiguity; the reasons for accepting an axiom or a postulate, although outside the sphere of formalized proven argument, may be cogent enough. However, in the case of construction during the early modern period, none of the arguments proposed to underscore the legitimacy and exactness of the various procedures proved lastingly convincing; as stated above, the issue ultimately vanished from the mathematical agenda.

Importance

Yet the early modern attempts to interpret exactness of constructions merits — and repays — historical study. The reasons for or against accepting procedures of construction, and the arguments in which mathematicians made them explicit, convincing or not, were very important in the development of mathematics. They determined directions in mathematical research, and they reflected the mental images that mathematicians had of the objects they studied. Moreover, they changed essentially during the sixteenth to eighteenth centuries, and these changes reveal much about the processes of development within mathematics in that period.

Proofs

It should be noted that in the early modern period there was much less concern about the rigor of proofs than there was about the legitimacy of constructions. This may seem remarkable because at present exactness in mathematics is related almost exclusively to proofs; indeed the essence of mathematics is usually located in the fact that its assertions are proven.

The relaxation of the classical Greek rigor of proof in mathematics has long been recognized as a characteristic of seventeenth- and eighteenth-century mathematics. It may be less generally realized that this attitude did not imply a lack of interest in exactness. Mathematicians were concerned about the foundation of their science, but they regarded the questions about construction, or in general about procedures to make objects known, as more critical than rigor of proof. The example of Kepler is illustrative in this connection. Kepler was willing to replace rigorous Archimedean "exhaustion" proofs by loose arguments on infinitesimals;[14] at the same time, as we will see in Chapter 11, he adopted an extremely purist position on construction, rejecting all other procedures than the orthodox Euclidean use of circles and straight lines.

1.3 Structure of the story

First period: geometrical problem solving

The structure of the story of construction and representation in early modern mathematics is basically simple. It comprises two slightly overlapping periods, c. 1590 – c. 1650, c. 1635 – c. 1750, and one central figure, Descartes. During

[14] Cf. [Baron 1969] pp. 108–116.

the first period questions about construction arose primarily in connection with geometrical problems which required a point or a line segment to be constructed and which admitted one or at most a finite number of solutions. Examples were: dividing an angle in two equal parts (solved essentially by constructing one point on the bisectrix) and finding two (or n) mean proportionals between two given line segments. If translated into algebra, problems of this type led to equations in one unknown.

Around these problems a considerable field of mathematical activity developed, which may be called the *early modern tradition of geometrical problem solving.* It was in this field that the use of algebra as an analytical tool for geometry was pioneered by Viète and his followers and fully realized later by Fermat and Descartes. Indeed, the adoption of algebraic methods of analysis provided the principal dynamics of the developments in the field. The tradition also gave rise to a rich and diversified debate about the proper, exact ways of solving construction problems, especially those which could not be solved by the standard Euclidean means of straight lines and circles.

Descartes

Descartes' *La Géométrie*[15] of 1637 derived its structure and program from this field of geometrical problem solving. The two main themes of Descartes' book were the use of algebra in geometry and the choice of appropriate means of construction. The approach to geometrical construction that he formulated soon eclipsed all other answers to the question of how to construct in geometry. Thus Descartes closed the first episode of the early modern story of construction by canonizing one special approach to the interpretation of exactness concerning geometrical constructions. As stated above, my present study will be confined to this first episode.

Construction of equations

Geometrical problem solving remained alive for some time after Descartes had provided it with a persuasive canon of construction and a standard algebraic approach. According to this approach a problem had to be translated into an equation in one unknown; thereafter, the roots of the equation had to be constructed by geometrically acceptable means. The latter procedure was called the "construction of equations." For some time after Descartes the "construction of equations" remained recognizable as a definite field of interest for mathematicians. Later, however, it fell into oblivion, and by 1750 it was extinct.[16]

Second period: the study of curves

Descartes' *Geometry* may also be seen as the opening of a second episode lasting until around 1750. Now the problems that gave rise to questions about construction and representation were primarily quadratures and inverse tangent problems. These belong to a class of problems in which it is required to find or construct a curve. If translated in terms of algebra, these problems lead

[15] [Descartes 1637]; I refer to this text as "the *Geometry*."

[16] I have described this process of rise and decline in [Bos 1984], cf. also Section 29.3.

to equations in two unknowns, either ordinary (finite) equations or differential equations.

Curve construction problems belonged to a field of mathematical activity, flourishing from c. 1635 to c. 1750, which can best be characterized as the investigation of curves by means of finite and infinitesimal analysis.[17] The field was loosely connected to the earlier tradition of geometrical problem solving; it also had strong roots in classical Greek studies on tangents, quadratures, centers of gravity, and the like.

De-geometrization of analysis

It was from this field that, in the period 1650 – 1750, infinitesimal analysis gradually emancipated itself as a separate mathematical discipline, independent of the geometrical imagery of coordinates, curves, quadratures, and tangents, and with its own subject matter, namely, analytical expressions and, later, functions. This process of emancipation, which might be called the de-geometrization of analysis, constituted the principal dynamics within the area of mathematical activities around the investigation of curves by means of finite and infinitesimal analysis. It was strongly interrelated with the changing ideas on the interpretation of exactness with respect to construction and representation.

Representation of curves

The solution of a curve construction problem required the representation (in the sense explained above) of the curve sought. It should be stressed that it was only by the end of the seventeenth century that analytical representation of a curve by an equation became a generally feasible option. For a long time such equations were available for algebraic curves only, and even for these curves mathematicians long felt hesitant about accepting an equation as a sufficient representation. To make a curve known, they argued, more was required than the algebraic expression of one particular property shared by the points on it; what was needed was a construction of the curve. In the case of non-algebraic curves such constructions should provide transcendental relations and for these the extant methods of construction were insufficient. Indeed, new methods were devised and they occasioned new debates about their acceptability, legitimacy, and effectiveness.

Similarities between the debates

The new debates on the appropriate construction and representation of curves were related to the earlier debates on point constructions. Curves were often represented by pointwise constructions, that is, procedures for constructing arbitrarily many points on a curve. In these pointwise constructions of curves

[17] After Viète's introduction of the analytical use of his new algebra in geometry and number theory, the term *analysis* was soon used as practically synonymous with the term *algebra*, both indicating letter-algebra. Later, when special symbols for objects or processes involving limits, indivisibles, or infinitesimals were introduced, it became customary to denote letter-algebra combined with these new symbols as *analysis infinitorum* ("analysis of infinite quantities") and, in contrast, letter-algebra itself as *analysis finitorum* ("analysis of finite quantities"). The "infinite quantities" included infinitely small ones. Remaining close to this terminology, I use *finite analysis* for letter-algebra and *infinitesimal analysis* for letter-algebra in combination with symbols for infinite or infinitesimal objects and processes.

mathematicians had to rely on the techniques and arguments developed earlier about constructions of points. Moreover, the representations of curves, especially of non-algebraic curves, raised the same questions of interpretation of exactness as had arisen in connection with point constructions. Indeed, in the later debate about curve construction we can distinguish the same attitudes, strategies, and arguments as in the earlier debate (cf. Section 29.4).

Role of Descartes' ideas

As mentioned above, Descartes' work constituted the conclusion of the first debate; it was also a central point of reference in the later discussions. In pointwise constructions mathematicians relied implicitly on the Cartesian canon for constructing points. Moreover, they used Descartes' analytic techniques of studying curves by means of their equations. On the other hand, they had to break through the conceptual and technical restrictions of the Cartesian approach to geometry, in particular the restriction to algebraic curves and relationships, which was essential to Descartes' view of geometry.

The issues vanished

Although the interpretation of exactness with respect to geometrical construction and representation was discussed with some intensity during the periods I have sketched above, no ultimately convincing answers were found. By 1750 most mathematicians had lost the concern for issues of geometrical exactness and construction; they found themselves working in the expanding field of infinitesimal analysis, which had by then outgrown its dependence of geometrical imagery and legitimation.

1.4 Motivation of the study

Interpretation of exactness

Although at present the concerns about the acceptability of geometrical constructions and representations are no longer part of the mathematical awareness and practice, they were very central in the early modern period. The vanishing of reasonable and understandable concerns is an intriguing process and to understand it is, therefore, a valid objective of historical study. Moreover, the study of these issues of construction and representation seems important to me because it contributes in several different ways to our understanding of the development of mathematics. In Section 1.2 I have introduced my subject as an instance of the endeavor to interpret mathematical exactness. The early modern attempts to create and legitimize methods of exact geometrical construction constitute a revealing example of the processes involved in the interpretation of mathematical exactness, especially so because ultimately they failed. Thereby, they offer an enlightening contrast with other, successful endeavors to interpret mathematical exactness, such as the Greek treatment of incommensurable magnitudes, the rigorization of analysis in the nineteenth century, and the developments following the foundations crisis in the early twentieth century.

Furthermore, the study of the concepts of construction and representation contributes to an understanding of early modern mathematics with respect to the terminology and imagery of the texts, the directions of research, and the processes of development.

Terminology and imagery Knowledge of the early modern concerns about construction and representation is essential in reading the mathematical texts of the period. The conception of a geometrical problem as a task to be performed within a canon of acceptable procedures of construction is omnipresent in early modern texts; it strongly determines their structure and terminology. To give one example, to "construct" an equation or a differential equation meant to solve it. Even when, by the mid eighteenth century, the mathematicians had transferred their interest almost entirely from the geometrical to the analytical properties of these solutions, the term "construction" remained in use.

Direction of research In several instances the concerns about construction and representation determined the direction of mathematical research. As I will show in Part II, the structure of the *Geometry* of 1637 was largely determined by Descartes' endeavor to develop a new conception of exact geometrical construction which extended beyond the Euclidean restriction to straight lines and circles, and which was compatible with the use of algebraic methods of analysis. Also several developments within the early infinitesimal analysis related directly to constructional concerns, for example, the earliest studies of elliptic integrals and the interest in tractional motion.[18] Thus the historical study of construction and representation offers essential insights into the early development of both analytic geometry and infinitesimal calculus.

Processes of development Early modern mathematics witnessed the creation and expansion of (finite and infinitesimal) analysis and its later emancipation from geometry. These developments brought deep and fundamental changes, not only concerning the problems that mathematicians solved and the methods they developed for that purpose, but also concerning the very conception of what it meant to solve a problem or to acquire new mathematical knowledge. Thus the rise and emancipation of analysis brought with it changes in interest, in directions of research, in canons of intelligibility[19] and in rules for acceptable procedure in mathematics. These changes were brought about by such processes as the habituation to new mathematical concepts and material, and the progressive shift of methodological restrictions. By habituation, a mathematical entity that was earlier seen as problematical (for instance, some transcendental curve) could later serve as solution of a problem, even though the mathematical knowledge about it had not changed essentially. Methodological restrictions were mitigated or lifted as the result of conflicts on legitimacy of procedures and because of the appeal of new mathematical material. These processes are not exclusive to early modern mathematics, they belong also to other periods in the development of mathematics. I think they are important and interesting, and I find that the study of

[18] Cf. [Bos 1974] and [Bos 1988].

[19] A term I borrow, with appreciation, from M. Mahoney, cf. [Mahoney 1984].

the changing ideas and procedures of construction and representation provides a valuable key to the understanding of such general processes.

1.5 Historiographical scheme

Methodology

Some remarks about my method are in order here. Basically I just want to tell about the interesting and exciting things I have found concerning the concept of construction in early modern mathematics, and the way they reflect the mathematicians' interpretation of the exactness of their science. However, I have experienced, to the cost of much time and bewilderment, that "interesting" and "exciting" are poor guides in structuring a report. So, in order to make the necessary choices and to plan strategies of writing, I found myself forced to define my "methodology." I think I should make that methodology explicit here. I do so with some reluctance because writing on "methodology" easily suggests directive and factiousness. Hence I use the term "historiographical scheme," and under that title I give a description of the method adopted in the present study and, more important, of its restrictions.

Model

The primary motivation of my study is to understand the processes that occur when mathematicians interpret or reinterpret the meaning and the criteria of mathematical exactness. I restrict myself to a single period, the early modern one. In that period the arguments and debates on the interpretation of mathematical exactness centered on one concept, that of construction. For that reason the central subject of my study is the concept of construction and the changes it underwent in the early modern period. The concept did not subsist and change in isolation, it functioned within mathematics and its changes interrelated with the large-scale developments within mathematics. So the underlying simple model of my investigation is that of one changing entity, the *subject*, within a broader domain, the *context*, in which global developments occur.

It will be useful at the outset to identify not only the subject but also the context, and the developments within it, with respect to which I investigate the subject. As to the developments within the context, I have tried to simplify the model further by identifying what I call the *principal dynamics* in the context, that is, the main stimulus or agent of change in the developments.

Below I describe the model in some more detail; I also state the methods I have used in gathering evidence and rendering results.

Elements of the model

The elements of the historiographical model adopted in this study are:

- The *subject* of my study, namely: the *concept of construction* in the early modern period.
- The subject is studied in its appropriate *contexts*, namely,

a in Parts I and II of the present study: the *early modern tradition of geometrical problem solving* c. 1590 – c. 1650; and

b in the later part of the story (not treated in the present study): the *investigation of curves by finite and infinitesimal analysis* c. 1635 – c. 1750.

- Within these contexts or fields various developments occurred that I consider as following a *principal dynamics.* I define these principal dynamics as,

a for Parts I and II: the *creation and adoption of (finite*[20]*) algebraic analysis as a tool for geometry,* and,

b for the later part: the *emancipation of* (finite and infinitesimal) *analysis from its geometrical context.*

I study the subject, the concept of construction, primarily with the aim to understand the processes involved in the interpretation of mathematical exactness. It will turn out, however, that this purpose requires a fairly broad covering of the mathematical-technical aspects of geometrical construction in the early modern period.

Occasionally I view my subject in relation with processes outside of these particular contexts. In Part II, for example, I touch upon philosophical issues related to construction. However, in these cases I am far from exhaustive and there will be much room for further research.

Other contexts than the above will generally be disregarded, notably the professional and personal contexts and the general cultural context. I disregard them for reasons of time and because I have the impression that these contexts will not provide essentially new insights in my subject. (Conversely, however, the developments concerning the interpretation of exactness may well shed new light on certain aspects of, for instance, the professional context.)

As the scheme makes clear, my aim is not to write a comprehensive history of the domains identified as contexts. Rather I present the necessary background information about context and principal dynamics by means of characteristic examples of procedures and arguments. T hus Chapters 3–6 of Part I illustrate by various examples the nature of the early modern tradition of geometrical problem solving and the creation and adoption of (finite) algebraic analysis as a tool for this activity.

Sources Although I do attempt a certain completeness in my study, it is neither possible nor desirable to report exhaustively on all incidents of interpretation of exactness that occurred with respect to the concept of construction within the contexts as indicated. Therefore, I concentrate on significant and characteristic statements of a restricted number of mathematicians who represent archetypal positions in the debates on exactness. In choosing these protagonists (notably

[20]Cf. Note 17 above.

the ones discussed in Chapters 9–13) I have used a classification of attitudes of mathematicians vis-à-vis the interpretation of exactness. I explain that classification in Section 1.6 below.

My investigation is based almost entirely on published primary sources. I have not searched for relevant but as yet unpublished archival material such as letters or manuscripts. For the pre-Descartes period such a search would probably bring new details to light, but I think it unlikely that they would affect the overall picture as it can now be sketched. For the Descartes episode one can hardly expect much new Cartesian archival material to turn up — although I can't help hoping that the "écrit" of 1631–1632 (cf. Section 19.1) will surface one day and provide an opportunity to test my conjectures on one stage in the development of Descartes' geometrical thought.

In rendering mathematical argument I have adopted some conventions, which I explain in Section 1.7 below.

Finally I should add that whenever I felt that it would be useful or entertaining to step outside the historiographical scheme detailed above I have not hesitated to do so.

1.6 Strategies in the interpretation of exactness

A classification

The early modern mathematical literature offers an at first bewildering variety of geometrical constructions and of arguments about their legitimacy. As I noted in Section 1.2 these arguments concern a meta- or extra-mathematical question; therefore, they cannot be classified according to their correctness. Yet a classification[21] is needed to bring order into the material and to understand the significance and influence of the arguments. I have found that such a classification emerges quite naturally from the source material if one views the arguments as reflecting certain attitudes, approaches, or strategies adopted by mathematicians when confronted with (ultimately undecidable) questions of interpretation of exactness. I distinguish a number of basic attitudes in this respect, namely: (1) appeal to authority and tradition, (2) idealization of practical methods, (3) philosophical analysis of the geometrical intuition, and (4) appreciation of the resulting mathematics. Two further, slightly different, categories are: (5) refusal, rejection of any rules, and (6) non-interest.

This classification is primarily based on the material collected for the purpose of the present study, concerning the early modern tradition of geometrical problem solving and Descartes' *Geometry*. I believe that the classification may be relevant as well for the study of other episodes of interpretation of exactness in the history of mathematics.

In the paragraphs below I give short descriptions of the categories, formulated primarily with respect to the early modern tradition of geometrical problem solving; for each I note the persons or episodes dealt with in Parts I and II, that belong, totally or partly, within the category.

---

[21] I have presented the classification explained in the present section in [Bos 1993].

(1) Appeal to authority and tradition

Early modern geometry continued the classical geometrical tradition. Within that tradition certain construction procedures had been developed and codified. Although classical works provided little explanation of why they were acceptable, the procedures themselves were clear enough. Therefore, early modern geometers could opt for working in the style of that tradition, restricting themselves to the procedures of construction they found in the classical works and arguing that these needed no other endorsement than the authority of great mathematicians such as Euclid, Archimedes, and Apollonius. Kepler's purist opinion on the proper ways of geometrical construction is an example; he rejected all other means of construction than straight lines and circles, warning that the whole fabric of classical geometry would falter otherwise (Chapter 11). The recurrent invocation in early modern geometrical texts of a passage of Pappus in which the geometer is warned against the "sin" of constructing by improper means (cf. Section 3.4) is also a clear case of the appeal to authority and tradition.

(2) Idealization of practical methods

Some mathematicians argued for or against the adoption of procedures of construction by referring to the practice of draftsmen who use rulers, compasses, and other instruments to construct figures with great precision, such as the divisions of scales or the curves on the plates of astrolabes or on sundials. The precision and reliability of an instrument or a procedure in practice was an argument in favor of adopting the idealization of the instrument or the procedure as legitimate in pure geometry. In this case the interpretation of exactness was based upon an idealization of a geometrical practice. Clavius' attitude toward construction is an example; he defended pointwise construction of curves as acceptable in pure mathematics because of its precision in practice (Chapter 9). Molther's arguments (Chapter 12) belong in this category as well.

(3) Philosophical analysis of the geometrical intuition

Construction and representation served to make objects known. Hence, behind any choice of procedures for construction lay the intuition of "known-ness," or, in general, the intuition of certainty in geometry. Some mathematicians analyzed this intuition philosophically in order to find arguments for the adoption of their particular interpretation of constructional exactness. The most important proponent of this approach in the early modern period was Descartes, whose whole vision of geometry was shaped by his philosophical concern for the certainty of the geometrical operations, in particular of the constructions (cf. Chapter 24). But others, too, considered the mental processes involved in geometrical procedures: Kepler for instance argued against other means of construction than the classical ones on the basis of their lack of certainty in principle.

(4) Appreciation of the resulting mathematics

The result of accepting certain procedures of construction as legitimate is a mathematical system in which certain problems are solvable (they may be hard to solve or easy) and others are not. Different choices result in various systems that geometers may consider as qualitatively different. The mathematics of one system may be experienced as richer, more interesting, more challenging than that of the other. The quality of the resulting mathematics, recognized implicitly or explicitly, played a role when mathematicians accepted or rejected certain methods of construction. In that case the reasoning did not primarily concern the merit of the postulated procedure itself, but rather the interest and the quality of the resulting mathematical system. Viète provides an example here; his choice of one extra postulate to supplement the classical Euclidean ones seems to have been inspired more by a wish to secure extant interesting practices than by the intrinsic persuasiveness of that postulate (Chapter 10).

(5) Refusal, rejection of any rules

In interpreting geometrical exactness mathematicians make choices, accepting certain construction and rejecting others. The attitudes or strategies characterized in the four preceding categories have in common that the necessity of a choice is accepted. However, occasionally I found geometers who seemed to reject the practice of prescribing certain procedures of construction and forbidding others. Characteristically they argued that any problem had its own procedure of solution regardless of general criteria of acceptability. Clear examples of such open refusal are not to be found in the period covered in Parts I and II of this study, but in the later seventeenth and eighteenth century there are a few, which I hope to discuss in later publications.

(6) Non-interest

Finally, there were the mathematicians who simply were not interested in the matter. This attitude was probably rather common, but in individual cases it is difficult to distinguish it from the attitude listed above under (1). One major figure in the category of non-interest is Fermat, who, despite his important contributions to methods for analyzing geometrical problems, seems to have had very little affinity to questions of the acceptability of the constructions found by these methods (Chapter 13).

1.7 Methods of exposition

Mathematical terms and arguments

At the beginning of Chapter 6 (Section 6.2) I specify some terminology, notably concerning concepts such as "arithmetic," "algebra," "unknowns" and "indeterminates." Rather than giving these here I have inserted them at the place where they become essential, namely in the chapter on the interrelations of arithmetic, algebra, geometry, and analysis during the beginning of the early modern period.

It is impossible to render mathematical arguments from earlier times exactly as they were. The modern reader does not have the mathematical training and background of those for whom the old texts were originally written; he or she[22] has both too much and too little mathematical knowledge. He knows modern theories which often readily reveal connections in the mathematical material that remained hidden for earlier mathematicians. On the other hand,

[22] I use "he" henceforth, also in the cases where I mean "she or he."

the modern reader lacks certain skills that earlier writers would assume their readers to posses as a matter of course: knowledge of Latin, familiarity with classical authors, and the skill to elicit procedures or arguments from long and typographically unstructured prose texts referring to complicated and randomly lettered figures.

The historian, wanting to convey the essence and the meaning of earlier mathematics to a modern reader, cannot avoid modifying the mathematical argument and adapting it to the purposes of his presentation; indeed, he should do so. He may shorten it where modern insight makes the result obvious and no lessons can be learned from going into detail through elaborate and unfamiliar notation and terminology, and he does well to unify the presentation where differences are unessential. On the other hand, he may need to expand certain now unfamiliar steps in the argument.

In accordance with these general remarks I have not hesitated to modify the presentation of mathematical argument with respect to notation, symbols, and figures, in such cases where this makes the essence more clear and does not change the meaning.

The mathematical arguments I study were thought or written in the past. Most of them are still valid at present, but a considerable number of the expressions, statements, and reasonings are no longer natural or acceptable in modern mathematics. This leads to a peculiar problem of representation: should the mathematical arguments from earlier times be presented in the past or in the present tense? I have adopted the following convention with respect to this question: In general, the arguments from past mathematics are rendered in the past tense; however, when they are presented in special typographical formats such as the one I use for constructions, they are given in the present tense. It may at first be somewhat strange to read that the equation $x^2 - ax = b^2$ *could* be written as $x(x - a) = b^2$, because it still can. However I have found that the past tense in these cases is a very useful reminder that past mathematics is at issue and that its validity at present is not the main interest of historical research.

Figures When taking over figures from primary sources, I have generally changed the letters. I have coordinated the lettering in figures that are related (as for instance the figures belonging to an analysis and a construction of the same problem). I indicate points and straight lines of indefinite length by capital letters and line segments by lower case letters. Without indication otherwise, a lower case letter indicates the line segment along the nearest straight line between the two nearest marked points. If one point in a figure has an obvious central role, I denote it as O. Axes will usually be OX and OY. Otherwise the letters marking points are chosen in alphabetical order from A; if possible, the alphabetical order follows that of the appearance of the points in the argument. Indices are used if functional (they do not occur in any of the original figures). The lay-out of the figure, however, will be the same as that of the original unless stated otherwise. In case line segments clearly have the function of generic abscissa and ordinate of a curve with respect to some vertex, axis, and ordinate angle, I often indicate these segments by the lower case letters that would suitably denote the coordinates in the equation of the curve, even if the original argument involves no analytic geometry. In the case of facsimiles of original figures the conventions above, of course, do not apply.

Geometrical operations

In classical geometrical reasoning line segments were combined to form parallelograms (rectangles in particular) or ratios. Later the operations of forming rectangles and ratios came to be seen as multiplication and division, respectively. As the acceptance of these later conceptions was an essential element of the gradual adoption of algebraic analysis as a tool for geometry, it would be confusing to use the algebraic symbols for multiplication and division in cases where the texts have rectangles or ratios. In such cases I use the notations $a : b$ for the ratio of a and b, $\text{sq}(a)$ for the square with side a, and $\text{rect}(a, b)$ for the rectangle with sides a and b. Proportionalities, that is, equalities between ratios, are denoted as $a : b = c : d$. In all cases where I use the common notations for multiplication ($a \times b$, $a \cdot b$, ab, or the exponent notation) and the quotient bar for division, this means that these notations were actually used in the original text or that the meaning in the text was clearly that of multiplication and division rather than forming rectangles or ratios.

Constructions

Constructions are stepwise procedures, each step being one of a canon of accepted simple standard constructions. I have stressed this unifying conception by presenting almost all constructions in the same format, indicating the given and required elements and numbering the separate steps in the procedure. If relevant, I also render the "analysis" — the argument by which the construction was found — in a stylized, stepwise format. Within this format, my presentations are as close as possible to the original texts. I add (between square brackets) proofs or proof sketches of the constructions. As the original proofs, if extant, are usually little informative, I have felt free to shorten and streamline these proofs considerably and to present them in modernized notation. However, unless noted otherwise, the proofs involve no arguments unknown at the time.

Algebra

For rendering the unknowns and indeterminates in algebraic arguments I use the common system whose appearance, in the *Geometry* of Descartes, is actually part of my story. That is: unknowns and indeterminates are indicated by lower case letters, those for the unknowns from the end of the alphabet (preferably x), those for indeterminates from the beginning. As in the case of figures, I have felt free to change the choice of letters, even if in the original document they were used in the Cartesian way just mentioned. However, if, as was mostly the case, the homogeneity or inhomogeneity of the equations was meaningful in the argument, I have kept these features in my rendering. I have used indices, summation signs, and the general function symbol wherever these modern expedients make it possible to express a generality which I think was meant by the original author.

A special remark has to be made about Vietean algebra. The algebra of Viète and his followers was written down in a beautifully consistent but somewhat cumbersome and not fully symbolic notation. In general, this notation can be transformed straightforwardly into the notation described above. Occasionally information about the dimension of the terms or factors is lost in that transformation. In cases where this is essential I note it. I have occasionally provided examples of the Vietean notation.

1.8 Survey

Structure of the study

In accordance with the structure of the story outlined in Section 1.3, my research project is naturally divided into three parts, of which the first two are dealt with in the present study. Part I (Chapters 1–14) is on construction and exactness in the early modern tradition of geometrical problem solving before Descartes; Part II (Chapters 15–28) is on Descartes' conception of construction and his related redefinition of geometrical exactness. The last part of the project, about the construction and representation of curves in the period c. 1650 – c. 1750, is briefly sketched in the Epilogue (Chapter 29), but not further dealt with in the present volume; I hope to return to it in a later publication.

Starting point: Pappus' Collection *1588*

The chapters of Part I naturally divide into four groups. In the first, Chapters 2–3, I sketch the starting point of the developments concerning construction and constructional exactness within the early modern tradition of geometrical problem solving. This starting point was located in time around 1590 because it was then that, through the publication in 1588 of Commandino's Latin translation of Pappus' *Collection*,[23] the discussions on the interpretation of exactness with respect to construction acquired substance and structure. Chapter 2 is devoted to the earlier and rather elusive stage of these discussions beginning c. 1500; it also deals with the relevant classical sources that became known during the sixteenth century. Chapter 3 concerns Pappus' *Collection* as it became available by 1590. Several constructions from this work were influential in the later development of the concept of construction. They are discussed, together with Pappus' classification of problems and his precept for geometrical construction. It was precisely this classification that, together with the related precept, provided the substance and structure alluded to above.

Background: the tradition

The next three chapters sketch the background of the debates on construction: the early modern tradition of geometrical problem solving in the period c. 1590 – c. 1635. Chapter 4 gives an overview of the problems that constituted the subject matter of the tradition and provides a representative set of examples illustrating both the nature of these problems and the main methods of construction that were used. In Chapter 5 I turn to the analytical methods, that

[23] [Pappus 1588].

is, the formalized procedures for finding solutions. Besides acquainting themselves with the classical Greek method of analysis from sources such as Pappus' *Collection*, mathematicians pioneered the use of algebra as an analytical tool in geometry. The use of algebra soon became so central that we may identify it as the principal dynamics within the tradition of geometrical problem solving. The chapter provides examples of both methods of analysis. Chapters 6 and 7 take up a number of general issues that arose within the tradition of geometrical problem solving, in particular the legitimacy of using in geometry such fundamentally ungeometrical sciences as arithmetic and algebra. In Chapter 8 I discuss how Viète dealt with the fundamental issues concerning the analytical use of algebraic techniques.

Interpreting constructional exactness

With the starting point, the background and the fundamental issues thus sketched I turn in Chapters 9–13 to my main theme, the debates on the exactness of constructions within the early modern tradition of geometrical problem solving. I discuss successively the opinions of five mathematicians who stated their position in the matter explicitly, namely Clavius (Ch. 9), Viète (Ch. 10), Kepler (Ch. 11), Molther (Ch. 12), and Fermat (Ch. 13). By way of a summary of Part I, I sketch in Chapter 14 the state of the art of geometrical problem solving around the year 1635.

Part II

Apart from being of interest in itself as the description of an important tradition within early modern mathematics, the material gathered in Part I also serves to sketch the background of Descartes' program for restructuring the art of geometrical problem solving that led to the *Geometry* of 1637.[24] The program of Descartes is the subject of Part II, which is organized chronologically. After a brief introductory Chapter (15), Chapters 16–19 deal with Descartes' studies before the publication of the *Geometry*, his first programmatic ideas on problem solving and construction, the adoption of algebra as a means for analysis of geometrical problems, his first geometrical interpretation of the algebraic operations, and the difficulties he encountered in generalizing this interpretation.

The next eight chapters (20–27) are about the *Geometry* itself, its contents, its underlying methodological problems, and the canon of geometrical construction which it presented. The concluding chapters 28 and 29 summarize the dynamics of the development of Descartes' geometrical thinking and sketch the lines of later mathematical development that emanated from the early modern tradition of geometrical problem solving and in particular from Descartes' *Geometry*.

1.9 Conclusion

Validity and value

With the scheme, the categories, the conventions, and the structure intro-

[24] [Descartes 1637].

duced in the preceding sections, I attempt to render and analyze the various approaches adopted by early modern mathematicians when they had to forge programs, answers, or tactics with respect to the exactness of geometrical construction and representation. As said above (Section 1.6), these endeavors cannot be classified according to their validity because they attempted to answer an extra-mathematical question. To conclude, let me stress that this does not imply that the arguments are without value. On the contrary. They are very informative about mathematical ways of thinking in an earlier period and some of them have an independent intellectual quality. Indeed, I find that the mathematicians' arguments on the interpretation of geometrical exactness were largely sincere and serious, and occasionally showed a depth and penetration that can be admired and enjoyed even by readers of a later age for whom the question of exactness of geometrical construction has lost its former urgency.

Chapter 2

The legitimation of geometrical procedures before 1590

2.1 Introduction

Antiquity

Sixteenth- and seventeenth-century mathematicians were not the first to struggle with the interpretation of exactness of geometrical constructions; also in the classical Greek period there existed a rich variety of constructional procedures and many opinions on their acceptability. It appears that these opinions never converged to a clear *communis opinio*. This lack of uniformity and the relative scarcity of sources meant that sixteenth-century mathematicians had no clear classical examples in developing their ideas about the exactness of geometrical constructions. Therefore, it is not necessary to survey the classical ideas on the legitimacy of constructions here; rather I can restrict myself to dealing with those few classical sources that actually influenced the early modern debates. Besides, and fortunately, I can refer to an excellent recent study on the classical tradition of geometrical problem solving, Wilbur Knorr's *The ancient tradition of geometric problems*.[1]

The sixteenth century

During the sixteenth century several writers participating in the Renaissance discovery and elaboration of classical mathematical knowledge commented on the geometrical status of constructions. Usually the occasion was provided by one of the three so-called "classical problems": the quadrature of the circle, the construction of two mean proportionals, and the trisection of the angle. These problems cannot be solved, i.e., constructed, by straight lines and circles only.

[1] [Knorr 1986].

This impossibility, which was formally proved only in the nineteenth century,[2] had been accepted as a matter of experience by classical Greek mathematicians, and most early modern mathematicians assumed, by experience or on authority, that the classical problems could not be constructed by straight lines and circles. In the course of history these problems inspired, because of their conditional insolvability, several new lines of mathematical argument and repeated rethinking of the rules concerning geometrical construction.

However, the opinions expressed during the sixteenth century about the solvability of the classical problems, and about the acceptability of the means used to solve them, largely remained inconclusive and very few of them were backed by clear arguments. In particular, I have found no explicit criteria for accepting or rejecting geometrical procedures in the relevant literature before c. 1590. Below I present a number of statements characteristic for the level of the discussion on the matter during most of the sixteenth century. The level increased markedly after the publication in 1588, of Commandino's Latin translation of Pappus' *Collection*, which contained enough material to instigate an ordered and effective discussion of both the techniques of construction and their acceptability.

2.2 The legitimacy of the Euclidean constructions

Circles and straight lines

No mathematician of the sixteenth century doubted the legitimacy of using circles and straight lines in geometrical constructions. This use was codified in the first three postulates of Euclid's *Elements*. Although the actual execution of these constructions required the use of instruments, namely, a ruler and a compass, no one objected to their geometrical legitimacy on the grounds that they involved mechanical instruments — as we will see, this objection was raised against other instruments and procedures.

Proclus' commentary on the first book of Euclid's *Elements*, available in print since 1533,[3] contained several comments on the special status of straight lines and circles in geometry. Proclus primarily invoked the authority of Plato. Thus in a passage early in the book he wrote:

> . . . Plato constructs the soul out of all the mathematical forms, divides her according to numbers, binds her together with proportions and harmonious ratios, deposits in her the primal principles of figures, the straight line and the circle, and sets the circles in her moving in intelligent fashion. All mathematics are thus present in the soul from the first. Before the numbers the self-moving numbers, before the visible figures the living figures, before the harmonised parts the ratios of harmony, before the bodies moving in a circle

[2] Cf. [Kline 1972] pp. 38–42, 764 and 981–982. Cf. also Section 26.6 for Descartes' explanation why problems like the trisection cannot be constructed by straight lines and circles.

[3] Greek text in [Euclid 1533], Latin translation in [Proclus 1560].

> the invisible circles are already constructed, and the soul is the full company of them.[4]

Commenting on Euclid's definition of the straight line in *Elements* I (Def. 4) Proclus explained:

> Plato assumes that the two simplest and most fundamental species of line are the straight and the circular and makes all other kinds mixtures of these two, both those called spiral, whether lying in planes or about solids, and curved lines that are produced by the sections of solids.[5]

In the comments on the first postulates, Proclus argued that drawing straight lines and circles was a supremely evident procedure.[6] The definition of the circle in *Elements* I (Def. 15) gave him the occasion to explain that straight lines and circles were the "first and simplest and most perfect of figures"[7] and to state some more profound analogies such as the one between the circle and the heavens, on the one hand, and the straight line and "the world of generation,"[8] on the other hand.

Sixteenth-century mathematicians would readily interpret Proclus' statements about circles and straight lines as arguments in favor of their use in geometrical constructions, or at any rate in support of their non-mechanical nature, as opposed to such means and procedures as applied in the constructions of two mean proportionals (cf. Sections 2.4 and 4.4).

However, Proclus did not state explicitly that geometry should restrict itself to circles and straight lines in constructions, nor did he formulate any other explicit criterion for demarcating the domain of legitimate geometrical procedures.

2.3 The quadrature of the circle

In 1559 Buteo published his *Two books on the quadrature of the circle, in which many quadratures are refuted and Archimedes is defended against the attacks from all and sundry.*[9] Apparently the squaring of the circle was widely discussed; Buteo had occasion to refute the attempts of no less than ten classical and later authors.[10] The book provides an illuminating picture of the discussions on the quadrature of the circle around the middle of the sixteenth century. A brief report on some of the arguments from Buteo's work may serve here to illustrate the nature and the level of these discussions. *Buteo*

[4] [Proclus 1992], p. 14.
[5] [Proclus 1992], p. 84.
[6] [Proclus 1992], pp. 140 sqq.
[7] [Proclus 1992], p. 117.
[8] [Proclus 1992], p. 117.
[9] [Buteo 1559].
[10] Namely Antiphon, Bryson, Hippocrates, an Arab writer which Buteo did not identify, Campanus, Cusanus, Dürer, Fortius (i.e., Joachim Fortius Ringelberg), Bouvelles, and Fine.

Buteo's main target was Oronce Fine, who had published a quadrature of the circle[11] in 1544. Fine's incorrect quadrature involved a construction of two mean proportionals, which was equally wrong. His work drew immediate fire; in 1546 Nonius published an elaborate reaction[12] pointing out all the errors in the book.

Buteo refuted most of the quadratures he discussed by showing that they implied a ratio between the diameter and the circumference of the circle that lay outside the limits proved by Archimedes. But he also dealt with more subtle matters, in particular, the distinction between existence and constructibility. The issue arose in connection with Bryson's quadrature,[13] which consisted merely in an argument to the effect that for any given circle there existed an equal square. Buteo agreed, because, as he wrote,

> the circle and the square belong to the same kind of magnitude [namely area][14]

so that, because clearly there was a square smaller than the given circle, and a larger one as well, there must also be a square whose area was equal to that of the circle. Apparently Buteo tacitly assumed that the concept of magnitude implied continuity and thereby a kind of intermediate-value theorem for magnitudes of the same kind; he gave no explicit arguments for this. However, he insisted that the quadrature of the circle required more than an argument for the existence of an equal square — this equal square must actually be found. Elsewhere, criticizing Cusanus, he repeated this argument, stressing

> the point that among geometers it is not considered impossible that something can be demonstrated although it is not given.[15]

Thus Buteo was competently able to refute false quadratures, and he was aware of the subtle distinction between existence and constructibility. Nevertheless, he did not formulate any positive criteria for a circle quadrature to be truly geometrical. This approach was characteristic for the discussions on the quadrature of the circle throughout the period until 1590: the problem was recognized as important and intriguing; it was also understood that merely proving the existence of the required square was not enough, the square had to be found or constructed in some geometrically acceptable way; but no positive criteria for acceptability were formulated. It should be noted that Buteo's prudent comments did not prevent later circle squarers from publishing mistaken quadratures; thus the 1690s witnessed a considerable row about Scaliger's failed attempts to solve this classical problem.[16]

---

[11] [Fine 1544].

[12] [Nonius 1546].

[13] On this quadrature, dating from the fourth century b.C., see [Knorr 1986] pp. 76–78.

[14] [Buteo 1559] p. 14: "circulus et quadratum sunt in eadem specie magnitudinis."

[15] [Buteo 1559] p. 121: "id quod non est apud Geometras impossibile, scilicet aliquid posse demonstrari, quamvis non detur id ipsum." Buteo's criticism of Cusanus concerned a passage in the latter's *Quadrature of the circle*, [Cusanus 1980] pp. 58–67, in particular p. 59.

[16] Cf. [Scaliger 1594] and [Scaliger 1594c].

2.4 Two mean proportionals

Preeminence among the classical problems

I now turn to the construction of two mean proportionals. Of the three classical problems, the construction of two mean proportionals attracted most notice in the sixteenth century. There were two related reasons for this preeminence. The first was that the list of 12 different constructions of two mean proportionals that Eutocius had included in his commentary to Archimedes' *Sphere and Cylinder*[17] became available in print, first in works of Valla[18] and Werner,[19] later in editions of Archimedes' works.[20] A similar, though smaller set of constructions of the trisection in Pappus' *Collection*[21] became known only much later. Secondly, mathematicians learned and found that several problems not solvable by straight lines and circles could be reduced to the problem of two mean proportionals, whereas fewer, if any, problems were found to be reducible to trisection; so the former problem acquired a central position among problems beyond the constructional power of straight lines and circles.

Two mean proportionals

It is generally assumed that the problem of two mean proportionals arose in antiquity as a useful generalization of the cube duplication problem[22] (see below). It is as follows:

Problem 2.1 (Two Mean Proportionals)
Given two line segments a and b, it is required to find their two mean proportionals, that is, two line segments x and y such that

$$a : x = x : y = y : b\,. \tag{2.1}$$

Equivalently the problem can be stated as requiring the construction of four line segments forming a geometrical sequence whose first and last terms are given. If in particular $b = 2a$, then the first of the two mean proportionals, x, solves the cube duplication problem, because in that case $x^3 = 2a^3$, that is, x is the side of a cube twice as large as the cube with side a.[23]

In another sense the problem is a generalization of a construction that can be performed by Euclidean means, namely the determination of one or *the* mean proportional of two given line segments, i.e., the segment x satisfying $a : x = x : b$. On the further generalization to any number of mean proportionals see Section 4.4.

[17] [Eutocius CommSphrCyl] pp. 588–620.
[18] [Valla 1501] book XIII, Caput II, pp. u v^v–x iv^r.
[19] [Werner 1522] pp. c iv^r–h iv^r.
[20] [Archimedes 1544], [Archimedes 1615].
[21] Cf. Construction 3.8.
[22] The insight that the duplication of the cube can be achieved if a general method is available for determining two mean proportionals is attributed to Hippocrates of Chios, cf., e.g., [Knorr 1986] pp. 23–24.
[23] It should be noted that the converse does not apply; a method for duplicating the cube cannot be used for determining two mean proportionals between an arbitrary pair of line segments. For this reason it is somewhat confusing to consider the duplication, rather than the determination of two mean proportionals, as one of the three "classical problems."

The occasion for Eutocius' list of constructions was Proposition I of Book II of *On the sphere and cylinder* in which Archimedes constructed a sphere equal to a given cone or cylinder. In his construction he assumed that it was possible to find two mean proportionals between two given line segments. As he did not explain how, Eutocius had good reason to provide constructions in his commentary; the fact that he could provide twelve different ones testified to the classical interest in the problem.

Three constructions As the problem cannot be constructed by straight lines and circles, each classical construction of two mean proportionals reported by Eutocius involved one step that was alien to the Euclidean corpus of geometrical procedures. These steps differed from construction to construction; some involved a procedure of trial-and-adjustment with rulers, others employed special instruments or curves traced by such instruments, again others used the intersection of conic sections. We will see examples of all these types in the course of this study. The most popular construction in the period before 1590 was the one which Eutocius ascribed to Heron; two other often quoted ones were those attributed to Plato and to Nicomedes. Those of Heron and Plato employed a ruler or a system of rulers, which was shifted over the figure until, by trial, adjustment, and trial again, the desired position was achieved. The third construction used a so-called "neusis" procedure for which Nicomedes had devised a special curve, the conchoid.

Heron's construction Before turning to the opinions about the legitimacy of these procedures I discuss the three constructions mentioned above.

Construction 2.2 (Two Mean Proportionals — Heron)[24]
Given: two line segments a and b (see Figure 2.1); it is required to find their two mean proportionals x and y.

Construction:
1. Construct a rectangle $OACB$ with sides $OA = a$ and $OB = b$; prolong the sides OA and OB; mark the middle D of the rectangle.
2. Place a ruler along point C; its points of intersection with OA and OB are E and F, respectively; turn the ruler around C until a position is reached in which $DE = DF$.
3. Then $x = BF$ and $y = AE$ are the required mean proportionals.
[**Proof:** We have $a : x = y : b = (a+y) : (b+x)$ by similar triangles. Further $DE^2 = DF^2$ yields $(y + \frac{1}{2}a)^2 + \frac{1}{4}b^2 = (x + \frac{1}{2}b)^2 + \frac{1}{4}a^2$, so $y(a + y) = x(b + x)$, whence $x : y = (a + y) : (b + x) = a : x$.]

Step **2** of the construction cannot be performed by straight lines and circles. It is an example of a construction by shifting rulers: a ruler is shifted until two line segments in the resulting figure (here DE and DF) are equal to each other. In such procedures sometimes (as here) both line segments were affected by the

[24] [Eutocius CommSphrCyl] pp. 590–592.

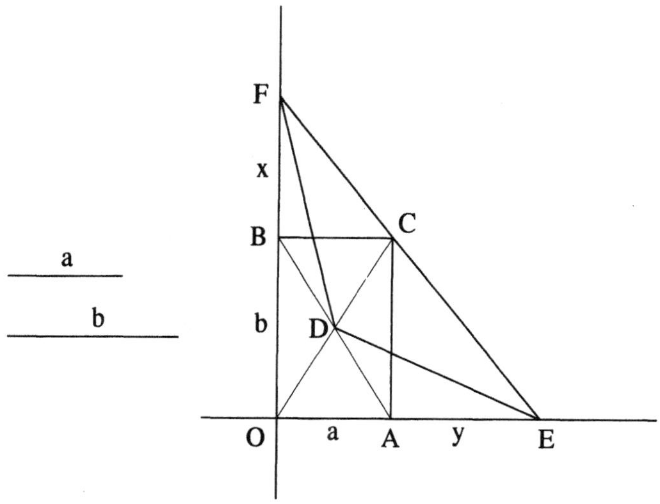

Figure 2.1: Heron's construction of two mean proportionals

motion of the ruler, sometimes only one of them, the other being fixed (as in the classical neusis construction, see below). The methods to check whether or not the line segments were equal might differ; in the case of Heron's construction a compass could be used.

Plato's construction

The construction which Eutocius ascribed to Plato[25] proceeded by shifting rulers, or rather shifting gnomons:

Construction 2.3 (Two Mean Proportionals — Plato)[26]
Given: two line segments a and b (see Figure 2.2); it is required to find their two mean proportionals x and y.

Construction:
1. Mark $OA = a$ and $OB = b$ along two lines AOC and BOD intersecting perpendicularly in O.
2. Place a gnomon with one arm along A and its vertex on OD; the vertex marks E on OD; the second arm intersects OC in F.
3. Place a second gnomon with its vertex in F and one arm along EF; its second arm intersects OB in G.
4. If G coincides with B we are done; if not, move E up or down along OD, keeping the one arm along A, and adjust the second

[25]On the attribution to Plato, which is doubtful because Plato would hardly have accepted the construction as geometrical, see [Heath 1921], vol. 1, p. 255 and [Knorr 1986] pp. 57–61.
[26][Eutocius CommSphrCyl] pp. 589–590.

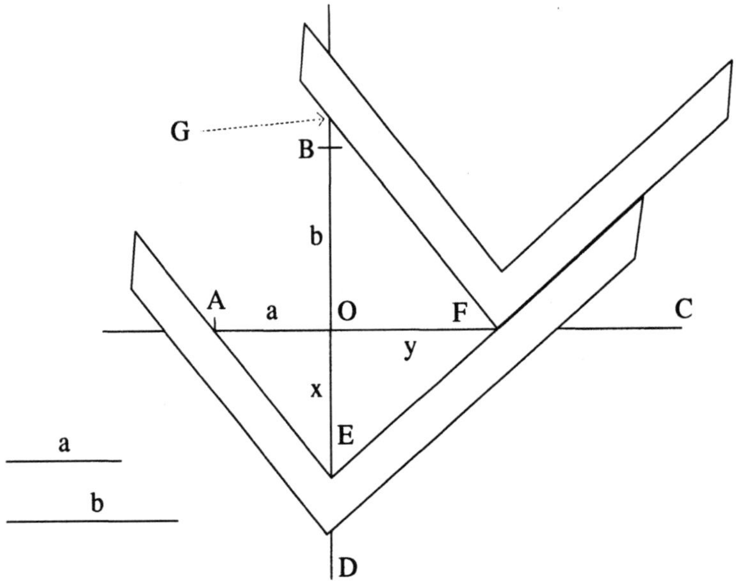

Figure 2.2: Plato's construction of two mean proportionals

gnomon along the first with its vertex on OC, until G coincides with B.
5. Now $G = B$, and $x = OE$ and $y = OF$ are the required mean proportionals.
[**Proof**: The triangles AOE, EOF, and FOB are similar, hence $a : x = x : y = y : b$.]

Step **4** of the construction cannot be performed by straight lines and circles.

The conchoid of Nicomedes The construction of two mean proportionals attributed by Eutocius to Nicomedes employed a procedure called "neusis" (see below), for which Nicomedes used a curve called the conchoid. Eutocius explained an instrument that Nicomedes had devised to trace this curve; it consisted (see Figure 2.3) of two perpendicularly connected rulers AB, CD and a movable ruler EG. The rulers CD and EG had slots along their central lines; at O on AB and at F on EG pins were fixed, which fell in the slots as shown. The distances $FG = a$ and $AO = b$ were constant. (Nicomedes probably envisaged an instrument in which these distances were adjustable.) When the ruler EG was moved, the point G described the conchoid. The pins and the slots ensured that in all its positions EG passed through the fixed point O while F remained on the line CD. Thus any point H on the conchoid had the property that its distance to the base line CD measured along the line HO was equal to a.

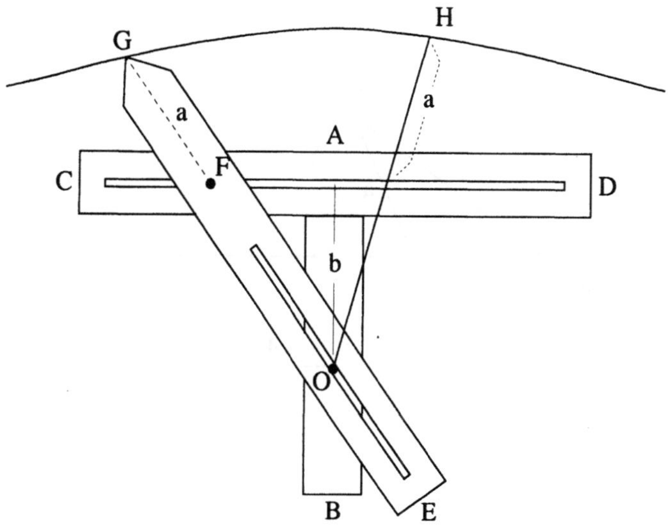

Figure 2.3: Nicomedes' instrument for tracing the conchoid

The neusis problem

Eutocius went on to show how Nicomedes' conchoid could be used for solving the so-called neusis problem:

Problem 2.4 (Neusis)
Given: two straight lines L and M, a point O (referred to as the "pole" of the neusis) and a segment a (see Figure 2.4); it is required to find a line through O, intersecting L and M in A and B, respectively, such that $AB = a$.

Literally "neusis" means "verging": the given segment is placed between the two given lines such that it verges or points toward the given pole. In general, a neusis cannot be constructed by straight lines and circles. The neusis problem played an important role in classical Greek constructional practice. However, the importance of this role became apparent to early modern mathematicians only after the publication of Pappus' *Collection* in 1588. Accordingly I discuss it in more detail later (cf. Sections 3.6 and 4.6, Problem 3.7 and Constructions 3.8, 4.16, 5.4, 5.6, and 12.1).

Nicomedes' solution of the neusis problem by means of the conchoid was as follows:

Construction 2.5 (Neusis by means of a conchoid — Nicomedes)[27]
Given and required: see Problem 2.4.

Construction:
1. Draw (see Figure 2.5) a conchoid with axis along L and pole

[27] [Eutocius CommSphrCyl] pp. 618.

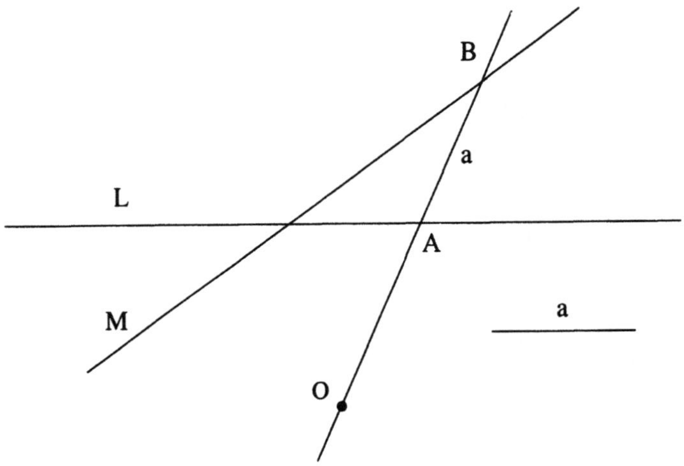

Figure 2.4: The neusis problem

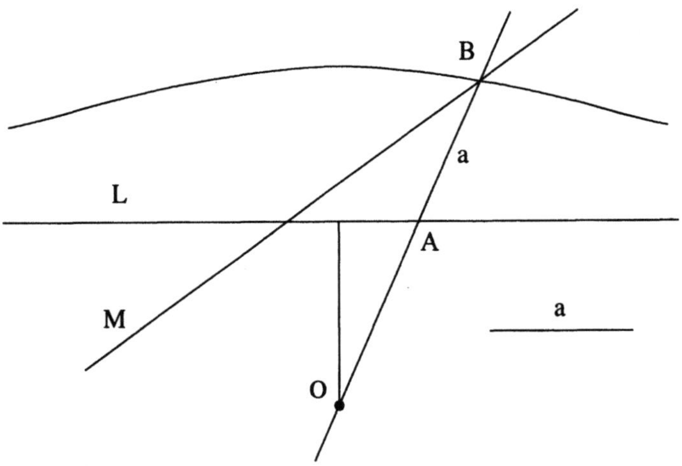

Figure 2.5: Neusis by means of a conchoid — Nicomedes

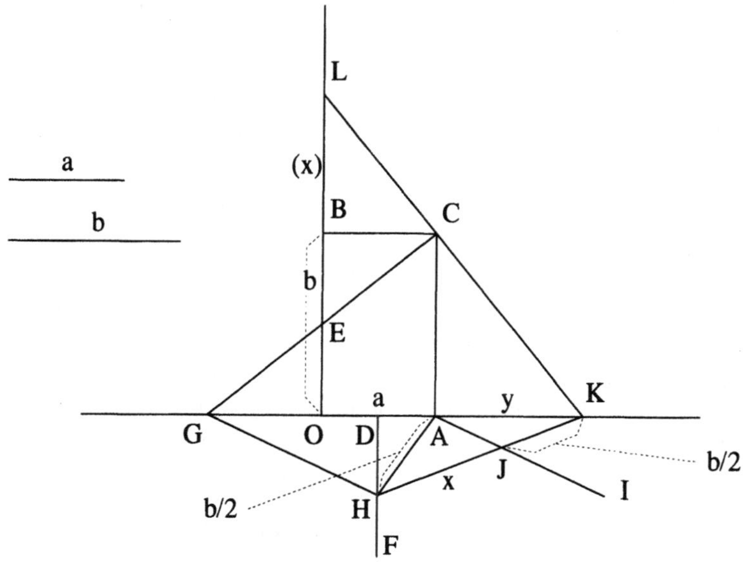

Figure 2.6: Nicomedes' construction of two mean proportionals

in O (this could be done with help of the instrument described in (Figure 2.3) above, by adjusting the pins O and F such that b is equal to the distance of O from L and a is equal to the given segment); the conchoid intersects M in B.
2. Draw OB; it intersects L in A.
3. OAB is the required line.
[**Proof**: Immediate from the definition of the conchoid.]

Nicomedes' construction

Nicomedes' construction of two mean proportionals, as reported by Eutocius, reduced the problem to a neusis. It was as follows:

Construction 2.6 (Two Mean Proportionals — Nicomedes)[28]
Given: Two line segments a and b, $a < b$ (see Figure 2.6); it is required to find their two mean proportionals x and y.

Construction:
1. Construct a rectangle $OACB$ with sides $OA = a$ and $OB = b$; prolong OA to both sides and OB upwards; bisect OA and OB in D and E, respectively; draw DF perpendicular to OA; draw CE intersecting AO prolonged in G and making $GO = OA = a$.
2. Take H on DF such that $AH = OE$ $(= \frac{1}{2}b)$; draw GH; draw AI parallel to GH.

[28] [Eutocius CommSphrCyl] pp. 615–620.

3. By "neusis" draw a line HJK through H, intersecting AI and OA prolonged in J and K, respectively, and such that $JK = OE$ $(= \frac{1}{2}b)$.
4. Draw and prolong KC, it intersects OB prolonged in L.
5. $x = HJ$ and $y = AK$ are the required proportionals.
6. Moreover, $LB = x$, so Nicomedes' construction yields the same configurations as Heron's.
[**Proof**: In the right triangles DHA and DHK we have $DH^2 = (\frac{1}{2}b)^2 - (\frac{1}{2}a)^2 = (\frac{1}{2}b + x)^2 - (\frac{1}{2}a + y)^2$, whence $ay + y^2 = bx + x^2$, or $(a + y) : (b + x) = x : y$ $(*)$. The triangles AJK and GHK are similar, so $JK : AK = HJ : GA$, whence $(\frac{1}{2}b) : y = x : 2a$, and $a : x = y : b$ $(**)$. From this proportionality it follows further that $a : y = x : b$, so $(a + y) : y = (x + b) : b$ and $(a + y) : (x + b) = y : b$ $(***)$. From $(*)$ and $(***)$ follows $x : y = y : b$, and together with $(**)$ this yields $a : x = x : y = y : b$. Finally, by step (4) of Nicomedes' construction $a : BL = y : b$, whence $BL = x$.]

Step **3** of the construction cannot be performed by straight lines and circles.

2.5 The status of the constructions

Mechanical nature Eutocius hardly discussed the question whether the constructions of two mean proportionals were geometrically acceptable; as far as he expressed preferences, they concerned the practical ease of the constructions, not their theoretical exactness. The general attitude of sixteenth-century writers to the problem seems to have been the same. Most of them stated explicitly that no truly geometrical solution had yet been found. The only explicit argument against the available constructions was that they involved the use of instruments and were therefore "mechanical." Ramus, for instance, wrote about Heron's construction:

> Until now no way has been found by which this mechanical procedure could be performed on the basis of certain geometrical principles which would ensure that the equality of the distances was caught right-on by direct action.[29]

In the 1615 edition of Archimedes' works, which included Eutocius' commentary, the editor, Rivault, surveyed the list of constructions of two mean proportionals and aptly remarked that

> nobody left this stone unturned,

but that all solutions used instruments and were therefore mechanical,

> so that this problem has remained unsolvable until the present day, which is one among many reasons why Geometry is defective and

[29] [Ramus 1569] p. 95: "et hoc mechanicum nullo principio geometriae certo adhuc inventum est, ut protinus actione prima aequidistantia deprehendatur." Similar statements can be found e.g. in [Benedetti 1599] p. 353, [Bombelli 1579] pp. 47–48, [Nonius 1546] pp. 1 and 8, [Stevin 1583] p. 1.

> imperfect. It is indeed not certain that anybody ever will provide this supplement to the art.[30]

Despite their insistence on the ungeometrical nature of the mechanical means and the procedures of trial and adjustment that were involved in the constructions of two mean proportionals, mathematicians apparently felt little reluctance in accepting and presenting them; indeed the problem of two mean proportionals engendered less controversy than the quadrature of the circle. And, as in the case of the circle quadrature, the problem of two mean proportionals did not induce mathematicians to formulate positive criteria for truly geometrical procedures.

Impossibility not proved

Besides the legitimacy of the various constructions there was of course the question whether it was really impossible to square the circle or to determine two mean proportionals by means of circles and straight lines only. Many mathematicians wrote that truly geometrical constructions were not yet found. But it is unclear whether they meant constructions by straight lines and circles or by other means that were to be considered geometrical. There was, as far as I could ascertain, no awareness that the possibility or impossibility might be or ought to be proved by a formal demonstration. Still the informed opinion among mathematicians seems to have been that it was useless to try to find a construction of two mean proportionals according to strict geometrical requirements — whatever these were.

Mesolabum

But, as there were circle squarers there were also those who did think they had found the true constructions of two mean proportionals. One of these was Salignac who in his *Exposition of the mesolabe*[31] claimed that he had found a solution.[32] His book revealed that he did not understand the question; he only showed that in the configuration of Heron's construction it may indeed happen that the ruler passes through the vertex of the rectangle, and that if so, the construction does give the required two mean proportionals. Salignac added a long discussion about the provability of theorems,[33] which only helped to increase the confusion. His work, however, was of some importance because apparently it was the first of a series of works, often featuring the word "Mesolabum"[34] in their titles, about the duplication of the cube and the construction of two mean proportionals. Some of these books showed ignorance similar to Salignac's

[30] [Archimedes 1615] p. 100: "nemo enim hunc lapidem non movit"; "ita ut insolubile huc usque manserit hoc problema, quo Geometria ut multis aliis manca et imperfecta est. Incertum an unquam quisquam arti supplementum istud addiderit."

[31] [Salignac 1574] pp. 9–16.

[32] Another was Fine who based his quadrature of the circle ([Fine 1544]) on a construction of two mean proportionals by circles and straight lines. Both the quadrature and the construction were wrong, as Nonius ([Nonius 1546]) and Buteo ([Buteo 1559]) took great care to explain.

[33] [Salignac 1574] pp. 16–19.

[34] The term "mesolabum" was taken from Eutocius' report on Eratosthenes' construction of two mean proportionals ([Eutocius CommSphrCyl] pp. 609–615), where the instrument devised by Eratosthenes for determining any number of mean proportionals was called ⟨mesolabos⟩, "taker of means"; cf. [Knorr 1986] pp. 17, 210–218.

(for instance [Scaliger 1594b]), but others, such as [Viète 1593] and [Sluse 1659] were serious mathematical studies.

2.6 Conclusion

No explicit criterion for exactness

The examples given above make clear that several sixteenth-century mathematicians showed concern about geometrical exactness, criticizing procedures and proofs for the quadrature of the circle and questioning the geometrical status of constructions beyond the Euclidean canon, in particular those for two mean proportionals. But there was little clarity about the distinctions between questions of proof, construction, and existence. We find no explicit formulation of criteria for acceptability of geometrical procedures, nor an insistence that geometry should restrict itself to constructions by circles and straight lines.

Thus the debates on the exactness of constructions were opaque and they remained inconclusive. That situation changed with the publication in 1588 of Pappus' *Collection*.

Chapter 3

1588: Pappus' "Collection"

3.1 Introduction

Themes

Commandino's Latin translation of Pappus' *Collection* became available in print in 1588.[1] Pappus' text, composed in the early fourth century AD, provided clear-cut statements on the aim and rules of geometrical problem solving, many examples of constructions, and suggestive information about analytical methods for finding the solutions of problems. At the same time Pappus' practice of problem solving was often inconsistent with the rules and aims he professed. As a result the influence of the book on the subsequent ideas about construction was strong but undogmatic.

After c. 1590, largely under the influence of Pappus' text, geometrical problem solving became a recognizable, well-defined subfield of geometry with a shared understanding of its first aims and principal methods. And, although at first there was no *communis opinio* on the legitimacy of means of construction beyond straight lines and circles, the discussions on this issue were much more clear and objective than they had been before. Three themes were of particular importance in this process of clarification: Pappus' classification of problems, the use of curves in constructions, and neusis constructions.

3.2 The classification of problems

Plane, solid and line-like problems

In two famous passages Pappus classified geometrical problems as either "plane," or "solid," or "line-like". Plane problems were those that could be constructed by circles and straight lines. Solid problems were non-plane problems that could be constructed by straight lines, circles, and conic sections. If

[1] [Pappus 1588]; there were two re-issues: [Pappus 1589] and [Pappus 1602]. The 1588 edition was posthumous; Commandino had died in 1575. By that time a considerable number of manuscript copies of the Greek text circulated among humanists and mathematicians, cf. [Pappus 1986] pp. 62–63, [Treweek 1957], and [Passalaqua 1994]. In 1569, for instance, Ramus referred to the treatment of two mean proportionals in book III of the *Collection* ([Ramus 1569b] p. 95); cf. also Note 4 of Chapter 9.

other curves than these had to be used in a problem's construction, it was called line-like.[2] The two passages[3] were almost literally the same. I quote the first one, keeping my translation close to Commandino's Latin version:

> The ancients stated that there are three kinds of geometrical problems, and that some of them are called plane, others solid, and others line-like; and those that can be solved by straight lines and the circumference of a circle are rightly called plane because the lines by means of which these problems are solved have their origin in the plane. But such problems that must be solved by assuming one or more conic sections in the construction, are called solid because for their construction it is necessary to use the surfaces of solid figures, namely cones. There remains a third kind that is called line-like. For in their construction other lines than the ones just mentioned are assumed, having an inconstant and changeable origin, such as spirals, and the curves that the Greeks call ⟨tetragonizousas⟩, and which we can call "quadrantes,"[4] and conchoids, and cissoids, which have many amazing properties.[5]

The classification evidently presupposed that all problems should be constructed by the intersection of curves, rather than, for instance, by instruments or by shifting rulers. We will see that Pappus himself was nevertheless quite interested in these alternative methods of construction.

Solid problems The first class of problems, the plane ones, comprised the common problems constructible by Euclidean means. The best known example of a solution of a solid problem by the intersection of conics was the construction of two mean proportionals that Eutocius attributed to Menaechmus. It was as follows:

[2]Commandino used the term "linearis" in translating the Greek "⟨grammike⟩"; I will use "line-like" rather than "linear" because in modern mathematics the latter term suggests straight lines and first-degree equations, which is the opposite of what Pappus meant.

[3][Pappus Collection] III, § 7, pp. 38–39 (introduction to Prop. 5), and IV, § 36, pp. 206–208 (introduction to Props 31–34).

[4]The term ⟨tetragonizousa⟩ means "square making" and indicates that the curve was important for a quadrature. Indeed, Pappus used the same term for the curve used in a quadrature of the circle, see below, Definition 3.3; Commandino translated this as "linea quadrans" ([Pappus 1660], p. 88). It seems that Clavius was the first to call this curve "quadratrix," cf. Section 9.2.

[5][Pappus Collection] III (§ 7, pp. 38–39), in Commandino' translation ([Pappus 1660] p. 7): "Problematum geometricorum antiqui tria genera esse statuerunt, et eorum alia quidem plana appellari, alia solida, alia linearia, que igitur per rectas lineas et circuli circumferentiam solvi possunt, merito plana dicuntur; etenim lineae, per quas eiusmodi problemata solvuntur, in plano ortum habent. Problemata vero quaecumque solvantur, assumpta in constructionem aliqua coni sectione, vel pluribus, solida appellantur namque ad constructionem necesse est solidarum figurarum superficiebus, nimirum conis, uti. Restat tertium genus, quod lineare appellatur. Lineae enim aliae praeter iam dictas in constructionem assumuntur, varium, et transmutabilem ortum habentes, quales sunt helices, et quas graeci ⟨tetragonizousas⟩ appellant, nos quadrantes dicere possumus, conchordes [sic; read: conchoides] et cissoides, quibus quidem multa, et admirabilia accidunt."

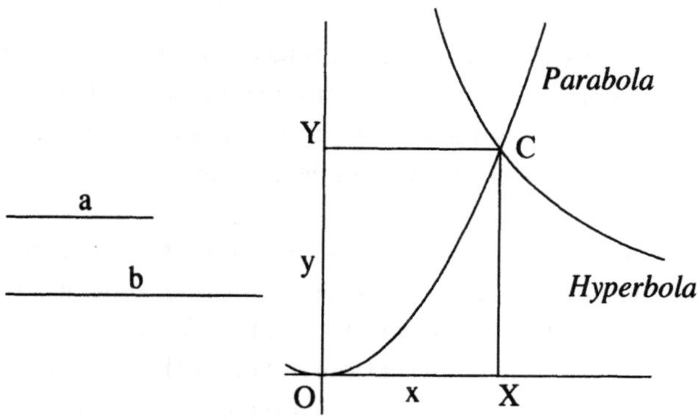

Figure 3.1: Menaechmus' construction of two mean proportionals

Construction 3.1 (Two mean proportionals — Menaechmus)[6]
Given: two line segments a and b (see Figure 3.1); it is required to find their two mean proportionals x and y.

Construction:
1. Mark a quadrant by a vertical and a horizontal half-axis through O. Draw a parabola with vertical axis, vertex O, and *latus rectum*[7] a.
2. Draw in the quadrant a (single branch of a) hyperbola, which has the two axes as asymptotes and whose abscissae and ordinates form rectangles equal to rect(a, b).
3. The two curves intersect in C; draw perpendiculars CX and CY through C to the axes, with X and Y on the horizontal and vertical axes, respectively.
4. Now $x = OX$ and $y = OY$ are the required two mean proportionals.

[6] [Eutocius CommSphrCyl] p. 603–605.

[7] *Latus rectum* and *latus transversum* are the Latin terms for certain line segments occurring in the defining properties of conic sections. If the vertex of the conic section is taken as origin and the X- and Y-direction are taken along the axis of the conic and perpendicular to the axis, respectively, then the *latus rectum* a and the *latus transversum* b occur in the analytical equations for the conics in the following way: $y^2 = ax$ (parabola), $y^2 = ax - \frac{a}{b}x^2$ (ellipse), and $y^2 = ax + \frac{a}{b}x^2$ (hyperbola). For an excellent explanation of Apollonius' definition of these line segments see [Hogendijk 1991] pp. 4–12.

[**Proof**: Because C is on the parabola with *latus rectum* a, we have $ay = x^2$; and because C is on the hyperbola we have $xy = ab$. Hence $a : x = x : y = y : b$.]

As the example shows, construction by the intersection of conics presupposed that, given their parameters (vertex, axes, *latus rectum*, *latus transversum*, or equivalent data) these curves could somehow be posited in the plane, and that thereby their points of intersection became known. I return to this presupposition in Section 3.5 below.

Only a few other constructions by conics were known from classical sources before c. 1590.[8] The method of constructing used in these solutions apparently made very little impression in the sixteenth century. As we will see, Pappus provided more examples of such constructions. Yet, although the possibility of constructing by the intersection of conics was recognized as important after 1590, the method was hardly ever used before Descartes and Fermat did so in the 1620s and 1630s.

Line-like problems

The line-like problems that attracted most interest after 1590 were those which made use of the spiral and the quadratrix. These curves were defined by specifying a procedure for generating them. I give the definitions and one example of a construction by means of the quadratrix.

Pappus devoted several sections of his fourth book to Archimedes' theory of the spirals.[9] In his book on spirals Archimedes defined the curve as follows:

Construction–Definition 3.2 (Spiral — Archimedes)[10]
Given: a point O in the plane (see Figure 3.2); a curve, starting in O, is traced, called the Archimedean spiral.

Construction:
1. Let a line L through O rotate one full turn at uniform speed; at the same time let a point A move uniformly along L starting in O.
2. The curve traced by the point A is the spiral.

Pappus assumed the definition of the spiral to be known. He was more explicit about the quadratrix. He mentioned that the curve was used by Dinostratos and Nicomedes.[11] He gave its definition in Book IV:

Construction–Definition 3.3 (Quadratrix — Pappus)[12]
Given a square $OBCA$ (see Figure 3.3) and a quadrant OBA, a curve is traced, called the quadratrix.

---

[8] Eutocius inserted in his list after the one by Menaechmus discussed above, a construction by means of two parabolas; this construction is usually also ascribed to Menaechmus. Eutocius also gave some constructions by conics in his commentary at *Sphere and Cylinder* II-4 [Eutocius CommSphrCyl] pp. 626–666.

[9] [Pappus Collection] IV-19-22 (§§ 21–25), pp. 177–185.

[10] [Archimedes Spirals] Def. 1; p. 165.

[11] The curve is often called "quadratrix of Dinostratos," but a more likely originator of the curve is Hippias.

[12] [Pappus Collection] IV-25-26 (§ 30), pp. 191–192.

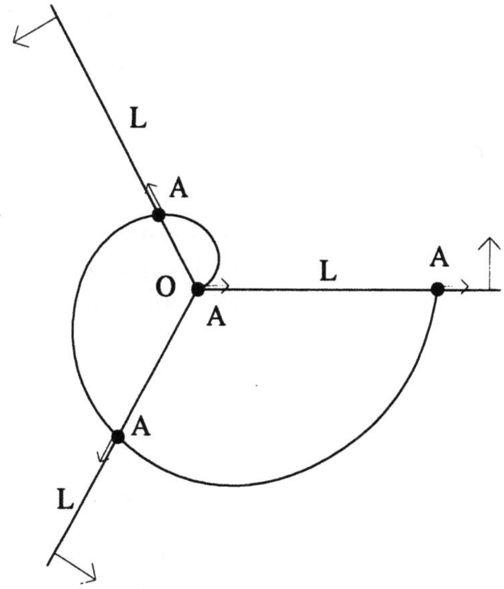

Figure 3.2: Definition of the Archimedean spiral

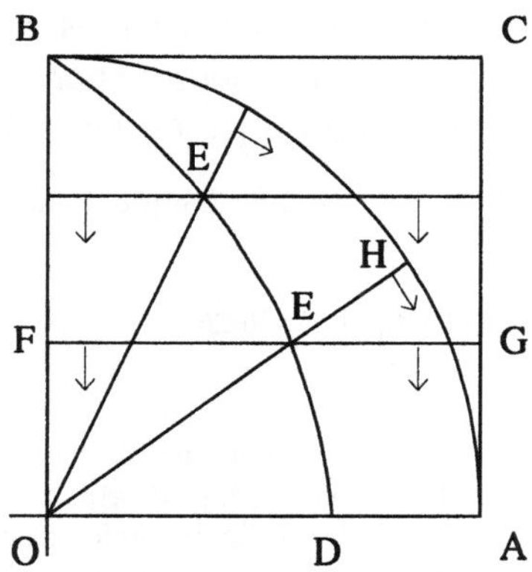

Figure 3.3: Definition of the quadratrix

Construction:
1. Let a line segment turn uniformly around O from position OB to position OA (its endpoint describes the arc BA); let during the same time interval another line segment move uniformly from position BC to position OA, keeping parallel to OA.
2. During the motion the point of intersection E of the two line segments traces the quadratrix $BEED$.

An immediate consequence of the definition, noted by Pappus, is that for any point E on the quadratrix, with corresponding positions FG of the moving horizontal line segment and OH of the moving radius, the following proportionality holds:

$$\text{arc}BH : \text{arc}BA = BF : BO\,. \tag{3.1}$$

Pappus then proved[13] that

$$OD : OA = OA : \text{arc}BA\,. \tag{3.2}$$

If a quadratrix was given, the lengths OD and OA were given too, and the proportionality of Equation 3.2 implied that arcBA could be determined. Thereby the circle was rectified and hence the circle could be squared as well, using Archimedes' proposition that the area of a circle is equal to half the area of the rectangle formed by its radius and its circumference. This application gave the curve its name.[14]

In the construction of the quadratrix it was assumed that the two uniform motions could be regulated such that the one line turned through the quadrant in exactly the same time as the other moved along the given square. Pappus reported objections to this assumption such as the one from Sporus, who noticed that in order to do so one should regulate the speeds in the ratio of the arc BA and the radius OB, hence that ratio should be known beforehand. But squaring the circle meant finding precisely that ratio; hence, Sporus objected, using the curve in solving that problem would imply a *petitio principii*.[15]

---

[13] [Pappus Collection] IV-26, §§ 31–32, pp. 194–196; the proof employed a *reductio ad absurdum* argument.

[14] Cf. Note 4 above.

[15] The relevant passage ([Pappus Collection] IV-25-26 (§ 31), p. 193) was translated by Commandino as follows ([Pappus 1660] p. 88):

> Hae autem linea spero [sic; read: Sporo] iure ac merito non satisfacit propter haec. Primum enim ad quod videtur utilis esse, hoc in suppositione assumit, quomodo, inquit, fieri potest, ut duo puncta ab ipso B principium motus capientia; hoc quidem in recta linea ad A, illud, vero in circumferentia ad D in aequali tempore simul restituantur, nisi prius proportio rectae lineae AB ad circumferentiam BED cognita sit. In hac enim proportione, et motuum velocitates sint, necesse est. Nam quo pacto arbitrantur ea simul restitui. Velocitatibus temere, et nulla ratione utentia? nisi forte quispiam dicat hoc casu evenire, quod est absurdum.

Few modern historians of mathematics have commented upon this objection. Heath accepts it ([Heath 1921] vol 1 p. 230); Van der Waerden considers it "only partially justified" ([Waerden 1961] p. 192) because, he claims, for practical purposes the quadratrix can be con-

Pappus also mentioned another objection. In using the quadratrix for squaring the circle by means of the proportionality of Equation 3.2, it was assumed that the intersection D of the quadratrix and the base was given. But that point was not covered by the procedure described in the definition. Indeed at the end of the procedure, the two moving lines coincided and so their intersection, which should be D, was not defined.[16]

Pappus qualified this construction of the quadratrix as "rather mechanical,"[17] but added that the curve could be generated in a more geometrical manner by the intersection of surface loci. He then described two ways in which the quadratrix could be considered as resulting from the intersection of surfaces.[18]

A construction using the quadratrix

A good example of the use of the quadratrix in constructing line-like problems is Pappus' solution of the problem to divide a given angle in a given ratio. As this construction could be used for any angular section problem (the trisection, for instance, by taking the given ratio to be 1 : 2) I refer to it as the "general angular section." It was as follows:

Construction 3.4 (General angular section — Pappus)[19]
Given: an (acute) angle φ and a ratio ρ (see Figure 3.4); it is required to divide φ in two angles φ_1 and φ_2 such that $\varphi_1 : \varphi_2 = \rho$.

Construction:
1. Draw a quadratrix BA with pertaining square $OBCA$ and BA.
2. Draw OH (with H on arcBA) such that $\angle HOA$ is equal to the given angle φ.
3. Mark the intersection E of OH and the quadratrix.
4. Draw $FEG \parallel OA$ with F and G on BO and CA, respectively.
5. Divide FO in F' such that $FF' : F'O = \rho$.

---

structed pointwise. Remarkably, it seems not to have been noticed that, contrary to what Sporus is reported to have objected, it is not necessary to pre-install a special ratio of velocities to draw a quadratrix. The ratio of the arc BA and the radius OB arises only because the square in which the quadratrix is to be drawn is supposed as given. For the curve's later use in constructions this supposition is unnecessary. One may define the quadratrix as the curve traced by the intersection of two lines, the one turning (cf. Figure 3.3) counterclockwise around O starting from position OA, the other moving parallel to itself upward, also starting from position OA, both motions being uniform. No preliminary supposition about the ratio of the motions need be made; once the curve is traced one can determine the pertaining square by determining the intersection B of the quadratrix with the perpendicular to OA through O, and completing the square. Note that in the case of the spiral (cf. Definition 3.2) the tracing point also moves from the center outward and no presupposition is made about the ratio of the speeds. No criticism similar to Sporus' was raised against the spiral.

[16]This objection remained valid in the case of the alternative tracing procedure for the quadratrix mentioned in Note 15; at the beginning of the motion the point of intersection of L and M is undetermined because the two lines coincide. Thus the second objection precluded the use of the curve in squaring the circle but not in performing angular sections by means of Construction 3.4 below.

[17][Pappus Collection] IV (§§ 31, 33), pp. 194, 197.

[18][Pappus Collection] IV-28, 29 (§§ 33–34), pp. 197–201; for further details on these stereometric constructions see [Knorr 1986] p. 129.

[19][Pappus Collection] IV-35 (§ 45) pp. 222–223.

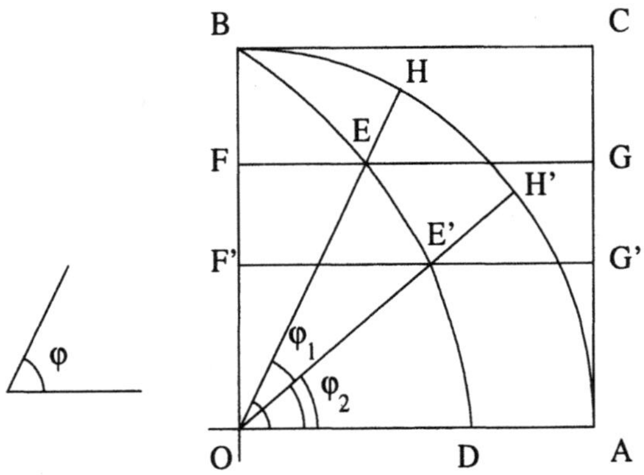

Figure 3.4: General angular section by means of the quadratrix — Pappus

> **6.** Draw $F'G' \parallel OA$ (with G' on CA) and mark its intersection E' with the quadratrix.
> **7.** Draw $OE'H'$ (with H' on arcBA); call $\angle HOH' = \varphi_1$ and $\angle H'OA = \varphi_2$.
> **8.** OH' divides $\angle HOA$ in the required manner, i.e. $\varphi_1 : \varphi_2 = \rho$.
> [**Proof**: The property of the quadratrix represented in Equation 3.1 implies that $\varphi_1 : \varphi_2 = \text{arc}HH' : \text{arc}H'A = FF' : F'O = \rho$.]

In his next proposition Pappus used the spiral to construct the same problem. In fact, although the name of the quadratrix refers to its use in squaring the circle, it seems likely that the curve, and similarly the spiral, were conceived precisely to solve angular section problems. In both curves the two motions are combined in such a way that the division of the angle is related to the division of the line, and this seems to be the only rationale to devise such a combination of motions.

A construction using the cissoid

An example of a line-like construction using one of the other curves mentioned by Pappus was the construction of two mean proportionals by means of the cissoid ascribed to Diocles. It occurred in Eutocius' list of constructions of that problem. The construction was based on a property of the configuration of lines in a semicircle illustrated in Figure 3.5. Let OAB be a semicircle with radius $OD = d$; let E and F be points on the diameter equidistant from D and let EG and FH be the corresponding ordinates with H and G on the semicircle; the

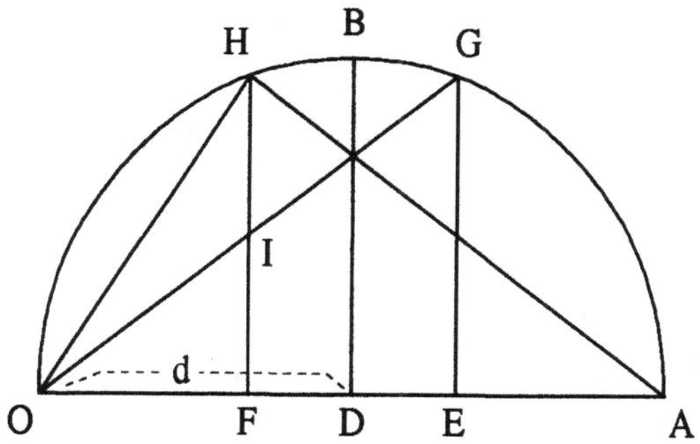

Figure 3.5: A property relevant to Diocles' cissoid

straight line OG intersects HF in I. Then

$$FI : FO = FO : FH = FH : FA\,; \tag{3.3}$$

in other words, FO and FH are the two mean proportionals between FI and FA. [**Proof**: Because $FD = DE$, the triangles FIO, FOH, and FHA are similar, from which the proportionality follows.]

The cissoid collects as it were all possible proportionalities 3.3 that occur within the semicircle in the way described above. Its construction, as explained by Eutocius, is as follows:

Construction–Definition 3.5 (Cissoid — Diocles)[20]
Given: a semicircle OAB with radius $OD = DA = d$ and vertex B (see Figure 3.6); a curve, called the cissoid, is constructed pointwise.

Construction:
1. Choose arbitrarily a point H on arc OB; draw a line through H perpendicular to OA intersecting the base in F.
2. Take G on arc BA such that $HB = BG$; draw GO, it intersects FH in I.
3. Proceed as in **1–2** starting with other points H_1, $H_2, \cdots$ on arc OB.
4. The points I, I_1, $I_2, \cdots$ thus found lie on a curve from O to B;

[20] [Eutocius CommSphrCyl] pp. 595–597.

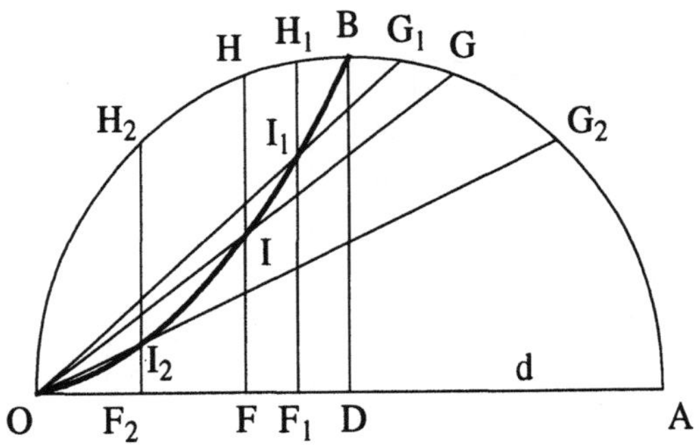

Figure 3.6: Construction of the cissoid — Diocles

this curve is the cissoid.
5. After having constructed sufficient points I_i, connect them by lines to get the curve itself.

I return below (Section 3.5) to the pointwise method of constructing the curve.

It followed from the definition and from the property in Equation 3.3 that a given cissoid could be used for constructing two mean proportionals. The construction, as explained by Eutocius, proceeded as follows:

Construction 3.6 (Two mean proportionals — Diocles)[21]
Given: two line segments a and b ($a < b$) (see Figure 3.7); it is required to find their two mean proportionals x and y.

Construction:
1. Draw a cissoid OB within a semicircle OAB with radius $OD = DA = d$.
2. Determine a line segment c such that $a : b = c : d$; take C on DB with $DC = c$.
3. Draw and prolong AC; it intersects the cissoid in I; draw FH through I perpendicular to OA with F on the base OA and H on the semicircle.
Note that by the defining property of the cissoid we now have four line segments e, u, v, f in continued proportion, namely, $e = FI$,

[21] [Eutocius CommSphrCyl] pp. 595–597.

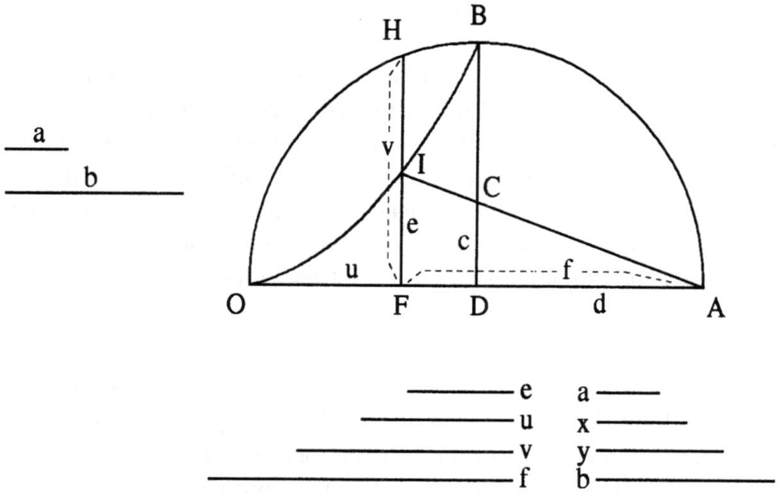

Figure 3.7: Diocles' construction of two mean proportionals by means of the cissoid

$u = FO$, $v = FH$ and $f = FA$. Moreover, because of similarity, $e : f = c : d = a : b$; so we have found two mean proportionals (u and v) between two line segments (e and f), which have the same ratio as a and b. Here Eutocius stopped his explanation of the construction; evidently he considered the further construction of x and y from a, b, e, u, v, f obvious. It could be done as follows:
4. Determine line segments x and y such that $e : a = u : x = v : y \ (= f : b)$.
5. The line segments x and y are the two required mean proportionals of a and b.
Note that the constructions in **2** and **4** are standard plane constructions (of fourth proportionals, *Elements* VI-12, cf. Construction 4.1).
[**Proof**: See note at **3**; the proportionality of a, x, y, b is proved by composition of ratios.]

This was the only construction by means of the cissoid known from classical sources, and Pappus probably had this construction in mind when he referred to the use of the cissoid in solving line-like problems. Yet the problem itself, as Pappus well knew, was solid because it could be constructed by conic sections. On the implication of this discrepancy see Section 3.4 below.

3.3 Terminology and context of the classification

Terminology

In the classification given by Pappus, problems were grouped according to the nature of the curves needed in their construction. The names of the classes also related to these curves: circle and straight line had their origin in the plane, hence problems constructible by them were called plane; conics had their origin in solid figures (namely, the cones from whose section with a plane they arose), hence problems constructible by conics were called solid; the remaining problems were called line-like because they were constructed by means of more intricate lines.

Context

The two passages in which Pappus explained the classification occurred in Book III and Book IV of the *Collection*.[22] The first related to the problem of two mean proportionals, the second to trisection. In both cases Pappus gave the classification of problems in order to explain the approaches of earlier geometers to these problems. These geometers, he said, first tried to solve the problems by plane means but did not succeed because the problems were solid. In Book III he reported that the geometers could not construct two mean proportionals in a geometrical way because it was not easy to draw conics in a plane; they therefore devised special instruments for finding mean proportionals and in that way succeeded admirably well. Pappus then discussed[23] instrumental constructions of mean proportionals by Eratosthenes, Nicomedes, Hero, and one by himself (all four also occur in Eutocius' list[24]). Those of Nicomedes and Hero were given in Section 2.4 above. The construction by Eratosthenes[25] employed a rather complicated instrument[26] which, by trial, readjustment and trial again, yielded any number of mean proportionals between two given line segments; Pappus' own construction was of the neusis type (cf. Problem 2.4, I explain this type of construction in Section 3.6). In Book IV Pappus explained that in the case of the trisection the ancient geometers were frustrated because they did not know about the conics. Later they succeeded in trisecting the angle by means of conics, based on a solution by means of a neusis (cf. Constructions 3.8 and 3.9 below).

According to Pappus, then, earlier mathematicians used special instruments or neusis constructions because they did not know the proper geometrical construction by conics, or because they found such a construction too difficult.

3.4 The precept

"A considerable sin"

If Pappus' text had only provided the classification of problems, its effect on

[22]Cf. Note 3 above.
[23][Pappus Collection] III-5 (§§ 7–10), pp. 40–50.
[24][Eutocius CommSphrCyl] pp. 588–620.
[25][Eutocius CommSphrCyl] pp. 609–615.
[26]The "mesolabum," cf. Note 34 of Chapter 2.

early modern mathematics would probably have been restricted to terminology and perhaps some increased interest in construction by means of curves. The reason why it had much more impact was that Pappus combined the classification with a strongly worded methodological precept: Problems should be constructed with the means appropriate to their class. It was not allowed in geometry to construct plane problems by solid or line-like means, nor to construct solid problems by line-like means. And it was unwise to attempt constructing solid problems by plane means or line-like problems by plane or solid means. Pappus formulated this precept only in Book IV. Commandino used strong words in the translation of the passage, notably the term "sin" ("peccatum"):

> Among geometers it is in a way considered to be a considerable sin when somebody finds a plane problem by conics or line-like curves and when, to put it briefly, the solution of the problem is of an inappropriate kind.[27]

In this connection Pappus referred to two examples of inappropriate constructions of problems, namely, "the problem in the case of the parabola in the fifth book of the conics of Apollonius and in the book on spirals an assumed solid neusis with respect to a circle."[28] Pappus gave no further reference but it is generally accepted[29] that he referred to Proposition 51 of Book V of Apollonius' *Conics* and Proposition 18 of Archimedes' *On spirals.* The Apollonian proposition concerned the construction of a normal to a parabola through a given point outside the parabola. Apollonius' construction employed the intersection of a hyperbola and the given parabola. The construction could also be performed by intersecting the given parabola with a circle. If the parabola was indeed considered as given, the latter construction used plane means only and was therefore, according to Pappus' precept, preferable over the one Apollonius gave. Probably this was Pappus' reason for criticizing Apollonius' result. Pappus' remark gave the problem of the perpendicular to the parabola a certain fame in the early modern period; I return to it in Section 4.10 below. Pappus' second example, the use of neusis in Proposition 18 of Archimedes' *On spirals*, concerned a more complicated matter which, it seems, was not taken up in the early modern period.[30]

Implications

Pappus' stern words made a strong impression; mathematicians quoted the passage on the "sin" or the "error" of geometers often in discussions and

[27][Pappus Collection] IV-30, (§ 36) p. 208; in Commandino' translation ([Pappus 1660] p. 95): "Videtur autem quodammodo peccatum non parum esse apud Geometras, cum problema planum per conica, vel linearia ab aliquo invenitur, et ut summatum [summatim?] dicam, cum ex improprio solvitur genere . . .."

[28][Pappus Collection] IV-30 (§ 36) p. 208, in Commandino' translation ([Pappus 1660] p. 95): "... in quinto libro conicorum Apollonii problema in parabola: et in libro de lineis spriralibus: assumpta solida inclinatio in circulo."

[29]Cf. e.g. [Zeuthen 1886] pp. 284–288, [Apollonius 1961] pp. cxxvii–cxxix, [Pappus 1986] pp. 529–530.

[30]Cf. [Knorr 1986] pp. 176–178.

polemics.[31] The precept could indeed be read as a strict directive. Understood in that sense it implied that the only legitimately geometrical constructions were those that employed the intersection of straight lines and curves — thus excluding the use of instruments or shifting rulers. Moreover, the only examples of line-like problems in the *Collection* were constructed by means of the spiral or the quadratrix, and Pappus reported reservations about their generation by motion. Thus the impression was given that beyond solid problems the extant construction procedures were somehow suspect.

Yet the practice of problem solving in the *Collection* implied a much more lenient attitude toward the legitimacy of constructions. Pappus himself freely used neusis, instruments, and intricate curves; indeed, he was quite interested in them. He gave many constructions by means of the quadratrix, and he discussed several instrumental constructions of two mean proportionals (a problem that, according to his precept, should be constructed by conics), even adding a neusis construction of his own. Moreover, Pappus did not report any explicit arguments in support of a preference for construction by means of curves as codified in the classification and the precept. In fact, he made clear that in practice construction by curves was not easy. He mentioned that earlier geometers had found it difficult to trace conic sections, and he did not give methods to do so himself. From his remarks on the construction of two mean proportionals it appeared that, as to practicability, he preferred instruments and neusis constructions over the use of conic sections.

Thus the stern formulation and the implications of the precept were mitigated by the practice throughout the *Collection*. The result was, as we will see in Chapters 9–13, that the precept had a marked but variegated influence on the later conceptions of geometrical construction. The issue was taken seriously, but mathematicians did not infer one uniform directive from Pappus' writings.

3.5 The constructing curves

How to construct the means of construction?

The doctrine about geometrical construction implied in Pappus' classification and precept left one obvious question unanswered. If geometers should construct by means of straight lines, circles, conics, or more intricate curves, how should they construct these lines and curves themselves? In other words: how to construct the means of construction? This question was to play a crucial role in the early modern discussions on construction. Pappus did not explicitly provide an answer, but early modern geometers could gather some implicit answers from

[31] The theme will frequently recur below; here three characteristic examples may suffice. Descartes wrote that it would be "an error in geometry" to construct with curves of a too high degree, as much as it was an error "to try in vain to construct some problem by a simpler kind of curves than the nature of the problem allows" ([Descartes 1637], p. 371). Fermat wrote: "for it has been often declared already, by Pappus and by more recent mathematicians, that it is a considerable error in geometry to solve a problem by means that are not proper to it" ([Fermat DissTrip] p. 121). Jakob Bernoulli wrote as late as 1688: "I can see nothing that could in this case acquit Descartes from the vice of acting ungeometrically which he mentions so often" ([Bernoulli 1688], p. 349).

Pappus' practice in the *Collection*.

In constructing plane problems Pappus used straight lines and circles without further comment; apparently he considered the first three postulates of the Euclidean *Elements* sufficient foundation for this practice.

Construction of conics

Pappus accepted the constructions of conics from Book I of Apollonius' *Conics*.[32] These constructions were indeed "solid" in the sense that they referred to solids in space. To construct a conic in a given plane Apollonius located the top and the base circle of a cone in the space surrounding the plane. The intersection of this cone with the plane then produced the required conic. Apollonius did not base this constructional practice on explicitly formulated postulates.

Proposition 33 of Book IV of the *Collection*[33] provides an interesting example of how Pappus dealt with the Apollonian constructions of conics. At that point in his text Pappus had shown (Propositions 31–32) that the trisection and the neusis problem (with respect to two straight lines, cf. Section 3.6) could be constructed by the intersection of a circle and a hyperbola with given asymptotes and passing through a given point (I discuss these constructions in detail below, Constructions 3.8 and 3.9). Now the construction of a hyperbola with given asymptotes and passing through a given point was not among the ones given by Apollonius in the first book of the *Conics*, but it did occur in Book II as Proposition 4, in which Apollonius reduced it to one of the constructions in book I. Pappus, however, did not refer to this construction but provided one himself, which involved a different reduction to one of the constructions of book I. This procedure suggests that Pappus considered the constructions from *Conics* I as postulates, but felt that in the case of a construction which Apollonius himself had reduced to one of these basic constructions, he (Pappus) could provide an alternative reduction.

Higher-order curves

As we have seen, Pappus mentioned the cissoid and the conchoid as curves used in constructing line-like problems.[34] In Section 3.2 above (Construction 3.6) I gave an example of the cissoid's use (by Diocles, as reported by Eutocius), in which the curve itself was constructed "pointwise"; Pappus himself did not discuss constructions by means of the cissoid.

Pappus did mention Nicomedes' instrumental generation of the conchoid and he explained how the curve could be used to perform a neusis.[35] He referred

[32]These constructions are: I-52: parabola with given vertex, axis, parameter, and ordinate angle of 90°; I-53: the same but with arbitrary ordinate angle; I-54: hyperbola with given vertex, diameter, *latus rectum*, and ordinate angle of 90°; I-55: the same with arbitrary ordinate angle; I-56: ellipse with given diameter, vertex, *latus rectum*, and ordinate angle of 90°, case that the *latus rectum* is smaller than the diameter; I-57: the same in the case that the *latus rectum* is larger than the diameter; I-58: as in 56 and 57 but with arbitrary ordinate angle; I-59: two branches of a hyperbola whose diameter, *latus rectum*, vertices, and ordinate angle are given; I-60: four branches of two opposite hyperbolas whose vertices and diameters are given.

[33][Pappus Collection] IV-33 (§§ 41–42) pp. 214–217.

[34][Pappus Collection] IV-30 (§ 36), pp. 206–208.

[35][Pappus Collection] IV-23 (§§ 26–28) pp. 185–188.

to this use of the curve in connection with Nicomedes' construction of two mean proportionals.[36] However, in discussing the neusis construction itself he did not refer to the conchoid and provided (cf. Construction 3.8) a proper solid construction by the intersection of a circle and a hyperbola.[37] Nor did he refer to the conchoid in connection with the trisection by neusis (cf. Construction 3.9).[38]

Thus the relevant passages from Pappus suggested that if the conchoid or the cissoid were used in constructing line-like problems, they were to be constructed pointwise or by instruments. Yet the only problem Pappus explicitly mentioned as solvable by means of these curves was a solid problem, namely, the finding of two mean proportionals, which according to Pappus' precept should be constructed by conics.

I have found no classical examples of non-solid problems constructed by means of the cissoid or the conchoid (or indeed any algebraic curve with degree higher than two). Hence it seems that Pappus' mention of the cissoid and the conchoid with respect to line-like problems was only relevant in so far as their use in constructing solid problems such as the determination of two mean proportionals or the neusis was inappropriate. However, Pappus did not spell out this consequence explicitly.

Spiral and quadratrix

Thus the only curves that were actually used in the classical sources as means to solve line-like problems were the spiral and the quadratrix. We have seen that Pappus presented these curves as traced by motion and that he noted objections to the feasibility or acceptability of this way of generating the curves.

The construction of the general angular section with the quadratrix (Construction 3.4) illustrated that, if accepted, the quadratrix was a very powerful means of construction. Bisection, trisection, or any division of the angle was possible and all were equally simple, because via the quadratrix any angle division was reduced to a division of a line segment in a given ratio. Similarly, the construction of regular polygons and squaring the circle were simple problems once the quadratrix was given. Pappus dealt with these quadratrix-based constructions extensively,[39] and it is noteworthy that he did not stipulate that in special cases (as the bisection or the trisection) one should prefer plane or solid constructions over the use of the quadratrix. We will see, however, that in the early modern period these constructions by means of the quadratrix and the spiral were considered somewhat suspect, among other things because of reported objections such as Sporus'.[40]

---

[36] [Pappus Collection] IV-24-25 (§§ 28–29) pp. 188–191, cf. Construction 2.6.

[37] [Pappus Collection] IV-31 (§§ 36–37) pp. 210–212.

[38] [Pappus Collection] IV-32 (§§ 38–40) pp. 212–214.

[39] [Pappus Collection] IV-36-41 (§§ 47–51), pp. 224–230; in particular: To cut equal arcs from unequal circles (IV-36); to construct an isosceles triangle in which the base angle has to the top angle a given ratio (IV-37); to construct regular polygons within a given circle (IV-38); to construct a circle with given circumference (IV-39); to construct on a given line segment an arc which has to the given segment a given ratio (IV-40); to divide an angle in two incommensurable angles (IV-41).

[40] Cf. Sections 9.3 and 16.5.

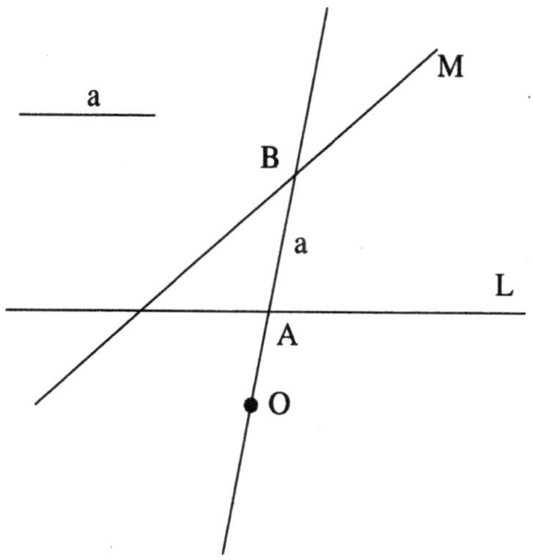

Figure 3.8: The neusis problem

3.6 Neusis constructions

The neusis problem

I now turn to the third important theme in Pappus' *Collection*: neusis constructions. As I briefly mentioned in the previous chapter, a neusis construction was one in which it was essentially assumed that the neusis problem could be solved. I repeat the formulation of the neusis problem given earlier (Problem 2.4):

Problem 3.7 (Neusis)
Given: two straight lines L and M, a point O (often referred to as the "pole" of the neusis) and a segment a (see Figure 3.8); it is required to find a line through O, intersecting L and M in A and B, respectively, and such that $AB = a$.

In particular cases (for instance when the distances of the pole O to L and M were equal) the problem could be solved by straight lines and circles, but not in general. There were variants of the problem in which one or both of the straight lines L and M were replaced by circles.

Neusis by intersection of conics

Problems unsolvable by straight lines and circles could often be solved by a neusis construction. Still, as we have seen, Pappus' precept could be understood as implying that neusis procedures did not supply properly geometrical constructions, and that geometers should try to find constructions by intersection of conics instead. Although apparently Pappus did not draw so strict a conclusion from the precept, he did explain how a neusis could be performed by

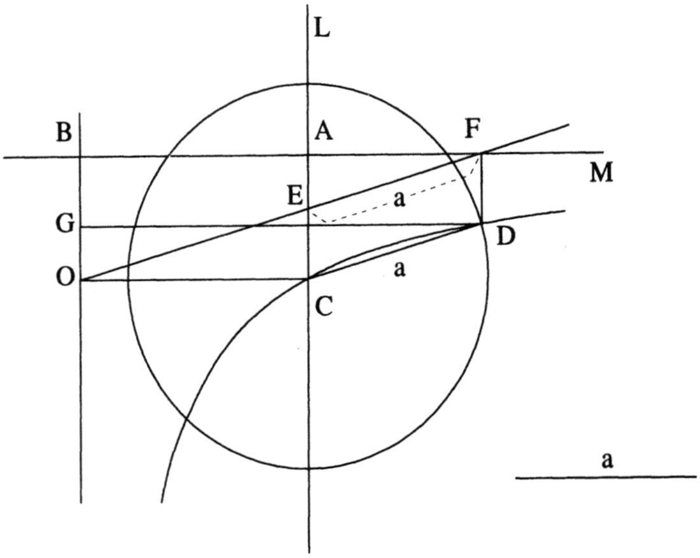

Figure 3.9: Neusis by intersection of a hyperbola and a circle — Pappus

the intersection of conics. He did so in Book IV of the *Collection*, in connection with the trisection. The relevant constructions were influential later and I present them here in the order in which they were treated in the *Collection*. Pappus started with the following:

Construction 3.8 (Neusis — Pappus)[41]
Given: two perpendicular[42] *lines L and M (see Figure 3.9), intersecting in A, a point O and a segment a; it is required to construct a line through O intersecting L and M in E and F, respectively, and such that $EF = a$.*

Construction:
1. Complete the rectangle $ABOC$; prolong BO.
2. Draw a hyperbola through C with asymptotes along BA prolonged ($= M$) and BO prolonged.[43]
3. Draw a circle with center C and radius a; it intersects the hyperbola in D.
4. Draw a line through D parallel to AC; it intersects BA prolonged in F; draw OF; OF intersects AC in E.
5. OEF is the required line; $EF = a$.
[**Proof**: Draw CD; draw GD parallel to AB. Then rect.$(FD, DG) =$

[41] [Pappus Collection] IV-31 (§§ 36–37), pp. 210–212.
[42] The construction can be easily adjusted to the case of non-perpendicular lines; Pappus, however, did not explicitly mention this.
[43] This step in the construction was criticized by Kepler, see Section 11.4.

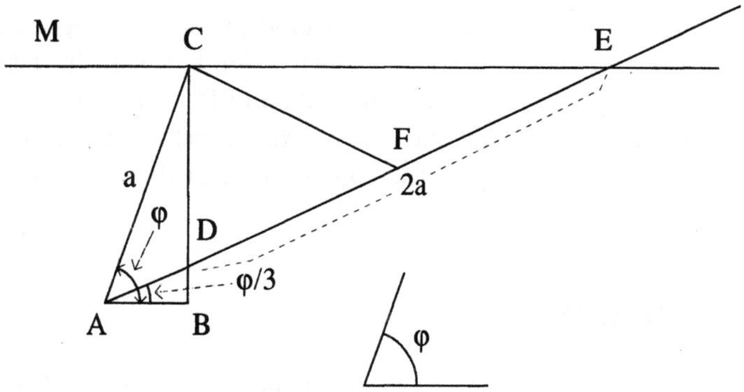

Figure 3.10: Trisection by neusis — Pappus

rect.(CA, CO) because D and C are on the hyperbola; hence rect. (BF, FD) = rect.(BA, AC), so $BF : BA = AC : FD$ ($\star$); furthermore, the triangles $\triangle BFO$ and $\triangle COE$ are similar, whence $BF : BO = CO : CE$, so $BF : CO = BO : CE$, i.e., $BF : BA = AC : CE$ ($\star\star$); comparing the proportionalities $\star$ and $\star\star$ yields $CE = FD$ so $CEFD$ is a parallelogram, so $EF = CD = a$ as required.]

Trisection by neusis

Having thus shown that a neusis between perpendicular lines could be performed by means proper to solid problems, Pappus went on to explain how to trisect an angle by this kind of neusis:

Construction 3.9 (Trisection — Pappus)[44]
Given: an angle φ (see Figure 3.10); it is required to construct an angle equal to $\frac{1}{3}\varphi$.

Construction:
1. Construct a right angled triangle ABC with $\angle CAB = \varphi$; call $AC = a$; draw a line M through C parallel to AB.
2. By neusis, draw ADE, intersecting BC in D and M in E and such that $DE = 2a$.
3. Then $\angle DAB$ is the required angle.

[44] [Pappus Collection] IV-32 (§§ 38–40), pp. 212–214.

[**Proof**: With F on DE such that $DF = FE$, we have $DF = CF = FE = CA = a$, hence, by isosceles triangles, $\angle CAF = \angle CFA = (\angle FCE + \angle FEC) = 2\angle FEC = 2\angle DAB$, so $\angle DAB = \frac{1}{3}\angle CAB$.]

Together the two constructions showed that the trisection could be performed by the intersection of conics; it was indeed a solid problem. Pappus added two further constructions of the trisection,[45] which I don't discuss in detail. Both used the intersection of a circle and a hyperbola and Pappus stressed that they did not use a neusis as an intermediary step. Apparently it was important for him to show that the neusis step in the construction of the trisection could be avoided.

3.7 Conclusion

Pappus' views on construction

I can now summarize Pappus' views on geometrical construction as they would appear to early modern readers of the *Collection*. Proper geometrical construction proceeded by the intersection of straight lines, circles, conics, or other more complicated curves. Other constructing procedures, such as neusis, shifting rulers or the use of special instruments, were useful if they were easy in practice, and they were interesting enough to be studied and proved; but the ultimate aim in solving geometrical problems was to find proper geometrical constructions in the sense above. These constructions induced a classification of problems according to their constructibility by plane, solid, or line-like curves. The geometer was obliged to construct problems by means proper to the problem's class. Although there was some doubt on the geometrical status of constructions by quadratrix and spiral, solid problems properly and legitimately belonged to geometry, and so did the conics by which they were to be constructed.

Exactness

The *Collection* offered little argument concerning the interpretation of exactness. Pappus' attitude to the question appeared ambivalent. He explained a strict interpretation of constructional exactness and based a classification of problems on it; yet he devoted considerable attention to constructions which were at variance with this interpretation. In so far as he argued explicitly about constructional exactness he took the position which in Section 1.6 I have characterized as appeal to authority and tradition. His remarks about the advantages of instrumental or other constructions seem to refer to actual rather than to idealized practice of construction.

Influence

Despite the absence of explicit arguments about exactness, the clear structure of the methods of properly geometrical construction that Pappus presented had a crucial influence on the early modern interpretation of geometrical exactness. Indeed the publication of the *Collection* in 1588 marked the starting point of an

[45] [Pappus Collection] IV-34 (§§ 43–44), pp. 217–221.

ordered debate on the geometrical legitimacy of construction procedures. Pappus' views on construction provided structure to this debate in several ways. First, the question of legitimate procedures in geometry became focused on construction rather than on methods of proof or questions of existence.[46] Second, the question of proper geometrical construction was split in two separate issues: the *demarcation* between geometrical and ungeometrical procedures of construction, and the *classification* of the geometrical procedures as to simplicity. Third, the fact that Pappus' classification of problems concerned the curves used in their construction led to a heightened interest in curves and the methods by which they could be generated.

As we will see in Chapters 9–14 these themes were indeed prominent in the debate on the interpretation of the exactness of geometrical procedures after c. 1590.

[46] Cf. Sections 1.2 and 2.6.

Chapter 4

The early modern tradition of geometrical problem solving; survey and examples

4.1 Introduction

Two classifications

In Section 1.5 I identified the early modern tradition of geometrical problem solving as the context of the debates on the interpretation of the exactness of construction during the period c. 1590 – c. 1650. The debates primarily concerned the solution of point construction problems, that is, problems that admitted one or a finite number of solutions only. Solving such problems was indeed seen as a major, if not the main, aim of geometry.[1]

The geometrical problems of the early modern tradition can be classified in two ways: by their class according to Pappus' distinction of plane, solid, and line-like problems, and by a distinction of types. The combination of both classifications yields an array as in Table 4.1. In the next sections I discuss the various types in the order of that table and illustrate them by examples.

Together with the constructions presented in the previous two chapters, the examples discussed below serve to illustrate the various methods of construction that were used. Table 4.2 gives a survey of these methods with references to the constructions. The examples are also chosen so as to be useful in explaining the early modern methods of analysis in Chapter 5.

---

[1]Thus in 1591 Viète's formulated the objective of his analytical program as "to leave no problem unsolved" (cf. Note 6 of Chapter 6) and Descartes opened his *Geometry* of 1637 with the words "All the problems of geometry..." (cf. [Descartes 1637] p. 297). Knorr notes a similar preeminence of problems over theorems in classical Greek geometry, [Knorr 1986] p. 300.

Class → Type ↓	Plane problems	Solid problems	Line-like problems
Standard problems and constructions (Section 4.2)	Euclidean standard constructions; later plane standard constructions	Vietean standard "constitutive problems"	
Angular sections (Section 4.3)	Bisection, 2^n-section; regular pentagon; "constructible" regular polygons	Trisection; regular heptagon and nonagon	General angular section; regular polygons
Mean proportionals (MP's) (Section 4.4)	One MP; three MP's; (2^n-1) MP's	Two MP's; doubling the cube; 11 MP's	Any number of means
Area and content problems (Section 4.5)	Quadrature of rectilinear figures; similar rectilinear figures with given areas	Cubature of polyhedra; addition of similar solids	Quadrature of the circle
Neusis problems (Section 4.6)	Special neusis problems	General neusis between straight lines	
Division of figures (Section 4.8)	Division of plane rectilinear figures	Division of a sphere	
Triangle problems (Section 4.9)	Plane triangle problems		
Varia (Section 4.10)		Normal to a parabola	

Table 4.1: The early modern tradition of geometrical problem solving: types and classes of problems

1. Construction of plane problems

By circles and straight lines and/or by reduction to standard plane problems: 4.1 (Fourth Proportional — Euclid), 4.2 (Mean proportional — Euclid), 4.3 (Scholium to *Elements* III-36 — Clavius), 4.4 ($x^2 - ax = b^2$ — Viète), 4.8 (Lines in continued proportion — Clavius), 4.16 (Special neusis between a circle and a line — Ghetaldi), 4.18 (Triangle division — Clavius), 4.21 (Triangle problem — Viète).

2. Construction of solid problems

2.1 By approximative procedures: 2.2 (Two Mean Proportionals by shifting a ruler — Hero), 2.3 (Two Mean Proportionals by shifting gnomons — Plato), 4.9 (Two mean proportionals by shifting a ruler — Clavius).

2.2 By the intersection of conics: 3.1 (Two mean proportionals by intersection of a parabola and a hyperbola — Menaechmus), 3.8 (Neusis by intersection of a hyperbola and a circle — Pappus).

2.3 By procedures using special constructing curves: 3.4 (Angular section by means of the quadratrix curve — Pappus), 2.5 (Neusis by means of a conchoid — Nicomedes), 3.6 (Two mean proportionals by means of the cissoid — Diocles), 4.11 (Two mean proportionals by means of a curve constructed pointwise earlier — Villalpando).

2.4 By reduction to standard solid problems: 4.6 (Root of a cubic equation by reduction to trisection — Viète), 4.19 (Division of a sphere by reduction to trisection — Huygens), 4.15 (Addition of similar solids by reduction to two mean proportionals — Stevin), 4.22 (Perpendicular to a parabola by reduction to a Vietean standard problem — Anderson), 2.6 (Two Mean Proportionals by neusis — Nicomedes), 3.9 (Trisection by neusis — Pappus).

3. Construction of curves

3.1 By motion: 3.2 (Spiral by combination of a rotating and a rectilinear motion — Archimedes), 3.3 (Quadratrix by combination of a rotating and a rectilinear motion — Pappus).

3.2 By "pointwise" procedure: 3.5 (Cissoid pointwise — Diocles), 4.10 ("First proportionatrix" pointwise — Villalpando).

Table 4.2: Methods of construction from Chapters 2–4

Methodological questions

Apart from illustrating the various aspects of the early modern tradition of geometrical problem solving mentioned above, the examples are meant to convey the urgency of the two main methodological questions faced by the practitioners of the art of geometrical problem solving. First, the diversity of the means chosen to construct non-plane problems illustrates that no *communis opinio* existed on the interpretation of exactness of constructions beyond the Euclidean means. And second, the apodictical and unenlightening manner in which the constructions were often presented (which my uniform format for rendering constructions enhances, but not much so) shows the need for a uniform and generally applicable method to find the constructions, that is, the need of an effective method of analysis.

4.2 Standard problems, standard constructions

Euclidean standard constructions

In solving problems geometers hardly ever wrote out their construction in complete detail. Rather than explicitly reducing the whole procedure to its smallest constituent steps, they assumed that their readers knew a range of standard constructions to which they merely referred by name or by reference to standard sources, notably to Euclid's *Elements*. The most important standard constructions concerned the fourth, the third, and the mean proportional. For given line segments a, b, and c (the terminology also applied to other magnitudes), a line segment x was called the *fourth proportional of a, b, and c* if

$$a : b = c : x \; ; \tag{4.1}$$

a line segment y was called the *third proportional of a and b* if

$$a : b = b : y \; ; \tag{4.2}$$

and a line segment z was called the *mean proportional of a and b* if

$$a : z = z : b \tag{4.3}$$

(cf. Problem 2.1). The Euclidean constructions of these proportionals were as follows:

Construction 4.1 (Fourth Proportional — Euclid)[2]
Given: three line segments a, b, and c (see Figure 4.1); it is required to find their fourth proportional.

Construction:
1. Draw two half lines through a point O under any angle.
2. Mark $OA = a$ and $AB = b$ on the one half line, and $OC = c$ on the other.
3. Connect AC and draw a line through B parallel to AC, it intersects OC prolonged in D.

[2] *Elements* VI-12.

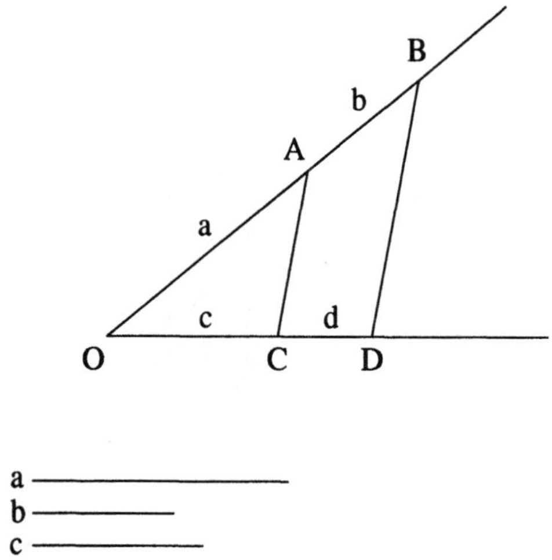

Figure 4.1: Construction of the fourth proportional — Euclid

4. $d = CD$ is the required fourth proportional, i.e., $a : b = c : d$.
[**Proof**: By similar triangles.]

The same construction yields the third proportional if one takes $c = b$.[3]

Construction 4.2 (Mean proportional — Euclid)[4]
Given: Two line segments a and b (see Figure 4.2); it is required to construct their mean proportional.

Construction:
1. Mark $AO = a$ and $OB = b$ along a straight line.
2. Draw a semicircle with diameter AB; draw a line through O perpendicular to AB, it intersects the semicircle in C.
3. $c = OC$ is the required mean proportional, i.e., $a : c = c : b$.
[**Proof**: By the similarity of the triangles AOC and COB.]

A standard construction by Clavius

Occasionally, geometers felt that some standard constructions were missing in Euclid and supplied them. An example is a construction which Clavius gave as a Scholium to *Elements* III-36 in his Euclid edition. *Elements* III-36 stated that (see Figure 4.3) if a line OA is tangent at A to a circle and if another line through O intersects the circle in B and C, then[5] sq.(OA) = rect.(OB, OC).

[3] *Elements* VI-11.
[4] *Elements* VI-13.
[5] See Section 1.7 for the notations sq. and rect.

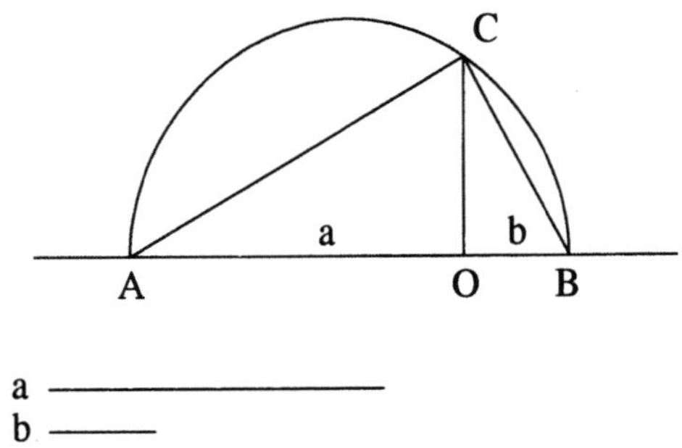

Figure 4.2: Construction of the mean proportional — Euclid

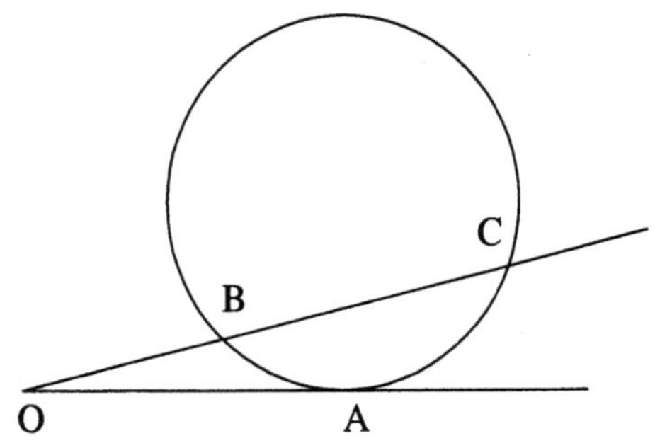

Figure 4.3: *Elements* III-36

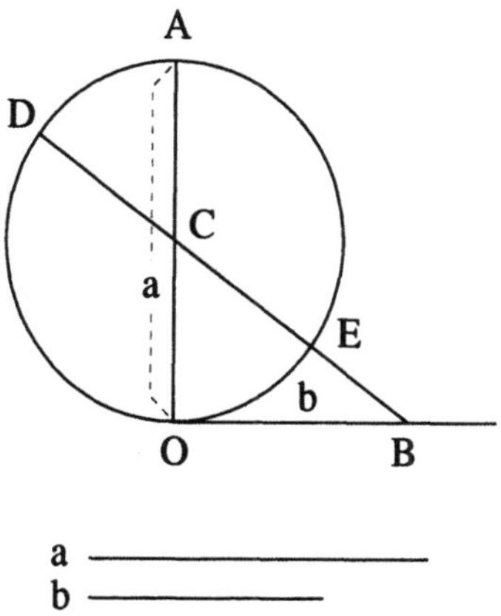

Figure 4.4: Clavius' construction in his Scholium to *Elements* III-36

Clavius used this result in the following construction (which he later used as standard construction[6]):

Construction 4.3 (Scholium to *Elements* III-36 — Clavius)[7]
Given: Two line segments a and b (see Figure 4.4). It is required to mark $DE = a$ *along a straight line and prolong it beyond E to B such that* rect.$(DB, EB) =$ sq.(b).

Construction:
1. Along perpendicular lines through O mark $OB = b$ to the right and $OA = a$ upward.
2. Draw a circle with diameter OA; its center is C.
3. Draw and prolong CB; it intersects the circle in D and E.
4. D, E, and B are points as required.
[**Proof**: $DE = a$ by construction, and by *Elements* III-36 rect.$(DB, EB) =$ sq.$(OB) =$ sq.(b).]

We will meet this construction later on (cf. Construction 22.1) because Descartes chose it as standard construction for the root of a quadratic equation of the form $x^2 = ax + b^2$. Indeed, if we choose $x = DB$, then $x(x - a) = b^2$, i.e. $x^2 = ax + b^2$.

A standard construction by Viète

As part of his program of reconstituting the ancient art of analysis (Sec-

[6]Cf. Construction 4.18
[7]Scholium to III-36 in [Euclid 1589] vol. 1, pp. 446–447.

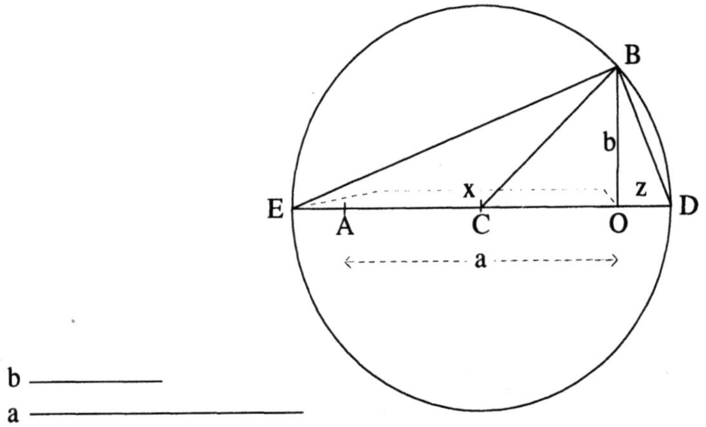

Figure 4.5: Viète's construction of the root x of $x^2 - ax = b^2$

tion 8.2) Viète gave standard constructions for the roots of quadratic and cubic equations. He explained how such equations could be reduced to a number of standard forms and related each of these forms to a standard problem. He called these the "constitutive problems" of the equations. Each of the constitutive problems related to a situation involving three or four line segments in continued proportion. Thus in the treatise in which he surveyed plane constructions[8] he explained that the equation

$$x^2 - ax = b^2 \tag{4.4}$$

could be written as $x(x-a) = b^2$ and therefore as a proportionality

$$x : b = b : (x - a) \, . \tag{4.5}$$

Hence solving Equation 4.4 was equivalent to determining three line segments x, y, z ($x > z$) in continued proportion, such that $y = b$ and $x - z = a$. On the basis of that interpretation of the equation, he gave the following standard construction:

Construction 4.4 ($x^2 - ax = b^2$ **— Viète**)[9]
Given: Two line segments a and b (see Figure 4.5); it is required to construct a

[8][Viète 1592].

[9][Viète 1592] Prop. 10, pp. 232–233 ([Viète 1983] p. 376). In illustration of Viète's terminology and notation I quote the original text, in which A, B and D correspond to x, a, and b above, respectively: "Itaque cum proponetur A quadratum, minus B in A aequari D

line segment x satisfying $x^2 - ax = b^2$; equivalently, it is required to construct three line segments x, y, z $(x > z)$ such that $x : y = y : z$, $y = b$ and $x - z = a$.

Construction:
1. Draw $AO = a$ and $OB = b$ intersecting perpendicularly in O; bisect AO in C.
2. Draw a circle with center C and radius CB.
3. The line AO, prolonged to both sides, intersects the circle in points D and E of which D is nearest to O.
4. $x = OE$, $y = OB = b$ and $z = OD$ satisfy the requirements.
[**Proof**: The triangles ΔEOB and ΔBOD are similar, hence $x : y = y : z$; $OD = AE$, so $x - z = a$.]

Proceeding in this way Viète set up the following correspondence between the standard forms of the quadratic equation and standard plane "constitutive problems" about three proportional line segments:[10]

$$\begin{aligned} x^2 + ax = b^2 &\iff x : y = y : z\ ,\ y = b\ ,\ z - x = a\ (1)\ , \\ x^2 - ax = b^2 &\iff x : y = y : z\ ,\ y = b\ ,\ x - z = a\ (2)\ , \\ ax - x^2 = b^2 &\iff x : y = y : z\ ,\ y = b\ ,\ x + z = a\ (3)\ . \end{aligned} \tag{4.6}$$

Solid standard problems

Solid problems were mostly reduced to one of a few standard problems that were assumed solved, solvable, or still to be solved. The most important standard solid problem was the construction of two mean proportionals between two given line segments (Problem 2.1). We have seen several methods to solve this problem (cf. Constructions 2.2, 2.3, 2.6, and 3.1) and we will meet more below. Viète showed, as we will see in detail in Chapters 8 and 10, that all solid problems could be reduced either to the construction of two mean proportionals or to the trisection of an angle.[11] With this result the trisection acquired as it were a status among the solid problems symmetrical to the determination of two mean proportionals. From Pappus' *Collection* mathematicians learned that the trisection of an angle could be performed by neusis (Construction 3.9), and Nicomedes' construction of two mean proportionals (Construction 2.6) showed that this problem could also be reduced to a neusis. Combined with Viète's result mentioned above, this implied that any solid problem could be reduced to a neusis problem. Thus the neusis occupied an even more central position among solid problems than the trisection of an angle and the determination of two mean proportionals. Viète's use of the neusis as standard solid problem will be discussed in detail in Section 10.1.

quadrato: intelligetur D media inter extremas, B differentia earundem. Et ex media et differentia extremarum quaerentur extremae, quarum major erit A, de qua quaeritur." ("Therefore, if it is proposed that the square of A minus A in D be equal to a square D: then consider a mean D between extremes, and B their difference. And from the mean and the difference of the extremes, the extremes are required, of which the larger is the A, which the question concerned.")

[10][Viète 1592] Props. 9, 11, pp. 232–233 ([Viète 1983] pp. 375–377).

[11]To be precise: he showed that any problem whose algebraic equivalent was a third- or fourth-degree equation could be so reduced, cf. Section 10.3.

Vietean solid "constitutive" problems

The three standard solid problems discussed above, determining two mean proportionals, trisecting an angle, and performing a neusis, were rather different in character, nor was there a clear analogy between the three equations pertaining to these problems. This may have been the reason why Viète felt the need for another set of standard solid problems and equations similar to the ones he singled out for plane problems (Equations 4.6). He proceeded as follows. He knew that by a transformation removing the quadratic term, any cubic equation could be written in one of the following forms (in which a and b were positive and only positive roots were considered):[12]

$$\begin{aligned} x^3 &= a^2b\,, \\ x^3 + a^2x &= a^2b\,,\ (1) \\ x^3 - a^2x &= a^2b\,,\ (2) \\ a^2x - x^3 &= a^2b\,.\ (3) \end{aligned} \tag{4.7}$$

He did not separately discuss the first form, whose one positive root was the first of two mean proportionals between a and b. He related each of the three others to a standard problem of the form:

Problem 4.5 (Standard solid problems — Viète)[13]
Given: Two line segments a and b; it is required to construct four line segments a, x, y, and z in continued proportion, i.e.

$$a : x = x : y = y : z\,, \tag{4.8}$$

and such that, in the different cases,

$$\begin{aligned} x + z &= b\,,\ (1) \\ z - x &= b\,,\ (2) \\ x - z &= b\,.\ (3) \end{aligned} \tag{4.9}$$

The correspondence of the equations and the problems is seen, for instance, in case (2), as follows: by 4.8 we have $a^2 : x^2 = x : z$, hence $x^3 = a^2z$, so by 4.9-2 $x^3 = a^2(x + b)$ or $x^3 = a^2x + a^2b$, as in 4.7-2. Note that the equation $x^3 = a^2b$, which he did not discuss, fits the same general formulation as a problem; it is the case in which the additional requirement is $z = b$.[14]

In his *Supplement to geometry* of 1593 Viète showed how in each of these cases a root of the equation could be found either by the determination of two mean proportionals or by the trisection of an angle. I discuss these results in more detail in Section 10.3; here it may suffice to give an example concerning case (2). Viète rewrote the equation as

$$x^3 - 3a^2x = 2a^2b \tag{4.10}$$

[12] [Viète 1615], pp. 86–87, [Viète 1983] pp. 164–167.

[13] [Viète 1615] pp. 86–87, [Viète 1983] pp. 164–167.

[14] Viète also knew that, by methods as the one by Ferrari (see Note 18 of Chapter 10) the solution of fourth-degree equations could be reduced to the solution of quadratic and third-degree ones. Thus his list of constitutive equations covered all problems that were reducible to third- or fourth-degree equations, cf. Note 11.

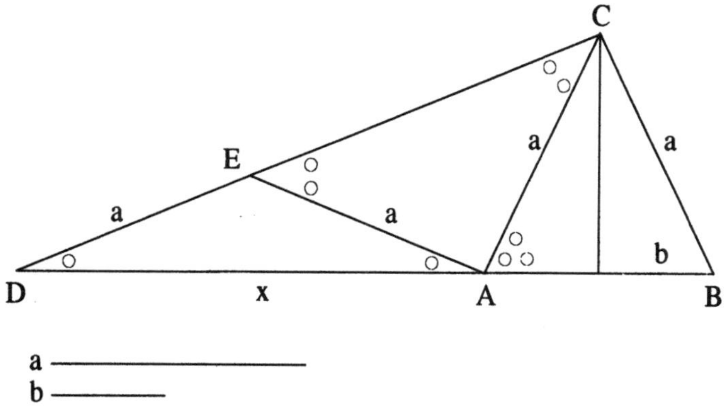

Figure 4.6: Construction of the root x of $x^3 - 3a^2x = 2a^2b$ — Viète

in order to distinguish more easily between reduction to trisection and reduction to two mean proportionals; the former case applied if $a > b$, the latter if $a < b$.[15] Viète's construction of the former case was as follows:

Construction 4.6 (Root of a cubic equation — Viète)[16]
Given: line segments a and b, $a > b$ (see Figure 4.6); it is required to find a root of the equation $x^3 - 3a^2x = 2a^2b$.

Construction:
1. Construct an isosceles triangle ABC with $AB = 2b$ and $AC = BC = a$; prolong AB to the left.
2. Assuming trisection possible, draw CD such that $\angle CDB = \frac{1}{3}\angle CAB$.
3. Then $x = AD$ is a root of the equation.
[**Proof**: Take E on CD such that $EA = ED$ and note that $DE = EA = AC = a$ and that the angles are as indicated in the figure. Then $x = 2a\cos(o)$, $b = a\cos(3o)$, and the familiar relation $\cos(3o) = 4\cos^3(o) - 3\cos(o)$ coincides with the given equation.]

The construction shows that Viète did not choose the constitutive problems because they were particularly easy to construct; indeed instances of the same constitutive problem (case (2) in Equations 4.8 and 4.9) required essentially

[15] The corresponding condition for the a and b as in Equations 4.7 – 4.9 is: $2a >$ or $< 3\sqrt{3}b$; see also Section 10.3.

[16] [Viète 1593] Props 25 and 16, pp. 403–404, 416–417.

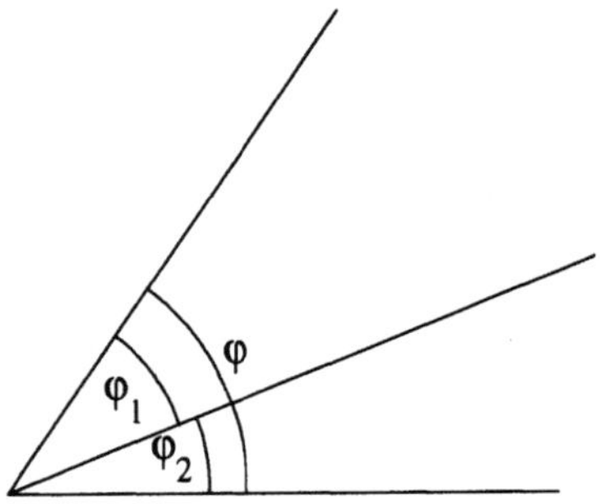

Figure 4.7: General angular section

different constructions depending on the ratio of a and b. Rather than relating to constructional aspects the "constitutive problems" reflect a wish to keep equation theory in close correspondence with proportion theory. This aspect remained characteristic in the Vietean school; as late as 1702 Ozanam mentioned the "constitutive problems" in a work on algebra.[17]

4.3 Angular sections

The problem The general problem of the section of an angle is:

Problem 4.7 (General angular section)
Given: An angle φ and a ratio ρ (see Figure 4.7); it is required to divide φ in two angles φ_1 and φ_2 such that $\varphi_1 : \varphi_2 = \rho$.

For the ratios $1:1,\ 1:2,\ 1:3,\cdots,1:k,\cdots$ the corresponding problems are the bisection, trisection, etc., or in general the multisection of the angle that is, the division of an angle in $2,\ 3,\ 4,\cdots,k+1,\cdots$ equal parts. A rational ratio $p:q$ leads to a division in aliquot parts, which can of course be performed by first multisecting the angle in $p+q$ parts. If no supposition is made about the rationality or irrationality of the given ratio, we have the general angular section. In the special case that the given angle φ is 360°, the multisection is the same as the construction of regular polygons.

[17] [Ozanam 1702] p. 224.

Plane angular sections

Bisecting an angle was a plane problem, the construction was given in Euclid's *Elements* I-9. Multisection of an angle in 2^k parts could be done by repeated bisection and was therefore plane as well.[18] Constructions by plane means of regular 3-, 4-, 5-, 6-, and 15-sided regular polygons were in Euclid's *Elements*.[19]

Solid angular sections

Classical Greek mathematicians recognized that trisecting an angle was a solid problem; it was indeed one of the three "classical problems." Also the construction of a regular heptagon was recognized as solid in antiquity.[20] We have seen a classical trisection above in Construction 3.9. In the early modern period the trisection attracted much less attention than the problem of two mean proportionals. One reason was that no treatment of the trisection existed comparable to Eutocius' list of 12 constructions of two mean proportionals. Also it seems that early modern geometers more often met solid problems reducible to two mean proportionals than problems reducible to trisection. On the other hand, the algebraic aspects of angular sections received considerable attention, especially through Viète's exploration of the equations for the successive multisections.[21]

Line-like angular sections

As we have seen above, Pappus classified the general angular section (Problem 4.7) as a line-like problem and solved it by means of the quadratrix (cf. Construction 3.4); it could also be solved by the Archimedean spiral.

4.4 Mean proportionals

Proportionals

If line segments a, x_1, $x_2, \cdots$, x_k, b were in continued proportion, i.e.,

$$a : x_1 = x_1 : x_2 = \cdots = x_{k-1} : x_k = x_k : b \,, \tag{4.11}$$

the $x_1, x_2, \cdots, x_k$ were called the k "mean proportionals" between a and b. Algebraically this implied

$$x_1 = \sqrt[k+1]{a^k b} \,. \tag{4.12}$$

Once the first, x_1, of k mean proportionals between two given line segments a and b was known, all others could be constructed by repeated application of the Euclidean construction 4.1 of the third proportional. Thus finding k mean proportionals was the geometrical equivalent of extracting the $(k+1)$-th root. Constructing one (or *the*) mean proportional of a and b was a standard plane construction (4.2), and, by inserting mean proportionals between previously

[18] By means of Galois theory it can be proved that all other rational angular sections of a general angle are non-plane.

[19] *Elements* IV-2, -6, -11, -15, -16, respectively; based on these any polygon with 2^n, 3.2^n, 5.2^n, and 15.2^n could be constructed by plane means. The possibility that other regular polygons than these might be so constructible seems not to have been considered before Gauss did so in 1796.

[20] Cf. [Hogendijk 1984].

[21] Cf. [Viète 1615b].

constructed proportionals one could construct 3, 7, 15 or in general $2^n - 1$ mean proportionals by straight lines and circles.[22]

Solid mean proportionals

Finding *two* mean proportionals, and its application in doubling the cube, were solid problems; in the previous chapters we have seen several constructions of two mean proportionals (Constructions 2.2, 2.3, 2.6, 3.1, 3.6). Another special solid mean proportionals problem enjoyed some renown in the early modern period, namely, the case in which $k = 11$ and $b = 2a$. If b is the length of a string, these mean proportionals are the string lengths of 11 equal semitones between the string's tone and its octave. Evidently, the determination of 11 mean proportionals can be reduced to that of two mean proportionals because if $x_1 \cdots x_{11}$ are the mean proportionals, then x_4 and x_8 are the two mean proportionals between a and b, and once these are known the others can be inserted by repeated construction of single mean proportionals. Barbour reports that the first mathematically precise definition of equal semitones (and thereby of the equal temperament) was given in 1577 by Salinas, who explained the connection with the geometrical problem of mean proportionals. He was aware that the problem required other than the Euclidean constructions and explained that instruments ("mesolabes"[23]) such as the one suggested by Eratosthenes should be used. After Salinas, mean proportionals and instruments to construct them were regularly mentioned in theoretical works on music in connection with problems of temperament.[24]

Stevin also studied the 11 mean proportionals between a and $2a$ in connection with equal temperament.[25]

Clavius' construction of proportionals

In one of the commentaries in his Euclid edition of 1589 Clavius gave an approximate construction of two mean proportionals. He had devised this construction by, in a sense, inverting a general procedure he had devised for constructing sequences of line segments in continued proportion. This procedure employed plane means only; it was as follows:

Construction 4.8 (Line segments in continued proportion — Clavius)[26]
Given: line segments a and b, $a > b$ (see Figure 4.8); it is required to find line segments u, v, w, ... and x, y, z, ... such that $b : a = a : x = x : y = y : z = \cdots$ and $a : b = b : u = u : v = v : w = \cdots$.

Construction:
1. Draw a circle with diameter $OA = a$; draw a chord $OB = b$ and

[22]By means of Galois theory it can be proved that for all $k \neq 2^n - 1$ the problem of constructing k mean proportionals is not plane.

[23]Cf. Chapter 2 Note 34.

[24]Cf. [Barbour 1961] in particular pp. 49–55. Barbour refers to [Salinas 1577] p. 173, and mentions among others [Zarlino 1571], [Zarlino 1588], [Mersenne 1636], and [Kircher 1650] as works in which the mean proportionals problem is discussed in connection with musical temperament.

[25][Stevin 1955–1966] vol. 5, pp. 440–443; cf. [Cohen 1984] pp. 55–57.

[26][Euclid 1589] pp. 778–779.

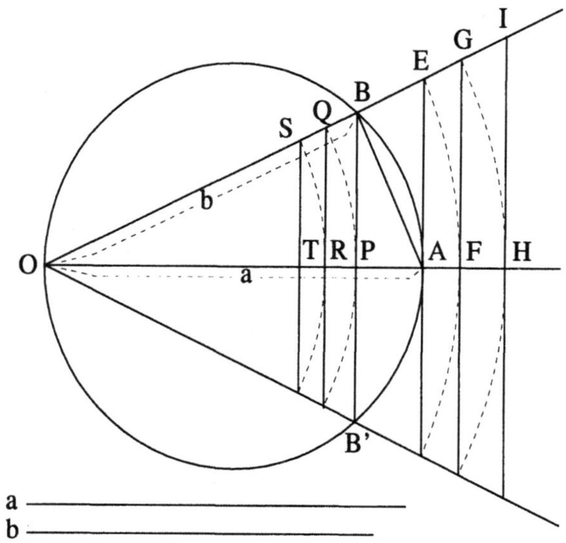

Figure 4.8: Clavius' construction of line segments in continued proportion

draw BA; prolong OB and OA.
2. Take P, R, T, $\cdots$ on OA and Q, S, $\cdots$ on OB such that $BP \perp OA$, $OQ = OP$, $QR \perp OA$, $OS = OR$, $ST \perp OA$, etc.
3. Take E, G, I, $\cdots$ on OB prolonged and F, H, $\cdots$ on OA prolonged such that $AE \perp OA$, $OF = OE$, $FG \perp OA$, $OH = OG$, $HI \perp OA$ etc. (Here Clavius used the symmetrical lower half of the figure to help drawing the perpendiculars more precisely.)
4. $u = OP$, $v = OR$, $w = OT$, $\cdots$ and $x = OE$, $y = OG$, $z = OI$ $\cdots$ are the required lines.
[**Proof**: By similarity of triangles and the constructed equalities.]

Clavius' use of a symmetrical figure illustrates his concern for precision in the practical execution of geometrical constructions; we will see this endeavor also in his construction of the quadratrix discussed in Section 9.2.

Clavius' approximate construction of two mean proportionals

By inverting this procedure Clavius arrived at a construction about which he wrote:

> From this we can find without much difficulty two mean proportionals between two given straight lines, not completely geometrical it is true, but as it were by trial and repetition of the same procedure again and again until we have reached what we want to find.[27]

[27][Euclid 1589] p. 780: "Ex his sine magno labore inter duas rectas datas reperiemus duas

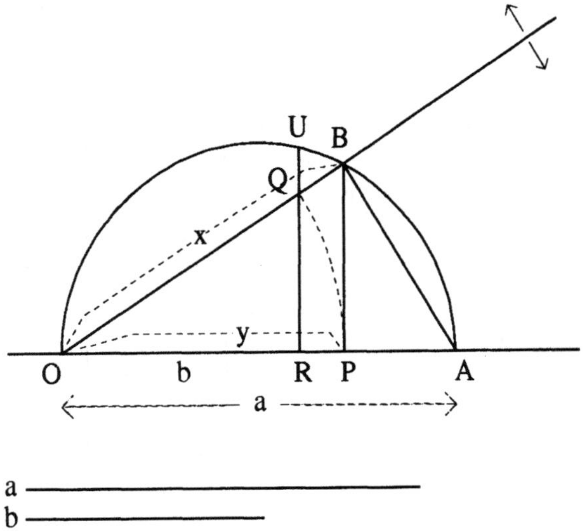

Figure 4.9: Clavius' construction of two mean proportionals

His construction employed a ruler which had to be turned until an appropriate configuration was reached. It was as follows:

Construction 4.9 (Two mean proportionals — Clavius)[28]
Given: two line segments a and b ($a > b$) (see Figure 4.9); it is required to find their two mean proportionals x and y.

Construction:
1. Draw a circle with diameter $OA = a$; take $OR = b$ on OA; take $RU \perp OA$ with U on the circle.
2. Apply a ruler to O; it intersects the circle and RU in B and Q, respectively; the perpendicular through B intersects OA in P.
3. Move the ruler around O, whereby Q, B, and P move as well, until $OQ = OP$.
4. In this position $x = OB$ and $y = OQ$ are the required mean proportionals.
[**Proof**: Draw BA; the triangles $\triangle OAB$, $\triangle OBP$, and $\triangle OQR$ are similar and $OP = OQ$, so $OA : OB = OB : OP = OQ : OR$, i.e., $a : x = x : y = y : b$.]

The two constructions are of particular interest because both as to configuration (the adjustable angle AOB, the series of perpendiculars to the sides of

medias proportionales, non quidem geometricè omnino, sed quasi attentando et praxim ipsam iterum atque iterum repetendo, donec id, quod quaerimus, assequamur."

[28] [Euclid 1589] p. 33.

the angle), and as to purpose (inverting the construction of proportionals to find mean proportionals) they are akin to the famous "Mesolabum" (Instrument 16.3) consisting of sliding rulers which Descartes later described in his *Geometry* and which was crucial in the development of his ideas about construction, cf. Sections 16.4 and 24.2.

Villalpando's "proportionatrix"

In Section 2.5 I quoted Rivault's remark of 1615 about the problem of two mean proportionals: "Nobody left this stone unturned." Indeed the problem was easily the most famous of the early modern tradition of geometrical problem solving. Over and again the classical constructions from Eutocius' list were presented and new ones were added. We find the results of this activity even at such unlikely places as the learned comments of the Jesuit fathers Prado and Villalpando on the prophet Ezechiel. The third volume of that work, written by Villalpando and published in 1604, contained a substantial chapter on proportionality and the determination of mean proportionals. The author considered the chapter of importance for the discussion of the system of weights and measures current at Ezechiel's time. He incorporated two constructions of two mean proportionals by means of specially constructed curves which he called "proportionatrices."[29] From a remark of Richard[30] we may surmise that in writing the mathematical parts Villalpando had received substantial help from another member of his Society, Christoph Grienberger.

Villalpando's (or Grienberger's) constructions provide examples of the use of special curves, which themselves were constructed pointwise. I discuss the first.[31] The construction of the curve was as follows:

Construction–Definition 4.10 ("First proportionatrix" — Villalpando)[32]
Given: a line segment OB with a point A on it such that $OB = 2OA$ (see Figure 4.10); upon the axis OB a curve, called the "first proportionatrix," is constructed pointwise.

1. Draw semicircles with diameters OA and OB.
2. Draw an arbitrary chord OC in the larger semicircle; it intersects the smaller one in D.
3. Take $OE = OD$ along OA; take F on OC such that $EF = OE$.
4. F will be on the "first proportionatrix."
5. Repeat steps **2–4** for other chords OC to find more points F on the curve.

---

[29] [Prado & Villalpando 1596–1605]; the mathematical part is in vol. 3, pp. 249–328; the constructions are on pp. 289–290.

[30] [Euclid 1645], last page of the "Argumentum librorum huius tomi," where Richard mentioned Grienberger, "cuius sunt omnia circa pondera et centrum gravitatis quae in suis de Templo Salomonico commentariis congessit Villalpandus etiam è nostra Societate." Richard supplemented his 1645 Euclid edition with a treatise of his own "Liber de inventione duarum rectarum linearum continue proportionalium inter duas rectas, ex antiquis geometris et recentioribus," pp. 545–563; in which he gave Villalpando's constructions.

[31] The second is discussed in [Ulivi 1985].

[32] [Prado & Villalpando 1596–1605] vol. 3, pp. 289–290.

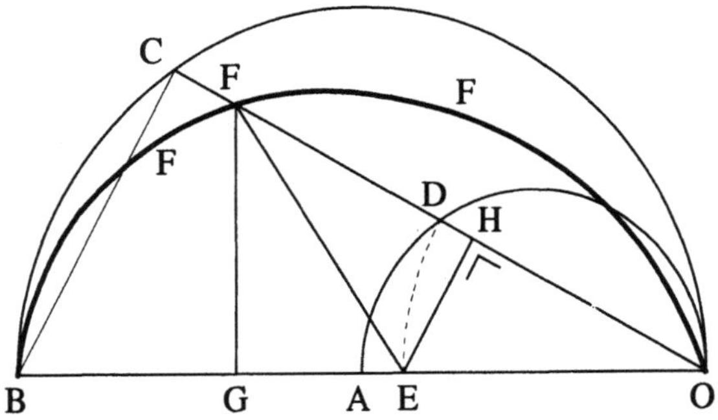

Figure 4.10: Construction of the "first proportionatrix" — Villalpando

6. Draw a smooth curve $BFFFO$ through all constructed points F; this curve is the "first proportionatrix."[33]

Villalpando's construction of two mean proportionals

The proportionatrix could be used for constructing two mean proportionals because it has the following property (cf. Figure 4.10): Let FG be drawn $\perp OB$, then

$$OB : OC = OC : OF = OF : OG\,, \tag{4.13}$$

that is, OC and OF are two mean proportionals between OB and OG. Indeed from the similarities of the triangles $\triangle OBC$, $\triangle OFG$, and $\triangle OEH$ (with H on OF such that $EH \perp OF$) it follows that $OB : OC = OF : OG = OE : OH$ ($\star$). Now, by construction, $OE = OD$, $OE = EF$ whence $OH = HF$, and $OD = DC$. Hence, $OE : OH = OD : OH = 2OD : 2OH = OC : OF$ ($\star\star$). Combining $\star$ and $\star\star$ yields the required proportionality 4.13.

Obviously, this proportionality was the reason for devising the proportionatrix; because of it the curve could serve for constructing two mean proportionals. Villalpando explained the procedure:

Construction 4.11 (Two mean proportionals — Villalpando)[34]
Given: two line segments a and b, $a < b$ (see Figure 4.11) and a "first propor-

[33]The equation of the curve in polar coordinates is $r = 2a\cos^2\varphi$ and in rectangular coordinates $(x^2 + y^2)^3 = 4a^2x^4$.
[34][Euclid 1645]. pp. 545–563

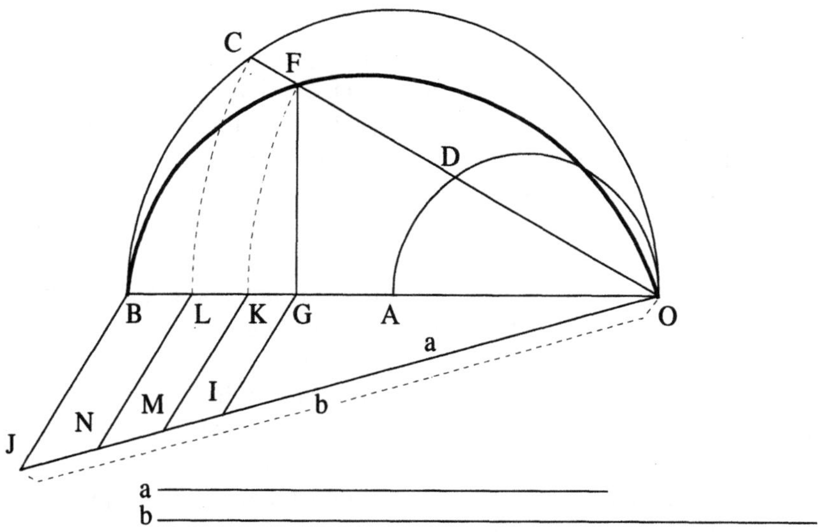

Figure 4.11: Construction of two mean proportionals by the "first proportionatrix" — Villalpando

tionatrix" drawn with respect to axis OAB; it is required to construct the two mean proportionals x and y of a and b.

Construction:
1. Draw a line through O making any angle with OB; mark $OI = a$ and $OJ = b$ along that line; draw JB; draw a line through I parallel to JB, it intersects OB in G.
2. Draw a line through G perpendicular to OB, it intersects the proportionatrix in F.
3. Mark K and L on GB with $OK = OF$ and $OL = OC$; draw lines through K and L parallel to JB, they intersect OJ in M and N, respectively.
4. $x = OM$ and $y = ON$ are the required mean proportionals.
[**Proof**: $a : x : y : b = OG : OK : OL : OB = OG : OF : OC : OB$; the last four line segments are in continued proportion (cf. Equation 4.13), hence so are a, x, y, and b.]

Curves devised for constructions

Any point F on the proportionatrix yields two mean proportionals between the corresponding OG and OB. If F moves along the curve, the ratio $OG : OB$ assumes all possible values (smaller than $1 : 1$). Thereby the curve as it were embodies the solutions of all possible two mean proportionals problems. This property was the basis of the curve's use and apparently the only reason for its introduction. Classical and early modern geometry presented several instances of curves introduced in a similar manner for the solution of one particular problem. The cissoid of Diocles (again for two mean proportionals, cf. Definition 3.5) and the conchoid of Nicomedes (for neusis, cf. Sections 3.6 and 4.6) are examples; as I noted earlier (Section 3.2), it seems that similarly the origin of the quadratrix lay in the endeavor to solve angular sections.

The introduction of curves for solving particular problems raises the question in how far such curves can be considered to achieve the required solution. The question is especially poignant in the example of Villalpando's proportionatrix; the curve was rather obvious (given the problem) but not simple. Only a finite number of its points were actually constructed, the others were obtained by drawing a line smoothly through the constructed ones. Villalpando did not discuss in how far these interpolated points could legitimately be used in the construction of two mean proportionals. Thus the use of the curve implied the assumption that by finding a finite number of quadruples of line segments in continued proportion one had in fact found all such quadruples. Such an assumption was practically a *petitio principii*.

It is clear that, in ways such as the one exemplified by Villalpando's construction, any problem gives rise to its own constructing curve (or curves), and thus the possibilities of introducing such curves are infinite; the activity could easily become vacuous. It appears indeed that prominent mathematicians of the early modern period required more of the constructing curves they introduced than that they encompassed all instances of the problem. Yet the exercise might be otherwise profitable, as is illustrated by Renaldini, who as late as 1670 proudly presented a set of curves, respectfully named after the Medici's. These curves were generated much in the same way as Villalpando's, for the special purpose of constructing roots of various types of equations.[35]

No special curve for general mean proportionals

Finding k mean proportionals between two line segments was also called dividing, or cutting, a ratio in $k+1$ equal parts. Thus if, e.g., a, x_1, x_2, x_3, x_4, x_5, x_6, b were in continued proportion,

$$a : x_1 = x_1 : x_2 = x_2 : x_3 = x_3 : x_4 = x_4 : x_5 = x_5 : x_6 = x_6 : b \, , \tag{4.14}$$

the ratio $a : x_1$ was called half the ratio $a : x_2$ and one seventh part of the ratio $a : b$; and x_3 was said to divide the ratio $a : b$ in aliquot parts of which the one, $a : x_3$ was 3/7-ths of $a : b$, and the other, $x_3 : b$, 4/7-ths.

The conception and terminology concerning the partition or section of ratios explained above suggests an analogy between mean proportionals and angular sections. Pappus presented the general section of an angle as a line-like problem, solvable by means of the quadratrix or the spiral (cf. Construction 3.4). Each of these curves related partitions of a straight line segment to corresponding partitions of an angle. It is noteworthy that, apparently, no classical mathematician pursued the analogy of angular sections and mean proportionals by introducing a curve, traced by some combination of motions, which similarly

[35] [Renaldini 1670]; the "Lineae Mediceae" are introduced pp. 12 sqq.

related partitions of a straight line segment to corresponding partitions of a ratio. A curve like that would have been logarithmic or exponential in nature.[36] Descartes seems to have been the first to consider such a curve; he called it the "linea proportionum" and he mentioned it in his notes of c. 1619,[37] but he did not pursue the subject; the logarithmic curve began to be discussed among mathematicians around 1640.[38]

4.5 Area and content problems

Euclidean area problems

With the term "area and content problems" I denote, for the sake of this overview and not because such a class was explicitly recognized by classical or early modern geometers, the problems that generalized some Euclidean transformations of rectilinear areas, notably:

Problem 4.12 (Parallellogrammic application — Euclid)[39]
Given: A rectilinear figure F, a line segment a, and an angle φ; it is required to construct a parallellogram with angle φ, one side equal to a, and equal in area to F.

Problem 4.13 (Quadrature of rectilinear figures — Euclid)[40]
Given: a rectilinear figure F; it is required to construct a square equal in area to F.

Problem 4.14 (Transformation into figure of given shape — Euclid)[41]
Given: two rectilinear figures F and G, it is required to construct a rectilinear figure H similar to F and equal in area to G.

Problem 4.12 asked to position a given rectilinear area as a parallellogram along a given line segment. Problem 4.13 was the quadrature of rectilinear areas, it asked to construct a square equal to the given area. Problem 4.14 was a generalization, reducing to 4.13 when F is taken to be a square.

Stevin and content problems

Stevin devoted two books (IV and V) of his *Geometrical problems* of 1583[42] to the exploration of the three-dimensional analogs of the area problems just mentioned. The first generalized Problem 4.14. He found that the generalization depended on the determination of two mean proportionals. Hence, at

[36] In the case of the exponential curve $y = e^x$, for instance, a partition of the segment $[a, b]$ by points, say, $x_1, \cdots x_4$ in five equal parts yields four mean proportionals $y_1, \cdots y_4$ between e^a and e^b.

[37] cf. Section 16.5.

[38] Cf. [Loria 1902] p. 542.

[39] [Euclid Elements] I-45; the enunciation in the *Elements* does not require the line segment a to be given beforehand; but from the use of the previous construction (I-44) it is clear that a can indeed be given and that Euclid intended the construction to be interpreted in that way.

[40] *Elements* II-14.

[41] *Elements* VI-25.

[42] [Stevin 1583].

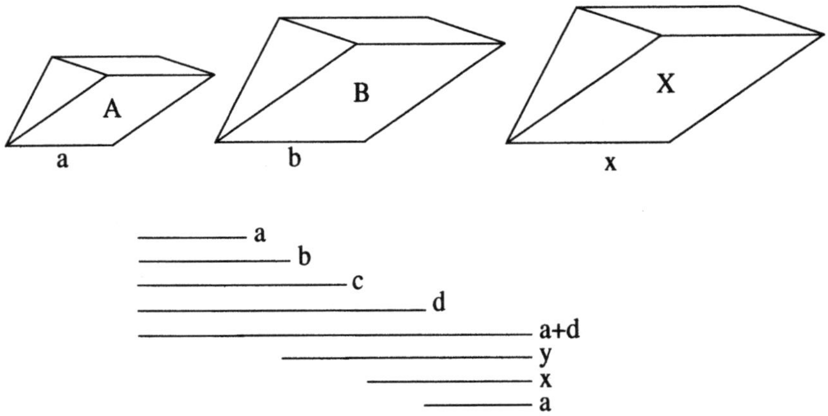

Figure 4.12: Addition of similar solids — Stevin

the beginning of the book he gave a construction of two mean proportionals, namely, Hero's (cf. Problem 2.2), mentioning that no "geometrical" construction was available.[43] In book V he proceeded to a further generalization, probably inspired by *Elements* VI-31. Euclid's proposition stated that if similar rectilinear figures were erected on the sides and the hypotenuse of a right-angled triangle, the one on the hypothenuse was equal in area to the two remaining figures together. The result could readily be interpreted as a construction problem, namely: given two similar rectilinear figures A and B, to construct a third similar figure C equal in area to A and B together. Stevin's problem was the three-dimensional analog. It may serve here as an example of how solid problems were reduced to the standard construction of two mean proportionals.

Construction 4.15 (Addition of similar solids — Stevin)[44]
Given: two similar solids A and B with homologous sides a and b (see Figure 4.12); it is required to construct a similar solid X equal in content to the sum of the contents of A and B.

Construction:
1. Construct a line segment c satisfying $a : b = b : c$.
2. Construct a line segment d satisfying $a : b = c : d$.
3. Construct (by Hero's construction 2.2 given previously) two mean proportionals x and y between a and $a + d$, so $a : x = x : y = y :$

[43] [Stevin 1583] Book IV, Problem 1, pp. 85–86.
[44] [Stevin 1583] Book V, Problem 2, pp. 108–114.

$(a+d)$.
4. With x as homologous side construct a solid X similar to the given solids; X is the required solid.
[**Proof**: The contents are proportional to the cubes a^3, b^3, and x^3. Now $x^3 = a^2(a+d) = a^3 + a \times ad = a^3 + abc = a^3 + b \times ac = a^3 + b^3$, hence the content of X is the sum of the contents of A and B.]

Problems of the same type can be found in Clavius' *Practical geometry* of 1604, for instance, a cubature generalizing Problem 4.13: given a parallelepiped, to construct a cube equal in content; Clavius reduced the construction to that of two mean proportionals.[45]

4.6 Neusis problems

The general neusis problem

The general neusis problem (cf. Problems 2.4 and 3.7) was solid. Its origin probably lay in an early period of classical Greek geometry during which the use of a marked ruler that could be shifted over the plane was accepted for constructional purposes, alongside the use of ruler and compass.[46] Two procedures of constructing the neusis problem have been mentioned in the previous chapters (Constructions 2.5 and 3.8). Its role as a solid standard problem has been discussed above (Section 4.2).

A plane neusis problem

Greek geometers had recognized that in special cases neusis problems were plane. Ghetaldi dealt with one such special case in his *Apollonius revived* of 1607:

Construction 4.16 (Special neusis between a circle and a line — Ghetaldi)[47]

Given: a semicircle with diameter $OA = a$ (see Figure 4.13), a line L intersecting the prolongation of OA perpendicularly in B, with $OB = b$, and a line segment c; it is required to find a line through O, intersecting the semicircle and the line L in F and G, respectively, and such that $FG = c$.

Construction:
1. Draw a semicircle with diameter OB; draw AC perpendicularly to OB with C on the semicircle; draw OC.
2. Draw BCD with $CD = \frac{1}{2}c$; draw OD.
3. Draw a circle with center D and radius DC; its intersection with OD is E.
4. Take F on the given semicircle such that $OF = OE$ (Ghetaldi proved that $OE < OA$ so that this can be done, I omit that proof);

[45] [Clavius 1604] Book VIII, Prop. 38, p. 416; cf. also the construction of solid figures enlarged in a given ratio, Book VI, Prop. 17, pp. 305–306.
[46] For the history of the neusis procedure in antiquity see a chapter by Heath in [Archimedes nd] pp. c–cxxii, and [Knorr 1986], index *s.v.* "neusis" (e.g. p. 34).
[47] [Ghetaldi 1607b] Probl. 2 Casus 1, pp. 5–6.

fig:GhetNeu1

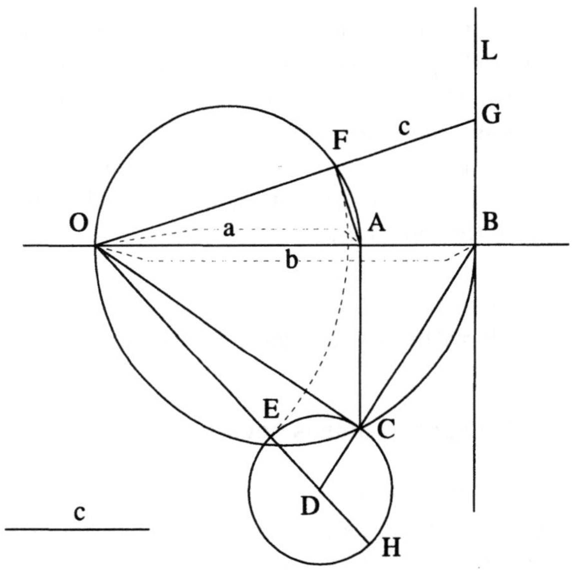

Figure 4.13: Special neusis between a circle and a line — Ghetaldi

prolong OF, it intersects L in G.
5. OFG is the required line, that is, $FG = c$.
[**Proof**: Draw FA and prolong OD, its second intersection with the circle around D is H. Because $\angle OCB$ is right, OC is tangent to the circle ECH. Hence sq.(OC) = rect.(OE, OH). Also sq.(OC) = rect.(OB, OA), so rect.(OB, OA) = rect.(OH, OE) $(*)$. Moreover $\triangle OFA \sim \triangle OBG$, so $OA : OF = OG : OB$, hence rect.(OB, OA) = rect.(OF, OG) $(**)$. Combining $(*)$ and $(**)$ yields rect.(OH, OE) = rect.(OF, OG); but $OE = OF$, so $OH = OG$, and $FG = EH = c$.]

In *Apollonius revived* Ghetaldi did not explain how he arrived at this construction. However, in his comprehensive work on analysis and synthesis *On mathematical resolution and composition*[48] of 1630 he went over many of the problems he had solved in earlier publications and added the relevant analyses, including the one for the present problem. I explain this analysis in the next chapter (Analysis 5.3).

4.7 Reconstructing classical texts

The "domain of analysis"

The example from Ghetaldi illustrates a particular interest shared by many early modern geometers: the reconstruction of lost classical mathematical texts. The main source of inspiration for these reconstructions was the seventh book of Pappus' *Collection*, which contained a list of twelve ancient works by Aristaeus,[49] Euclid,[50] Apollonius,[51] and Eratosthenes,[52] together forming what Pappus called the "domain of analysis."[53] Only two of these works[54] were extant around 1600, but Pappus gave enough information about the six lost treatises of Apollonius for mathematicians to attempt to reconstruct their contents. Thus in the seventeenth century Snellius, Ghetaldi, Anderson, Viète, and Fermat published reconstructions of the following Apollonian treatises: *On the cutting off of a ratio*,[55] *On the cutting off of an area*,[56] *On determinate section*,[57] *On contacts*,[58] *Vergings*,[59] and *Plane loci*.[60]

An example: Apollonius' Vergings

Ghetaldi's *Apollonius revived* was a characteristic example of this reconstruc-

[48] [Ghetaldi 1630]
[49] *Solid loci*, five books.
[50] *Data*, one book, *Porisms*, three books, *Surface loci*, two books.
[51] *Conics*, *On the cutting off of a ratio*, *On the cutting off of an area*, *On determinate section*, *On contacts*, *Vergings* and *Plane loci*.
[52] *On means*, two books.
[53] [Pappus Collection] opening sections of book VII, pp. 477 sqq., cf. [Pappus 1986] pp. 82 sqq. and 66–70.
[54] Euclid's *Data*, cf. Chapter 5 Note 8 and Apollonius' *Conics*, cf. Note 86.
[55] [Snellius 1607].
[56] [Snellius 1607].
[57] [Snellius 1608].
[58] [Viète 1600].
[59] [Ghetaldi 1607], [Ghetaldi 1613], [Anderson 1612]
[60] [Fermat Isagoge].

tional activity. He published its first volume in 1607. Ghetaldi concluded from Pappus' information that Apollonius' *Vergings* had dealt with particular cases of neusis between two circles and between a circle and a straight line.[61] For the neusis between two circles he considered the special cases in which the pole of the neusis was in one of the points of intersection of the circles with their common diameter. For the neusis between a circle and a straight line he took the pole to be in one of the intersections of the circle with its diameter perpendicular to the line (this was Construction 4.16 discussed above). In these special cases the neusis problem turned out to be plane. The problems called for the further distinction of many different cases, some of which were solvable only when the length to be inserted (c in the example) satisfied certain conditions. Ghetaldi set himself the task to deal with all the separate cases, to determine the conditions of solvability, and to provide the constructions. The volume published in 1607[62] was not a full reconstruction; in the case of the neusis between two circles Ghetaldi only gave a sketch of the solution, claiming lack of time and pressure of friends as an excuse to publish an incomplete treatment. The missing parts were then supplied by Anderson in 1612,[63] but they did not satisfy Ghetaldi and so in 1613 he published the second volume of *Apollonius revived*[64] containing his own treatment of the case he had merely sketched earlier; his treatment was indeed more complete than Anderson's.

4.8 Division of figures

The general problem

In its general form the figure division problem is as follows:

Problem 4.17 (Division of figures)
Given: a rectilinear plane figure F, a point P, or a line l, and a ratio ρ; it is required to draw a line through P or parallel to L, dividing F in two parts F_1 and F_2 such that $F_1 : F_2 = \rho$.

This problem is plane.

Division problems were a popular theme in books on geometry in the sixteenth and seventeenth century. Their study even acquired a special name: "geodesics."[65] The tradition went back to a work by Euclid *On divisions*, which was mentioned by Proclus. For the complicated history of that text I refer to Archibald's introduction to his reconstruction of it.[66] The sources for the early modern activity were a translation of a partial Arabic version published by Dee and Commandino 1570,[67] and a section in Leonardo of Pisa's *The practice of*

[61] Cf. [Knorr 1986] pp. 298–302 and [Pappus 1986] pp. 527–534.
[62] [Ghetaldi 1607].
[63] [Anderson 1612].
[64] [Ghetaldi 1613].
[65] Cf. [Clavius 1604] p. 263 where Clavius explains the term.
[66] [Archibald 1915] pp. 1–18.
[67] [Dee & Commandino 1570].

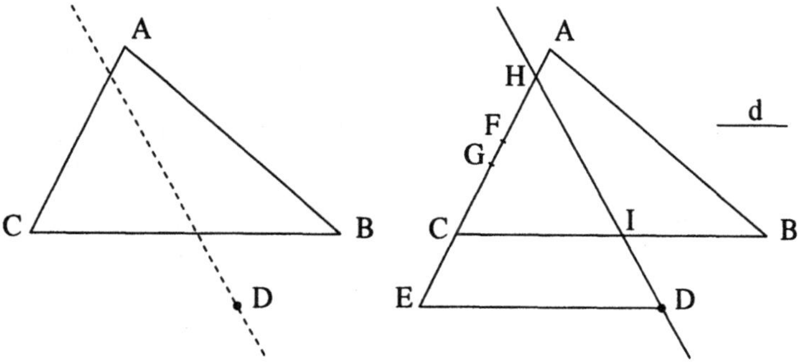

Figure 4.14: Division of a triangle — Clavius

geometry of 1220 most probably based on manuscripts of the Euclidean treatise of Greek or Arabic origin.[68]

Being plane, division problems were less difficult than their analogs, the general angular section and the division of ratios. Nevertheless, the constructions could be rather involved because many distinctions had to be made according to the particular shape of F and the position of P or L with respect to F.

An example from Clavius

My example of an area division problem is the relatively simple case where F is a triangle, P lies outside it and $\rho = 1:1$. It was treated by Clavius in his *Practical geometry* of 1604, where he gave the following construction:

Construction 4.18 (Triangle division — Clavius)[69]
Given: a triangle ABC and a point D (see Figure 4.14); it is required to construct a line through D that divides the triangle in equal parts. (Clavius dealt with the case in which the line DA divides the triangle in parts of which the left hand one is larger than the right hand one.)

Construction:
1. Draw a line through D parallel to CB; it intersects AC prolonged in E; bisect AC in F.

[68] Leonardo's treatise was first published in print in the nineteenth century: [Fibonacci 1857–1862] vol. 2, pp. 1–224; the section on division of figures is on pp. 110–148.
[69] [Clavius 1604] Book VI, Prop. 12, pp. 294–295.

2. Take G on AC such that CG is the fourth proportional of ED, CB and CF (i.e. $ED : CB = CF : CG$).
3. Determine the mean proportional d of CE and CG (i.e. $CE : d = d : CG$).
4. Find H on CA such that rect.(CH, GH) = sq.(d) (Clavius referred here to his Construction 4.3).
5. DH is the required line, i.e. $\triangle CIH = \frac{1}{2}\triangle ABC$.
[**Proof**: Draw DH, it intersects CB in I. Because of **3** and **4** we have rect.(CH, GH) = sq.(d) = rect.(CE, CG). Hence $HG : CG = CE : CH$; so $(HG + CG) : CG = (CE + CH) : CH$, that is $CH : CG = HE : CH = ED : CI$ (by similar triangles). So, using **2**, rect.(CI, CH) = rect.(ED, CG) = rect.(CF, CB) = rect.$(\frac{1}{2}AC, CB)$. Therefore, by a direct consequence of *Elements* VI-23 which Clavius had derived in an earlier proposition, $\triangle CIH = \frac{1}{2}\triangle ABC$.]

I use this example in Section 22.2 to illustrate Descartes' procedure of algebraic analysis and geometrical construction.[70]

A solid division problem

In Proposition 4 of the second book of his *Sphere and Cylinder* Archimedes dealt with a solid division problem: to cut a sphere by a plane into parts with a prescribed ratio. Archimedes did not provide a full construction but reduced the problem to one about a certain division of a line segment. His commentator Eutocius supplied three constructions, one which he thought could be Archimedes' own, one by Dionysidorus, and one by Diocles. All three constructions employed the intersection of conics — a parabola and a hyperbola in the first two cases, a hyperbola and an ellipse in the last.[71] The problem seems not to have attracted much attention in the early modern period, but at a rather late date it inspired Huygens to work out a construction in which he reduced the problem to a trisection. It was as follows:

Construction 4.19 (Division of a sphere — Huygens)[72]
Given: a sphere with center O and radius r, and a ratio $a : b$, $a > b$ (see Figure 4.15 which shows a great circle of the sphere); it is required to construct a point H on a diameter AB of the sphere such that the plane through H perpendicular to the diameter divides the sphere in two parts which have the given ratio.

Construction:
1. Extend AB to both sides and mark points C and D such that $CA = AO = OB = BD = r$.

[70] A variant of the triangle division problem, the so-called quadrisection (division of a triangle by two perpendicular lines into four equal parts) attracted interest around 1700, cf. [Hofmann 1960].
[71] [Eutocius CommSphrCyl] pp. 626–666; cf. [Archimedes nd] pp. 62–79.
[72] [Huygens 1654] Problem 1 of the *Problematum quorundam illustrium constructiones*, pp. 181–190, cf. p. 102 for the genesis of Huygens' construction.

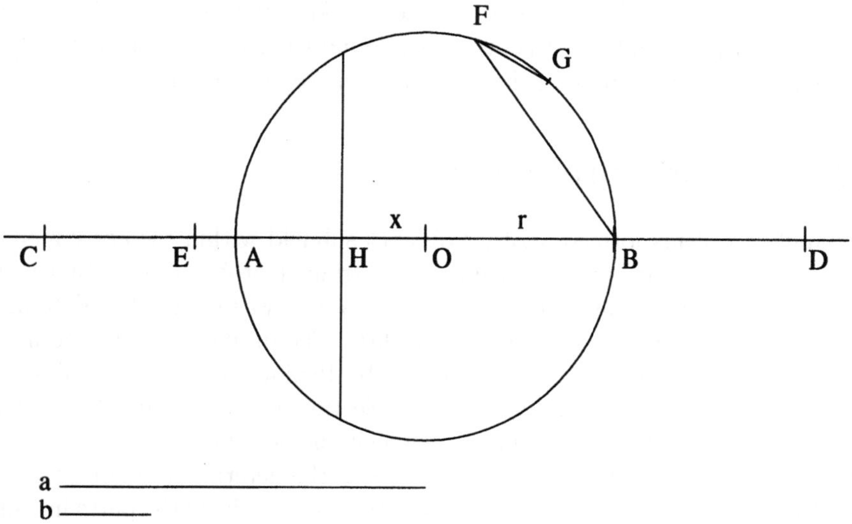

Figure 4.15: Division of a sphere — Huygens

2. Mark a point E on CO such that $DE : CE = a : b$; draw through B a chord BF of the circle with $BF = OE$.
3. Trisect arc BF and draw chord FG subtending a third part of the arc.
4. Mark H on CO such that $OH = FG$.
5. The plane through H perpendicular to AB divides the sphere in the required ratio.
[**Proof**: Call $OH = x$. From Prop. II-2 of Archimedes' *Sphere and Cylinder*, which expresses the volume of a segment of a sphere in terms of its base circle, its height, and the radius of the sphere, it follows that the volumes of the two parts of the sphere in this case are to each other as $(2r^3 + 3r^2x - x^3) : (2r^3 - 3r^2x + x^3)$. This ratio should be equal to $a : b$. Hence x should satisfy the equation $x^3 - 3r^2x + 2r^3\frac{a-b}{a+b} = 0$. This is the equation for the chord of $\frac{1}{3}$ of an arch whose chord is $2r\frac{a-b}{a+b}$ (cf., e.g., Construction 4.6). Now it follows from **2** that $OE = 2r\frac{a-b}{a+b}$. The construction of $x = OH = FG$ as the chord of one third of the arch spanned by $BF = OE$ ensures that x indeed satisfies the equation.[73]]

In presenting this construction Huygens mentioned the classical solutions by the intersection of conics and made it clear that he preferred the one with

[73] Huygens formulated his proof entirely in geometrical language, involving a separate lemma equivalent to the trisection equation; but the main steps in his proof are the same as here.

the trisection: "And this method of construction for solid problems seems in a way the simplest and best suited for use."[74] Although presented in classical geometrical style, it is likely that Huygens found the construction with help of Descartes' algebraic results on the trisection (cf. Section 26.6).[75]

4.9 Triangle problems

Regiomontanus' On triangles

Triangle problems formed another distinct subfield within the early modern tradition of geometrical problem solving. In a triangle problem, some (in general three) elements of a plane triangle were given and it was required to determine the triangle itself (that is, its sides and angles). The problems originated in the context of trigonometry; they were treated by Regiomontanus in the first two books of his *On triangles* of 1533 (cf. also Section 7.2). Regiomontanus dealt with many different triangle problems, from simple ones like the case that three sides were given,[76] to such complicated cases as the determination of a triangle given the lengths of a bisectrix of one angle and the lengths of the parts in which that bisectrix divides the opposite side.[77]

A triangle problem solved by algebra

Two of these problems acquired a certain notoriety in the later tradition because Regiomontanus had been unable to find their geometrical construction and had instead calculated the sides and angles of the triangle for specific numerical examples, deriving and solving the algebraic equation for one of the unknown line segments. In the first problem[78] he proceeded as follows:

Analysis 4.20 (Triangle problem — Regiomontanus)[79]
Given: line segments c, h, and a ratio $d : e$; it is required to find a triangle ABC with base $AB = c$, height $CD = h$ and $a : b = d : e$, where a and b are the sides BC and AC, respectively.

> **Analysis**: "Until now this problem has withstood solution in geometrical manner, but we shall try to solve it by the art of "res" and "census.""[80] Take, for example, $d : e = 5 : 3$ (hence $DB > AD$), $h = 5$ and $c = 20$.
> **1**: Take E on DB with $AD = DE$, call $EB = 2x$.
> **2.** Then $AD = 10 - x$ and $DB = 10 + x$.

[74][Huygens 1654] p. 45: "Atque haec construendi ratio in solidis problematibus quodammodo simplicissima videtur atque ad usum maximè accomodata."

[75]Cf. the related manuscripts in [Huygens 1888–1950] vol. 12 pp. 9–12, 16–18.

[76][Regiomontanus 1533] Book I Problems 42–47.

[77][Regiomontanus 1533] Book II Prop. 30 pp. 136–139.

[78]The second problem was: given the height h, the difference $b - a$ of the sides, and the difference $c_2 - c_1$ of the parts c_2 and c_1 in which the height divides the base c, to construct the triangle (Book II Problem 23).

[79][Regiomontanus 1533] Book II Prop. 12 p. 51.

[80][Regiomontanus 1533] p. 51: "Hoc problema geometrico more absolvere non licuit hactenus sed per artem rei et census id efficere conabimur." The terms "res" and "census" denoted the unknown and its square.

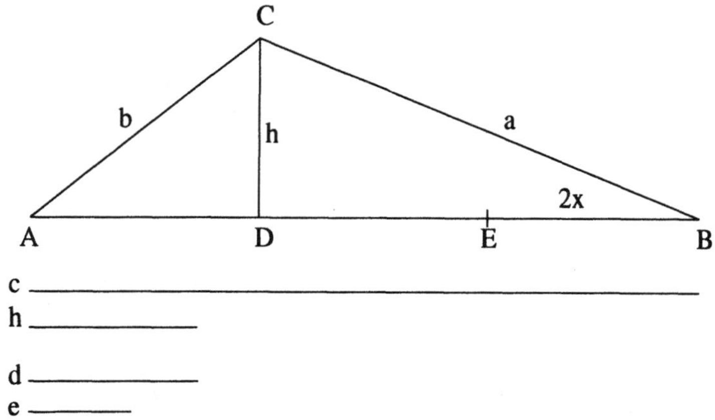

Figure 4.16: A triangle problem — Regiomontanus

3. $b^2 = AC^2 = AD^2 + CD^2 = (10 - x)^2 + h^2 = x^2 - 20x + 125$.
4. Similarly $a^2 = BC^2 = x^2 + 20x + 125$.
5. $b^2 : a^2 = 9 : 25$ yields $16x^2 + 2000 = 680x$.
6. "Hence the rules of the art will show us what remains to be done. The line BE which I posed to be 2 "res" will emerge as known. . . ."[81]

Regiomontanus assumed his readers knew how to solve the quadratic equation in step **5**. He was aware that in general such a solution in numbers would yield an approximate solution, not a geometrical one and from his words it appears that he felt defeated by the lack of a geometrical solution in the form of a general construction. However, as we will see (Section 7.2), he was not impressed by the advantages of exact over approximate solutions.

The result makes clear that for Regiomontanus algebra was still linked primarily with numbers and that general algebraic procedures could only be stated in terms of specific numerical examples. These two features of the "art of res and census" prevented him from translating the procedure of solving the equation (especially the root extraction it involved) into a construction by straight lines and circles. Regiomontanus' solution thus illustrates an obstacle against an easy merging of algebra and geometry; I discuss such obstacles in general in Section 6.4.

[81] "Quamobrem quod restat, praecepta artis docebunt. Linea ergo GE quam posui 2 res nota redundabit . . ." [Regiomontanus 1533] p. 51.

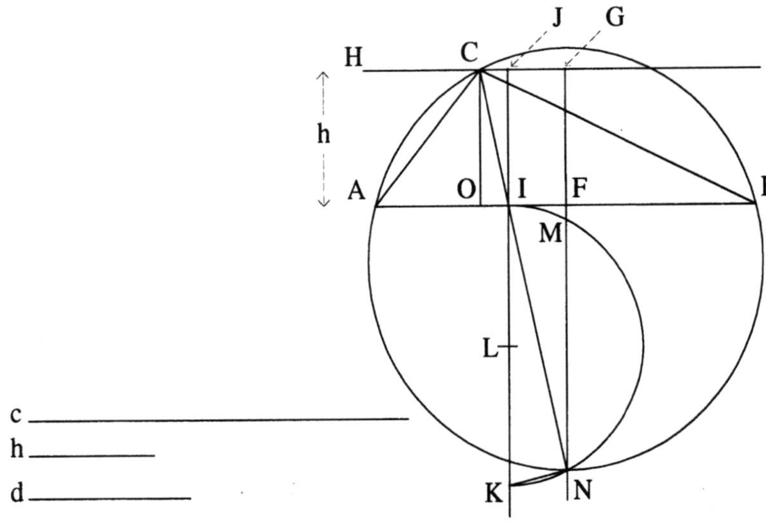

Figure 4.17: Construction of Regiomontanus' triangle problem — Viète

Viète's construction The two problems whose geometrical constructions eluded Regiomontanus were plane, as appears from the fact that Regiomontanus found quadratic equations. Several writers dealt with the two problems later,[82] some praising the decision to use algebra, others providing constructions and analyses. Viète, for instance, added an appendix to his *The French Apollonius* of 1600, entitled "On the problems whose geometrical construction Regiomontanus said he did not know."[83] Viète gave constructions and proofs but we don't have the analyses by which he reached them. His construction of Regiomantanus' first problem was as follows:

Construction 4.21 (Triangle problem — Viète)[84]
Given: two line segments c and h, and a ratio $d : e$ (see Figure 4.17); it is required to construct a triangle with base c, height h, and sides a, b, such that $a : b = d : e$. [Viète notes a condition: If c is divided into two segments c_1 and c_2 such that $c_2 : c_1 = d : e$, then c_1c_2 should be $\geq h(c_2 - c_1)$, where $c = c_1 + c_2$.)

Construction:

[82] [Nonius 1567] pp. 269^r–271^v (they are Nonius' problems nrs 46 and 51; in a passage introducing the first of these he mentions Regiomontanus' recourse to Algebra in support of the scientific stature of "esta subtilissima arte de Algebra"); [Viète 1600] pp. 339–342, "Appendicula I De problematis quorum geometricam constructionem se nescire ait Regiomontanus:" [Ghetaldi 1607] pp. 7–9; [Cyriacus 1616]; [Anderson 1617] (cf. [Grisard 1969]); [Anderson 1619] pp. 9–15; cf. also [Garibaldi 1992] pp. 183–186.

[83] Cf. Note 82.

[84] [Viète 1600] p. 340.

1. Take $AB = c$; bisect AB in F; draw GF perpendicular to AB with $GF = h$; draw $HG \parallel AB$.
2. Take I on AB such that $AI : IB = e : d$ (thus AI and IB are the c_1 and c_2 of the condition); draw JI perpendicular to AB with J on HG; prolong JI.
3. Take K on JI prolonged such that $JI \times IK = AI \times IB$ (hence $IK = \frac{c_1 c_2}{h}$).
4. Bisect IK in L; draw a semicircle with diameter IK and center L.
5. Prolong GF; it intersects the semicircle in two points M and N or it touches the circle in one point N [because $IF = \frac{1}{2}(c_2 - c_1) \leq \frac{1}{2}\frac{c_1 c_2}{h} = \frac{1}{2}IK = IL$].
6. Draw NI; its prolongation intersects HG in C; draw AC and CB.
7. ACB is the required triangle.
[**Proof**: $AB = c$ and $CO = h$ by **1**. Draw KN; now $CI : JI = IK : IN$ because of similarity of triangles, hence $CI \times IN = JI \times IK = AI \times IB$ (by **3**). So the points C, N, A, and B are on one circle; draw that circle; its centre is on GF (because GF bisects chord AB perpendicularly); arcAN = arcNB, so $\angle ACI = \angle ICB$ and therefore (by *Elements* VI-3) $AC : CB = AI : IB$. Now $AI : IB = e : d$ by **2**, so $AC : CB = e : d$ as required.]

From the construction and proof it seems likely that Viète arrived at this construction by a classical rather than an algebraic analysis of the problem.[85]

4.10 Varia

The normal to a parabola

Under "Varia" I add one example of a construction by reduction to one of Viète's standard solid problems discussed above in Section 4.2. It is the problem of constructing a normal to a given parabola through a given point outside the parabola. In the *Collection* Pappus had criticized Apollonius' solution of "the problem about the parabola in the fifth book of the *Conics*" for using solid means although a plane construction was possible (cf. Section 3.4). The first four books of the *Conics* were available in print since the middle of the sixteenth century.[86] Books V–VII, however, which came to the West via an Arabic translation, remained inaccessible until well into the seventeenth century.[87] Hence Apollonius' own construction of the problem was not known. But from Pappus' statement and from the information about book V that was available, it

[85] On the distinction between these two kinds of analysis see the next chapter. Viète may well have arrived at his construction by realizing that the bissectrix of the angle at C cuts the base in segments AI and IB such that $AI : IB = b : a = e : d$ (*Elements* VI-3), hence the point I can be constructed. Moreover, $OC : CI = IK : CN$, hence $h \times IK = CI \times IN$. By *Elements* III-35 $CI \times IN$ is equal to $AI \times IB$ and is therefore known; consequently, IK can be constructed. The rest of the construction then follows easily.

[86] [Apollonius 1537] and [Apollonius 1566].

[87] For the complicated history of their publication see [Apollonius 1990] vol. 1 pp. xiv–xxvii.

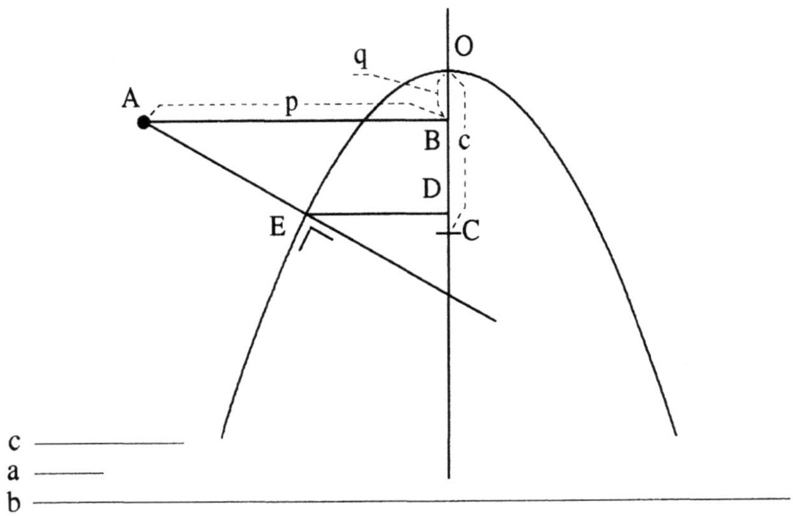

Figure 4.18: Construction of a perpendicular to a parabola — Anderson

was clear that Pappus' criticism concerned precisely the problem of finding a perpendicular to a parabola through a given point outside the parabola.

Anderson's construction In 1619 Anderson published a solution of the problem. He did not see the possibility of using the given parabola as means of construction[88] and concluded that Pappus had been wrong in claiming that the problem was plane. He provided a construction by reduction to one of Viète's solid "constitutive" problems:

Construction 4.22 (Perpendicular to a parabola — Anderson)[89]
Given: a point A, a parabola, not through A, with vertex O, vertical axis and latus rectum $c = OC$*; it is required to draw a line through A intersecting the parabola perpendicularly.* [Anderson distinguished several cases depending on the position of A with respect to O; I give only the first case he treated, which assumed that A was below O and that the vertical distance of A to O was smaller than $\frac{1}{2}c$; there are some calculation errors in his argument, which I have corrected.]

Construction:

1. Draw AB horizontally, with B on the axis; call $AB = p$ and

[88]Cf. Section 3.4; it seems that Huygens was the first to suggest this possibility and by it to justify, or at least explain, Pappus' criticism. He did so in a passage added by Van Schooten to the 1659 edition of Descartes *Geometry*, cf. [Huygens 1888–1950] vol. 1 pp. 242–243 and vol. 14 pp. 420–422.

[89][Anderson 1619] pp. 25–27.

$OB = q$.
2. Construct the mean proportional a of c and $\frac{1}{2}c - q$ (i.e., $c : a = a : (\frac{1}{2}c - q)$).
3. Construct the fourth proportional b of $\frac{1}{2}c - q$, $\frac{1}{2}c$ and p [i.e., $(\frac{1}{2}c - q) : \frac{1}{2}c = p : b$].
4. Now solve the following (Vietean) standard problem (cf. Problem 4.5 above): Find three line segments x, y, and z such that a, x, y, and z are in continued proportion and $x + z = b$.
5. The segment x is the ordinate DE of the point E in which the required line intersects the parabola; determine E and draw the perpendicular AE to the parabola.
[**Proof**: Anderson's proof was based on his analysis of the problem which I discuss in the next chapter (cf. Analysis 5.9 and Construction 5.10), so I omit a proof here.]

Anderson did not give the construction of the Vietean standard problem to which he had now reduced the parabola problem. He merely stated[90] that it was of the same kind as the problem of two mean proportionals and that it could be constructed by neusis or by motion, similar to Plato's construction of two mean proportionals by means of gnomons (cf. Construction 2.3). He also suggested that the reader might prefer reducing the cubic equation by a reduction analogous to Cardano's formulas[91] to the determination of cubic roots, that is, the determination of two mean proportionals (which can indeed be done in this case).

4.11 Heuristics and analysis

Heuristics often not explained

In most of the examples discussed in this chapter I have not explained how the construction and its proof were found. In leaving out this information I kept close to the spirit of the texts; in many cases the authors themselves did not discuss how they had found the constructions. Sometimes their readers could plausibly guess from the proof or from the construction itself how the author had found it. But for the more complicated constructions, such as Clavius' division of the triangle, Viète's solution of Regiomontanus' problem, and Ghetaldi's neusis,

90 [Anderson 1619] p. 27.

91 I use the term "Cardano's formulas" for formulas which represent the method for solving cubic equations published by Cardano in his *The great art* of 1545 ([Cardano 1545] Ch. 11, cf. [Cardano 1966] pp. 249–251). Cardano attributed the formula to Scipione Ferro. In modern terms the method may be summarized as follows: Any cubic equation $x^3 + ax^2 + bx + c = 0$ can be reduced (by the substitution $x \to x - a/3$) to a form without quadratic term: $x^3 + px + q = 0$; one root of the latter equation is given by

$$z = \sqrt[3]{-\frac{q}{2} + \sqrt{(\frac{q}{2})^2 - (\frac{p}{3})^3}} + \sqrt[3]{-\frac{q}{2} - \sqrt{(\frac{q}{2})^2 - (\frac{p}{3})^3}}\,. \tag{4.15}$$

Cardano's method became generally known among mathematicians in the second half of the sixteenth century, cf., e.g., [Kline 1972] pp. 263–266.

only the expert insider would be able to reconstruct the reasoning that led the author to the construction.

Reasons for not explaining

Mathematical writers could have various reasons for omitting explanations of how they arrived at their constructions. The paradigm classical works as Euclid's *Elements* and Apollonius' *Conics* were written in the same apodictical style. In fact several early modern mathematicians concluded from these works that the classical style of solving problems actually required to conceal the methods by which the solutions were found.[92] Not infrequently it would suit an author's purpose to conform to that classical style and to show his skill without sharing it. In other cases geometers might find it difficult to formulate how they arrived at constructions because their heuristic procedures were insufficiently formalized.

Interest in analysis

Yet, although the usual style of presentation did not call for explicit heuristics, there was a strong interest in it. From classical sources early modern geometers knew that the ancients had developed analytical methods for finding constructions of problems. But it was believed that, partly because of the synthetic style of presenting mathematics and partly because of a deliberate wish to keep methods secret, most knowledge about classical analytical methods had been lost.[93] The lack of good analytic methods was recognized and new methods were devised. These attempts culminated in Descartes' analytical use of algebra. In the next chapter I discuss various pre-Cartesian methods of analysis and give examples.

---

[92] Thus Viète wrote in his *Isagoge*: "So a skillful geometer, although thoroughly versed in analysis, conceals the fact, and, concentrating on the accomplishment of the task, proposes and explains his problem as synthetic." ([Viète 1591] p. 10: "Itaque artifex Geometra, quanquam Analyticum edoctus, illud dissimulat, et tanquam de opere efficiundo cogitans profert suum syntheticum problema, et explicat.")

[93] Descartes, for instance, remarked in the *Rules*: "we are well aware that the geometers of antiquity employed a sort of analysis which they went on to apply to the solution of every problem, though they begrudged revealing it to posterity." ([Descartes Rules] p. 373, translation quoted from [Descartes 1985–1991] vol. 1, p. 17).

Chapter 5

Early modern methods of analysis

5.1 Introduction

Pappus' account of the method of analysis

Two kinds of analysis were distinguished in early modern geometry: the classical and the algebraic.[1] The former method was known from examples in classical mathematical texts[2] in which the constructions of problems were preceded by an argument referred to as "analysis;" in those cases the constructions were called "synthesis." Moreover, a few classical sources[3] spoke in general about this arrangement. The most important of these texts was the opening of the seventh book of Pappus' *Collection*; I quote this passage here in full:

> Now, analysis is the path from what one is seeking, as if it were established, by way of its consequences, to something that is established by synthesis. That is to say, in analysis we assume what is sought as if it has been achieved, and look for the thing from which it follows, and again what comes before that, until by regressing in this way we come upon some one of the things that are already known, or that occupy the rank of a first principle. We call this kind of method "analysis," as if to say *anapalin lysis* (reduction backward). In synthesis, by reversal, we assume what was obtained last in the

---

[1]Thus in Book V, Ch. IV, pp. 330–343 of [Ghetaldi 1630] Ghetaldi treated a number of problems that he considered beyond the force of algebra and that he therefore solved "by the method which the ancients used in analysing and synthesizing all problems" ("Methodo, qua veteres in resolvendis et componendis omnibus problematibus utebantur" p. 330).

[2]Notably: Book II of Archimedes' *Sphere and Cylinder* together with the commentaries of Eutocius on Propositions II-1 and II-4, and Pappus' *Collection*. Furthermore, Euclid's *Data* was recognized as a collection of theorems useful in the analysis of plane problems.

[3]Primarily the passages in Pappus' *Collection* discussed below; also a scholium to *Elements* XIII-1–5, which in the Renaissance was attributed to Theon (cf. [Euclid 1956] vol. 3 pp. 442–443). In his *Isagoge* Viète referred to the definition of analysis in the scholium and attributed it to Theon ([Viète 1591] p. 1, [Viète 1983] p. 11).

> analysis to have been achieved already, and, setting now in natural order, as precedents, what before were following, and fitting them to each other, we attain the end of the construction of what was sought. This is what we call "synthesis."
>
> There are two kinds of analysis: one of them seeks after the truth, and is called "theorematic." In the case of the theorematic kind, we assume what is sought as a fact and true, then, advancing through its consequences, as if they are true facts according to the hypothesis, to something established, if this thing that has been established is a truth, then that which was sought will also be true, and its proof the reverse of the analysis; but if we should meet with something established to be false, then the thing that was sought too will be false. In the case of the problematic kind, we assume the proposition as something we know, then, proceeding through its consequences, as if true, to something established, if the established thing is possible and obtainable, which is what mathematicians call "given," the required thing will also be possible, and again the proof will be the reverse of the analysis; but should we meet with something established to be impossible, then the problem too will be impossible.[4]

The passage is difficult to understand and has led to a discussion on its meaning, which, as far as I can see, remains inconclusive.[5] Unsurprisingly, early modern mathematicians did not develop a unique interpretation of it either. However, one may say that, in combination with the extant classical examples of analyses, the passage in Pappus led to a view, generally shared by seventeenth-century mathematicians, according to which classical analysis was a procedure for finding constructions of problems or proofs of theorems in which the concept of "given" played a central role. This analysis was to be completed by another procedure, called synthesis, differing from the analysis with respect to the direction of the argument. I come back to the difference of direction of argument in analysis and synthesis below in connection with some of the examples (Section 5.3).

In the quoted passage Pappus made a distinction between theoretical and problematical analysis, the one leading to a proof of a theorem, the other leading to the construction of a problem. It should be remarked, however, that virtually all examples of analyses in the classical literature were analyses of problems, and the corresponding syntheses were constructions; there were no clear cut examples of the analysis of theorems.[6] It appears that early modern mathematicians

[4][Pappus Collection] opening sections of book VII, pp. 477–478 (vol. 2). The translation above is from [Pappus 1986], vol. 1, pp. 82–85; for variant translations see [Heath 1921] vol. 2, pp. 400–401 and [Knorr 1986] pp. 354–357.

[5]See in particular [Gulley 1958], [Mahoney 1968], [Hintikka & Remes 1974], [Knorr 1986], and [Behboud 1994].

[6]Knorr argues convincingly ([Knorr 1986]) that Pappus' notion of an analysis of theorems is in fact "entirely gratuitous" (p. 358) and that the extant examples of analyses of theorems are "trivial" (p. 378, note 102) apart from two examples in Pappus, which still are unconvincing

were not interested in formal analyses of theorems.[7]

Classical analysis

For plane problems the method of analysis by means of the concept "given" was codified in Euclid's *Data* (literally "givens," I refer to the work as *Data*), of which a Latin translation was available in print since 1505.[8] I explain the procedure below in connection with Construction 5.2.

However, most classical examples of analyses concerned solid problems. These analyses served either to find a standard solid problem to which the construction of the problem at hand could be reduced, or to find the conics by whose intersection the construction was to proceed. Instances of the first kind were in the propositions II-1 and II-4 of the *Sphere and cylinder*[9] in which Archimedes reduced the (solid) problems he was dealing with to others which he assumed solved (in the case II-1: the determination of two mean proportionals) or promised to deal with later (in the case II-4). Instances of analyses serving to find constructing conics occurred in Eutocius' commentary to Proposition II-1, containing the famous list of 12 constructions of two mean proportionals.[10] Two of these constructions (those atributed to Menaechmus) used the intersection of conics; they were preceded by an analysis in which these conics were found as loci.[11] Other examples of such analyses were in Eutocius' comment on Proposition II-4 of *Sphere and Cylinder*.[12]

Early modern geometers showed considerable interest in the classical method of analysis and were in general familiar with its terminology and techniques, at least as far as necessary for analyzing plane problems.

Analysis by algebra

Algebra[13] entered geometry through its use in the analysis of problems and from c. 1590 the development of this analytical use of algebra can be identified as the principal dynamics (cf. Section 1.5) within the early modern tradition of geometrical problem solving.

From 1591 onward Viète consciously and explicitly advocated the use of algebra as an alternative method of analysis, applicable in geometry as well as in arithmetic. The method consisted in reducing the problem to an equation, which

as examples of theorematic analysis.

[7]I am aware of only one example of such an interest, namely Van Schootens *Treatise on devising geometrical proofs from algebraic calculations* ([Schooten 1661]). Most of the propositions in this tract concern problems and constructions, but there are some geometrical theorems (typically one equivalent to the equality $\tan 2\alpha = \frac{2\tan\alpha}{1-\tan^2\alpha}$) whose proofs Van Schooten derived algebraically.

[8][Euclid 1505]; the Greek text was first published in 1625, [Euclid 1625].

[9][Archimedes SphrCyl] pp. 57–58, 62–65.

[10][Eutocius CommSphrCyl] pp. 588–620.

[11]In my presentation of one of Menaechmus' constructions, Construction 3.1 above, I have omitted the analysis.

[12][Eutocius CommSphrCyl], pp. 626–666; Eutocius gave three solutions of the problem by the intersection of conics; one, he suggested, might be Archimedes' own, the two others were by Dionysidorus and by Diocles; the first and third of these also contained an analysis.

[13]To be precise, algebra involving unknowns and indeterminates, cf. the explanation of my use of the terms "algebra" "geometry," and "analysis" in Section 6.3.

for the usual geometrical problems would be an equation in one unknown.[14] This equation was to be reduced to some standard form, corresponding to a standard construction. If the equation was quadratic the standard construction was a plane one; if its degree was higher the construction was a solid or an even more complex one.[15] This use of algebra in geometry had been pioneered by some Renaissance mathematicians before 1590, but it was Viète's conscious identification of this method with analysis that brought it into the center of attention. Descartes' method for geometry, although considerably different and more extensive than Viète's, was nevertheless based on the same conception of the use of algebra: reduce problems to equations and seek their construction from these equations, rewritten in standard form.

There is one part of classical analysis that on retrospect seems the most amenable to algebraic formulation; this is the determination of two constructing conics as loci for two properties shared by the required point. In the classical examples of such analyses the properties were relations involving the ordinates and abscissae of the conics; these relations were equivalent to quadratic equations in two unknowns. Thus the material for an algebraic characterization of curves was as it were readily available. In fact, however, this type of analytical procedure played a limited role in the creation of the early modern algebraic method of analysis. Viète was not interested in constructing solid problems by intersection of conics and therefore did not explore the possibility of formulating an algebraic equivalent. Even Descartes, who did develop the method of characterizing loci by equations in two unknowns, did not arrive at this conception via the classical analytical method of finding the constucting conics for solid problems. The only mathematician who elaborated an algebraic analysis based on the classical analysis by loci was Fermat, but his method had less influence than Descartes'. I return to this aspect of early modern analytical methods in Section 13.1.

Table 5.1 gives an overview of the methods of analysis mentioned above and discussed in the next sections.

5.2 Classical analysis of plane problems

Analysis of a triangle problem

The classical method of analysis is best introduced by a simple example of an analysis with the pertaining synthesis, for which I choose Ghetaldi's solution of a triangle problem in his *Collection of various problems* of 1607.

Analysis 5.1 (Triangle problem — Ghetaldi)[16]

Given: two line segments a and d, and an angle γ (see Figure 5.1); it is required to construct a triangle with base c, vertex angle γ, sides a and b, and such that $c - b = d$.

[14] Viète did not apply algebra to locus problems.

[15] Viète did not explore the possibility for higher-degree equations to be reducible; the importance of irreducibility for construction was first realized by Descartes, cf. Section 27.3.

[16] [Ghetaldi 1607] Problem XVI, pp. 41–42.

	Classical analysis	Algebraic analysis
Plane problems	By the concept of "given" — Section 5.2	By reducing the problem to an equation in one unknown — Sections 5.3 and 5.4
Solid problems	By the concept of "given:" reducing the problem to a standard one or determining the constructing conics as loci — Section 5.5	By reducing the problem to an equation in one unknown and then either (a) reducing the equation to a standard solid problem or (b) using equations in two unknowns to find the constructing conics — Section 5.6

Table 5.1: Early modern methods of analysis

Analysis:
1. "Factum iam sit" — "Let it be done" and let the triangle be ABC; let AC be prolonged to D such that $AD = c$, then $CD = c - b = d$; Let DB be drawn.
2. The side $BC = a$ is given in position and in magnitude;[17] $\angle BCA$ is given, so $\angle BCD$ is given too.
3. Since BC is given in position and magnitude, so (*Data* 29) CD is given in position.
4. The line segment CD is also given in magnitude (for it is equal to d), so (*Data* 27) the point D is given.
5. The point B is also given, so (*Data* 26) BD is given in position and magnitude.
6. Hence triangle CDB is given, so $\angle CDB$ is given.
7. Now $\angle CDB = \angle DBA$ because $AD = AB$; hence (*Data* 29) AB is given in position and therefore (*Data* 25) the point A is given in position.
8. Hence (*Data* 26) AB and AC are also given in magnitude.
9. Thus the triangle ABC is given.
[**Summary** The point A is found to be "given" in **7**, the sides b and c in **8**; the side a and the points B and C are given from the outset; hence the triangle is given.]

Construction 5.2 (Triangle problem — Ghetaldi)
Given etc.: see Analysis 5.1.

Construction (Synthesis) (the numbers of the steps correspond to those in the analysis.):

[17] See below for an explanation of the terms "given in position" and "given in magnitude," and for the references to the *Data*.

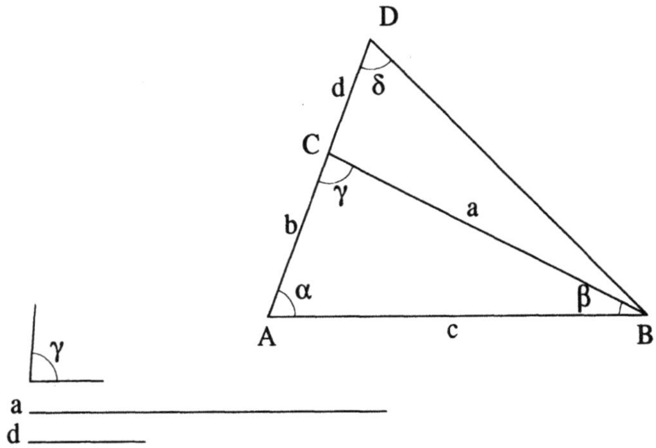

Figure 5.1: Analysis and synthesis of a triangle problem — Ghetaldi

1–2. Draw $BC = a$.
3. Draw angle γ at C, prolong the other side.
4. Take $CD = d$ along the other side.
5. Draw BD.
6–7. At B, draw an angle equal to $\angle CDB$ with one side along BD; the other side intersects DC prolonged in A.
8. ABC is the required triangle.
[**Proof**: By construction $CB = a$ and $\angle BCA = \gamma$; $BA = DA$ so $c = b + d$ or $c - b = d$.]

Role of the concept "given"

The example illustrates two characteristic features of a classical analysis of a problem: it proceeded by means of a concept "given," and it was performed with respect to a figure in which the required elements were supposed to be drawn already. The latter was indicated by such phrases as "factum jam sit," "let it be done," which served as a standard reminder that the subsequent argument was an analysis. The at-first-sight contradictory approach, namely to assume a problem solved in order to find its solution, was seen as the essential feature of analytical reasoning. In the supposed figure some elements were given at the outset; some were directly constructible from those originally given, and some required more steps. The analysis used a kind of shorthand, codified largely in Euclid's *Data*, for finding the constructible ("given") elements in the figure. The geometer used that shorthand as it were to plot a path from the primary given elements to the elements he ultimately wanted to construct.

Euclid's Data

In the *Data* Euclid distinguished between three modes of being given: given in magnitude, given in position, and given in kind. The first two modes occur in the example above. Geometrical entities (line segments, angles, rectilinear figures) were *"given in magnitude"* if, as Euclid phrased it: "we can assign equals to them,"[18] [19] The third mode applied to rectilinear figures (triangles, polygons); such a figure could be *"given in kind"*, which meant that its angles and the ratios of its sides were given, but not its size. Thus if a figure was given in kind, this meant that another figure similar to it could be placed anywhere in the plane. For ratios there was only one mode of being given: a ratio was *given* if a ratio equal to it could be obtained,[20] which meant in effect that two magnitudes could be produced whose ratio was equal to the given ratio.

Euclid's *Data* contained some 100 propositions of the form "If A is given in mode α, then B is given in mode β." In all these propositions the consequent (B) was constructible by straight lines and circles from the antecedent (A). These propositions, then, provided the steps by which the geometer planned the route from the given elements to the required ones. Once that route was planned, the construction could be written out along the same path, as we saw in Ghetaldi's construction.

In illustration I quote the propositions from the *Data*[21] to which Ghetaldi referred in his analysis:

> *Data* 25: If two lines[22] given in position intersect, the point at which they intersect each other is also given in position.
> *Data* 26: If the extremities of a straight line will have been given in position, the straight line will be given in position and in magnitude.
> *Data* 27: If one extremity of a straight line given in position and magnitude will have been given, the other extremity will also be given.
> *Data* 29: If from a straight line given in position and from a point given therein there is drawn a straight line making a given angle, the line so drawn is given in position.

One notes that in Ghetaldi's problem the three given elements (a, d, γ) were given in magnitude. Nevertheless, Ghetaldi started his analysis (cf. step **2**) with the assumption that one of them, namely the side a, was given in position as well. He could do so because it was required to construct any triangle with the given properties, so he was free to assume an arbitrary position in the plane for one of the elements given in magnitude.

Direction of the argument in analysis and synthesis

Once a route from the given to the required elements was planned in the anal-

[18] [Euclid Data] Def. 1 (in Ito's translation, [Euclid 1980] p. 55): "Spaces, lines and angles are said to be given in magnitude, when we can assign equals to them."

[19] [Euclid Data] Def. 4.

[20] [Euclid Data] Def. 2.

[21] [Euclid Data] 25–27, 29, translation quoted from [Euclid 1980] pp. 94–99.

[22] It appears from later items in the *Data* (e.g. 39) that both straight lines and circles are meant here.

ysis by means of the *Data*, the geometer could proceed to the synthesis, i.e., the actual construction, by writing out the separate construction steps along that route. In the case of our example the synthesis proceeded along the same steps and in the same direction as the analysis. Nevertheless, as we have seen at the beginning of this chapter, the *locus classicus* on analysis, Pappus' explanation at the beginning of *Collection* VII, stressed that the arguments in analysis and synthesis were each other's reverse. Modern commentators have also seen this reversal of argument as crucial.[23] It appears, however, that the actual practice of analysis as we find it in classical and early modern sources calls for a more nuanced view. Beyond the fact that analysis started with the assumption of the required, its direction of argument was not so definite. In Ghetaldi's case the directions in both analysis and synthesis were exactly the same. In the actual analytical argument (as opposed to the formal analysis written down after success) probably both directions occurred. Characteristically, the heuristic procedure of a "problematical" analysis would start arguing at any place that looked promising as a link in the chain which ultimately was to connect the "given" and the "required." Afterward the geometer might well choose to present the analysis thus found in a consistent order. In the case of analysis on the basis of the *Data* the obvious order was the same as that of the synthesis, namely from the given to the required, and that is what we find in Ghetaldi's case. In the analysis by means of algebra the situation was different. There at least the first part of the analysis, which argued from the problem to the equation, had a direction opposite to that of the synthesis. This will become clear in the next example.

5.3 Algebraic analysis of plane problems

A plane neusis problem analysed by algebra

In his *On mathematical resolution and composition*[24] Ghetaldi discussed, at great length, both the classical and the algebraic analysis of problems. He took the occasion to explain the analyses of many problems whose constructions he had published earlier without analysis. Among these was the special neusis problem whose construction, from *Apollonius revived* of 1607, I discussed in the previous chapter (Construction 4.16). It is a good example of a simple algebraic analysis in Vietean style.

[23]Cf. [Heath 1921] vol. 2, p. 401, [Klein 1968] pp. 154–155, 259–260, [Mahoney 1968] pp. 326–327, [Hintikka & Remes 1974], pp. 7–21, [Knorr 1986] pp. 348–360.

[24][Ghetaldi 1630]; *resolutio* and *compositio* were the Latin terms for analysis and synthesis.

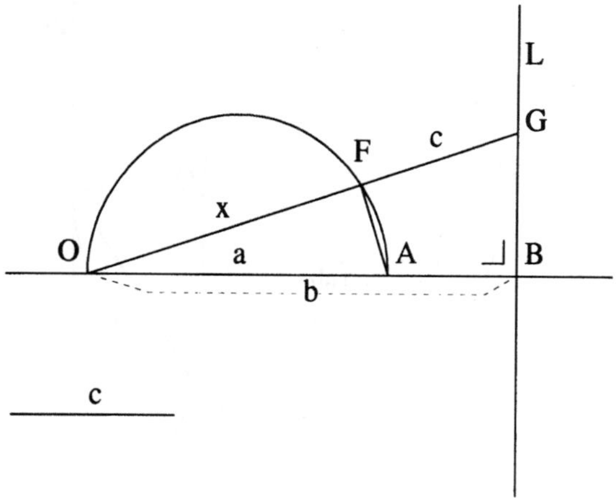

Figure 5.2: Special neusis between a circle and a line — Ghetaldi

Analysis 5.3 (Special neusis between a circle and a line — Ghetaldi)[25]

Given: a semicircle with diameter $OA = a$, a line L intersecting the prolongation of OA perpendicularly in B with $OB = b$, and a line segment c (see Figure 5.2, I use letters corresponding to those of Figure 4.13); it is required to find a line through O, intersecting the semicircle and the line L in F and G, respectively, and such that $FG = c$.

Analysis:
1. "Sit iam factum:" let OF be called x, let AF be drawn.
2. By similar triangles: $x : a = b : (x + c)$, hence $x^2 + xc = ab$.
3. Therefore $x = \sqrt{\frac{c^2}{4} + ab} - \frac{1}{2}c$, hence x is "given."

Construction 5.4 (Special neusis between a circle and a line — Ghetaldi)[26]
Given etc.: see Analysis 5.3.

[25][Ghetaldi 1630] Book III, Problem VII, Casus 1, pp. 118–120. Because the notational aspects of Ghetaldi's Vietean style are lost in my presentation, I give one example of it; the result in step **3** of the analysis is expressed as follows (the letters correspond by x = A, a = D, b = G, c = B): "Et explicata aequatione L.V.(BQ $\frac{1}{4}$ + D in G) - B $\frac{1}{2}$ aequabitur A." (L.V. stands for "latus universalis" indicating that the root has to be extracted from the whole of what follows; Q stands for "quadratus"). Ghetaldi formulated the result also in prose with reference to a figure: "Recta cuius quadratum aequale est quadrato dimidiae CD & rectangulo AOB, contracta dimidia CD aequalis est recta OC. Datur ergo OC quaesita." (I have changed the letters so as to fit the figure.)

[26][Ghetaldi 1607b], Problem II, Casus 1, pp. 5–6, also in [Ghetaldi 1630] p. 120.

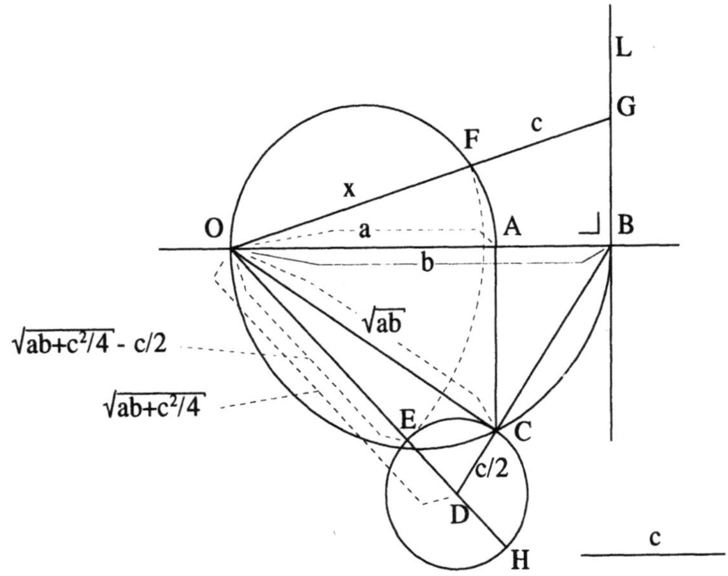

Figure 5.3: Special neusis between a circle and a line — Ghetaldi

Construction: (The construction is the same as Construction 4.16; the numbers do not correspond to those of the analysis; I note between brackets the algebraic interpretation of each step.)
1. Draw (see Figure 5.3) a semicircle with diameter OB; draw AC perpendicularly to OB with C on the semicircle; draw OC [construction of $OC = \sqrt{ab}$].
2. Draw BCD with $CD = \frac{1}{2}c$ [construction of $CD = \frac{1}{2}c$ perpendicular to OC]; draw OD [$OD = \sqrt{CD^2 + OC^2} = \sqrt{(\frac{1}{2}c)^2 + ab}$].
3. Draw a circle with centre D and radius DC; its intersection with OD is E [subtraction of $ED = \frac{1}{2}c$ from $OD = \sqrt{(\frac{1}{2}c)^2 + ab}$; hence $OE = x$].
4. Take F on the given semicircle such that $OF = OE$ [placing $OF = x$ in position] (here I omit Ghetaldi's proof that $OE < OA$); prolong OF, it intersects L in G.
5. OFG is the required line, that is, $FG = c$.
[**Proof**: See Construction 4.16.]

Ghetaldi also dealt with the case in which the line L intersects the given circle; its analysis led to the same equation (now with $b < a$), the construction was, *mutatis mutandis*, the same.[27] Ghetaldi used this construction in solving

[27] [Ghetaldi 1607b] Problem 2, Casus 4 p. 14; also in [Ghetaldi 1630] Book III, Problem VI, Casus 4, pp. 126–127.

another problem which I discuss below (Analysis 5.5 and Construction 5.6).

The example illustrates the procedure of the algebraic analysis of plane problems as codified by Viète: The problem was assumed solved; one unknown element was chosen, which, if known, would provide the solution; it was called x (or rather A, in the Vietean choice of letters); the required relations were translated into algebraic formalism and transformed until an equation for x was reached; the equation was solved, that is, an explicit expression for x was derived; finally this expression was translated into a construction. In the translation square roots of sums of second degree terms were constructed by means of appropriate right triangles. In later examples we will see several variants of this method.

Direction of the arguments in algebraic analysis

Unlike in the first example, the present construction did not retrace the steps of the analysis; in fact these steps did not occur in the construction at all, so one could speak neither of the same nor of opposite directions in analysis and synthesis. In the case of analysis by algebra the question of the directions of the argument is more complicated than simple equality or reversal. The whole procedure consisted of three or four distinct parts:

A: The derivation of the algebraic equation from the problem.

A′: If possible, the algebraic solution of the equation, that is, finding an explicit algebraic expression for the unknown. Otherwise, if necessary, a rewriting of the equation into some standard form.

B: The construction of the problem based on the expression found in **A′**, or, if such an expression was lacking or uninformative, on the basis of the equation found in **A** (if necessary reduced to some standard form).

C: The proof that that construction was correct.

Items **A** and **A′** constituted the analysis. Here one started with the assumption that the problem was already solved, and so in a sense one argued from the unknown, the x, toward the given elements, namely the constants and parameters that occur in the final expression for x in **A′**, or failing that, in the coefficients of the equation in **A**. Part **B**, the construction, started from the given elements and operated on them in the manner indicated by the explicit expression in **A′** or by the structure of the coefficients in the equation found in **A** (for an example of an analysis in which part **A′** is lacking see Analysis 5.9 below). Thus the direction of the synthesis was opposite to that of the analysis, but it did not retrace the steps of the analysis. The construction used the equation or its solution, but not the way it was derived in the analysis. In the case of the proof, part **C**, the elements of the analysis did recur, but not necessarily in their original order, and in combination with arguments from the construction.

From the equation to the construction

It is worth noting that in Ghetaldi's solution part **B** of the procedure constituted the whole, fairly complicated construction. Thus the example illustrates a crucial point, namely that the algebraic analysis of a problem (parts **A** and

A′) did not directly provide its geometrical solution, the construction. The step from **A** or **A′** to **B**, from the algebraic solution, or the equation, to the construction, was neither simple nor trivial. This point will be important in my interpretation of Descartes' mathematical achievements below (cf. Section 20.2).

Esthetics

Ghetaldi's works contained hundreds (counting all the separate cases he distinguished) of constructions like the one just discussed. As a result the modern reader experiences his mathematics as enormously laborious and boring. Yet there is an attempt at elegance to be discerned in Ghetaldi's constructions; he tried as much as possible to combine the different steps of a construction in one figure, using the minimum number of auxiliary lines and circles. If he had made the separate steps in separate drawings, the matter would have been structurally more clear, but the construction would have been less direct and simple. The use of algebra in the way later introduced by Descartes actually induced such a separation in different steps, performed, if at all, in different figures. In such an approach no place was left for elegance of the sort attempted by Ghetaldi.

5.4 Ghetaldi: algebraic analysis limited

A plane problem defying algebraic analysis

My next example, a classical analysis of a plane neusis problem, is again from Ghetaldi's *On mathematical resolution and composition* of 1630. Ghetaldi chose the classical style in this case because he considered the problem not amenable to algebraic treatment.[28] Since he knew the Vietean algebraic analysis, the example helps us to see where the limitations of this approach lay. Ghetaldi reduced the problem to a plane neusis that he had already solved, namely the neusis presented in previous example (Analysis 5.3), but for the alternative case in which the line L intersected the circle. Ghetaldi proceeded as follows:

Analysis 5.5 (Special neusis between two lines — Ghetaldi)[29]
Given: a line segment b and a rhombus $OACB$ with $\angle BOA = \varphi$ (see Figure 5.4), $AB = a$; side OA is prolonged; it is required to construct a line through B, intersecting AC in D, and OA prolonged in E such that $DE = b$.

Analysis:
1. Assume the problem solved. Consider the circle through A, D and E; BA prolonged intersects the circle in G. Consider the bisectrix of $\angle EAD$, it intersects the circle in F; line GF intersects DE in H. The diagonals AB and OC of the rhombus are perpendicular,

[28] Cf. Note 1.

[29] [Ghetaldi 1630] Book V, Ch. IV Prob. I, pp. 330–333. Ghetaldi had published the construction without the analysis in his *Apollonius revived* ([Ghetaldi 1607b] Probl. III, pp. 17–19). This work contained Ghetaldi's reconstruction of Apollonius' lost work on neusis. It was based on Pappus' remarks about Apollonius' treatise in [Pappus Collection] VII Props 70–72 pp. 603–608 (in [Pappus 1986] pp. 202–205). Pappus mentioned the present problem there, provided a lemma on which its solution could be based, and gave an explicit construction, attributed to one Heraclitus, for the special case of the problem in which the rhombus is a square. See also Note 33 below.

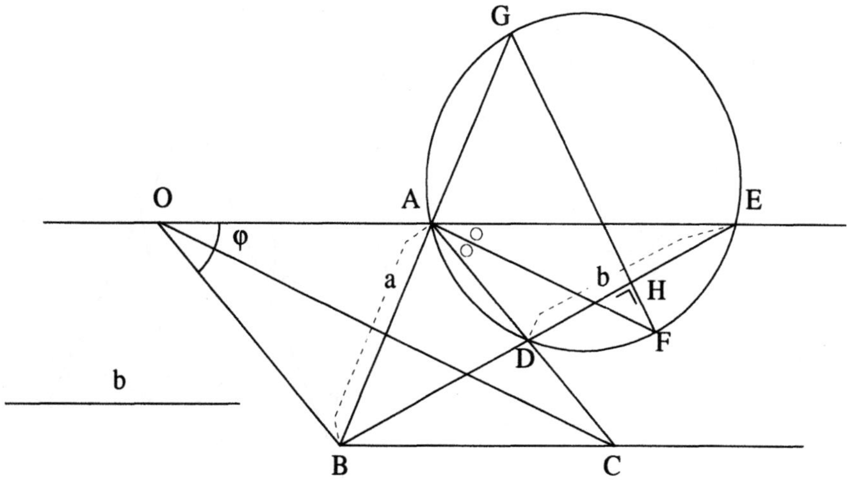

Figure 5.4: Special neusis between two lines — Ghetaldi

AF is parallel to OC, so $AF \perp BAG$; hence GF is a diameter of the circle. $\angle DAF = \angle FAE$, so arcDF = arcFE; and because FG is a diameter, $DH = HE$ and $DE \perp GF$.

2. Now DE is given in magnitude (it is equal to b) and we may consider DE given in position as well if we consider the rhomb given in magnitude only.[30] (Note how Ghetaldi used the relativity of the concepts "given in position" and "given in magnitude" in order to change the problem from one in which a line segment of given magnitude has to be positioned with respect to a rhombus given in position, to one in which a rhombus given in magnitude has to be positioned with respect to a line segment given in position.)

3. $\angle DAE$ is given in magnitude ($= \varphi$) because the rhombus is given.

4. So (*Data* Def. 8[31]) the arc DE is given in magnitude and position, and hence the whole circle as well.

5. The line that perpendicularly bisects DE is given in position because DE is given in position (*Data* 29).

6. So its intersections G and F with the circle are given (*Data* 25).

---

[30] "Et quoniam recta DE data est magnitudine, cum sit aequalis datae b; intelligatur ipsa DE, positione quoque data, nulla positionis rombi dati habita ratione, tanquam non esset positione datus; hoc modo liberum est datae rectae lineae b aequalem alteram DE positione et magnitudine datam exponere" p. 331 (letters changed).

[31] In Ito's translation, [Euclid 1980] p. 57): "Segments [of circles] are said to be given in position and in magnitude, when the angles in them are given in magnitude and the bases of the segments are given in position and magnitude."

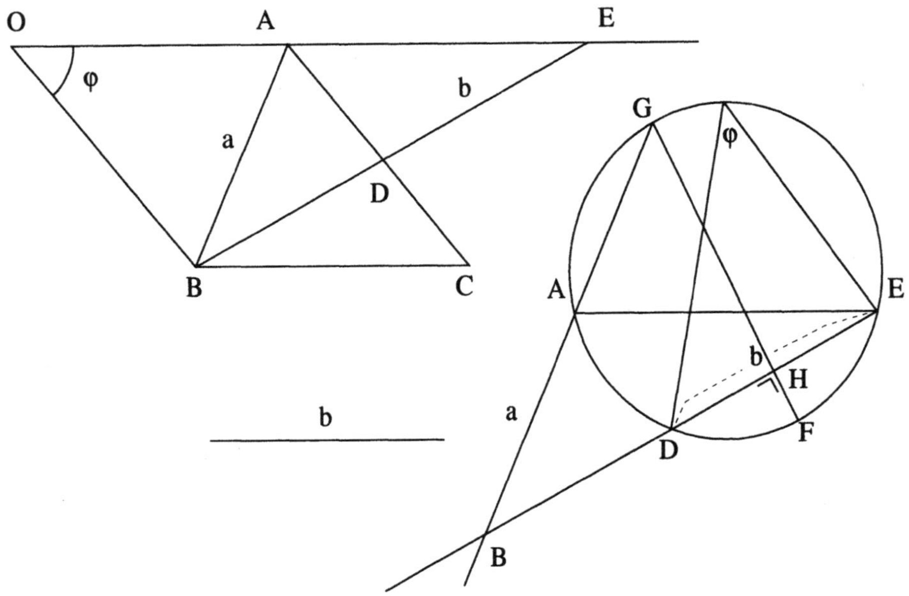

Figure 5.5: Special neusis between two lines — Ghetaldi

7. Consider GAB; $AB = a$ so AB is given in magnitude; A is on the circle, B is on the line DE, DE is perpendicular to the diameter through G. So GAB can be constructed by plane means as the special neusis (between a circle and a line) of the previous example (namely the neusis of a line segment a between the circle $GDFE$ and the line ED, with pole G on the diameter perpendicular to DE — note that a similar situation occurred in Analysis 5.3, although in that case the given line did not intersect the circle). So GAB is given in position too.
8. The rhomb can be completed from AB so the whole figure is given in position and magnitude.

Construction 5.6 (Special neusis between two lines — Ghetaldi)
Given etc.: see Analysis 5.5.

Construction (The numbers of the steps correspond to those in the analysis.):
1. Let (see Figure 5.5) $OACB$ be the given rhombus.
2. In an auxiliary figure, draw $DE = b$.
3–4. Draw an arc on DE such that the peripheral angle on DE is equal to φ (by *Elements* III-33); complete the circle; prolong ED.
5–6. Draw the diameter that bisects DE; it intersects the circle in G and F.

7. By the special neusis between a circle and a straight line (variant of Analysis 5.3), draw GAB with A on the circle, B on DE prolonged and $AB = a$; draw AE.
8. In the given rhombus prolong OA to E such that AE is equal to the segment AE in the auxiliary figure; draw BE; it intersects AC in D.
9. BDE is the required line.
[**Proof**: the proof follows from the analysis.]

Algebraic analysis

Ghetaldi did not explain why he thought that the algebraic method could not be applied in this case. In the Vietean approach, which Ghetaldi used, one would select a suitable unknown line segment (in this case AD would be an obvious candidate) and derive an equation for it. It may well be that Ghetaldi tried to do so and failed because that approach leads to a fourth-degree equation and thus suggests that the problem is solid, whereas it was known from Pappus that the problem was plane. In fact the quartic polynomial involved is decomposable into two quadratic factors,[32] but in the Vietean approach there was little interest in the decomposition of equations and so we may well suppose that Ghetaldi gave up the algebraic approach. It should be noted that Descartes chose precisely the present neusis problem (for the case that the rhombus is a square) as an example for explaining the necessity of and the techniques for decomposing reducible fourth-degree equations.[33]

[32]Setting $CD = x$, $DE = b$, $OA = c$, and $c\cos\varphi = d$, the equation can be found as follows: The triangles ΔADE and ΔCDB are similar, hence $b : (c - x) = x : BD$ ($\star$). In triangle ΔCDB $BD^2 = x^2 + c^2 - 2xc\cos\varphi = x^2 + c^2 - 2dx$ ($\star\star$). Eliminating BD from ($\star$) and ($\star\star$) and reordering yields the equation

$$x^4 - 2(c+d)x^3 + (2c^2 - b^2 + 4cd)x^2 - 2c^2(c+d)x + c^4 = 0\,. \qquad (5.1)$$

The fourth-degree polynomial on the left-hand side can be written as the product of two quadratic polynomials, and correspondingly the equation reduced to the following two quadratic equations:

$$\begin{aligned} x^2 - (c+d+\sqrt{(c-d)^2+b^2})x + c^2 &= 0\,, \qquad (5.2)\\ x^2 - (c+d-\sqrt{(c-d)^2+b^2})x + c^2 &= 0\,. \end{aligned}$$

As the coefficients in the two quadratic equations involve square roots only, they can be constructed by plane means from the given b, c, and d and hence all solutions of Equation 5.1 can be constructed by plane means. (Note that this construction-relevant reduction of the fourth-degree equation involves a concept of reducibility, which is different from the present-day concept. In the modern sense of the term Equation 5.1 is not reducible because the decomposition in Equation 5.2 involves $\sqrt{(c-d)^2+b^2}$ and hence requires a quadratic extension of the ground field. See Section 27.3 for a more detailed discussion of reducibility in the context of geometrical problem solving.)

[33]Cf. Note 29 of the present Chapter, Construction 27.1, and Sections 27.3 and 27.4. See also [Brigaglia & Nastasi 1986] pp. 112–115, 120–131 for further particulars about the history of this problem.

5.5 Classical analysis of solid problems

Two procedures

As in the case of a plane problem, classical analysis of a solid problem (cf. Table 5.1) started from the assumption that the problem was solved, located the given elements of the figure, and identified the point whose construction was required. Then there were basically two procedures. The first consisted of finding elements in the figure which, would be known if some standard solid problem were solvable (for instance, two mean proportionals between two given line segments). If the required point could be constructed by plane means from these auxiliary known elements, then the problem was reduced to the standard solid problem. Archimedes' analyses to propositions II-1 and II-4 of the *Sphere and cylinder* were of this kind.[34] The other analytical procedure consisted in identifying two properties of the required point, and showing that the points satisfying any one of these properties constituted two loci. If these loci were conics and if they could be considered known, then their point or points of intersection, among which the required point, were known as well, and the construction could be performed by the intersection of the conics thus found. Pappus' *Collection* contained several analyses of this kind.[35]

Scarcity of examples

There are very few, if any, formal analyses of either kind in the early modern literature. We have seen many examples of solutions of solid problems by reduction to standard problems in the previous chapters, but in these cases no formal analyses were given. It appears that the construction of solid problems by the intersection of conics, although recognized as a method of classical standing, was very little practiced before Fermat and Descartes. The only examples I know are in Commandino's notes to his edition of the *Collection*, where he elaborated on Pappus' trisection by a hyperbola and a parabola,[36] and Van Roomen's solution of the Apollonian tangency problem. In his note Commandino stayed very near to Pappus' text and did not clearly separate analysis and construction. Van Roomen's argument is more illustrative.

Van Roomen's analysis of Apollonius' problem

Van Roomen did not present his argument as a formal analysis, but in structure it clearly was an analysis. The principal idea of his argument was:

Analysis 5.7 (Tangency problem — Van Roomen)[37]
Given: three circles C_1, C_2, C_3 with centers O_1, O_2, O_3 and radii r_1, r_2, r_3 (see Figure 5.6); it is required to construct a circle C tangent to all three given circles.

Analysis:
1. Assume the problem solved and consider C externally tangent to all three circles. (One of the two cases of external tangency is

[34] Cf. Section 5.1 Note 9.

[35] Pappus' neusis construction (Construction 3.8), for instance, was preceded by such an analysis, cf. [Pappus Collection] IV-31 (§36) p. 210.

[36] [Pappus 1660], pp. 102–104; commentary to IV-34.

[37] [Roomen 1596] pp. 11–13.

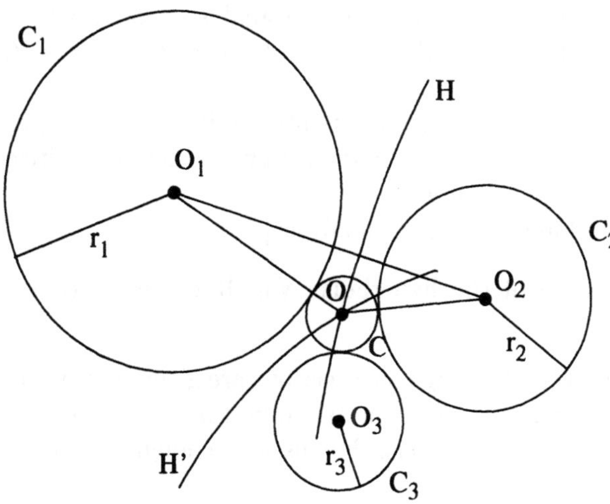

Figure 5.6: Van Roomen's analysis of Apollonius' tangency problem

illustrated in the figure; Van Roomen listed and treated all possible cases of internal and external tangency and relative position of the given circles.)
2. As C is tangent to C_1 and C_2, $OO_1 - OO_2 = r_1 - r_2$; call $O_1O_2 = d$.
3. By Apollonius' *Conics* III-51 all points whose distances to the two given points O_1 and O_2 have a constant difference $r_1 - r_2$ are on (one branch of) a hyperbola H whose center is on the middle of O_1O_2, whose *latus transversum*[38] b is equal to $r_1 - r_2$ and is situated along O_1O_2, and whose *latus rectum* a is equal to the fourth proportional of b, $(d-b)$, and $(d+a)$ (the other branch occurs when C_1 and C_2 touch C internally);[39] H is therefore known.
4. Similarly the centers of circles that touch C_1 and C_3 are on a known hyperbola H'.
5. The center O of the required circle C is in one of the points of intersection of the two hyperbolas and is therefore known too; its radius r satisfies $r = OO_1 - r_1$ and is also known.

Construction 5.8 (Tangency problem — Van Roomen)[40]
Given etc.: see Analysis 5.7.

[38]Cf. Note 7 of Chapter 3 for the meaning of the terms *latus transversum* and *latus rectum*.
[39]Prop. III-51 of Apollonius' *Conics* states a property of the hyperbola from which Van Roomen's statement follows by inversion.
[40][Roomen 1596] pp. 11–13.

Construction:
1. With parameters as derived in the analysis, construct the hyperbolas H and H' related to the circle pairs C_1, C_2 and C_2, C_3 respectively.
2. With the appropriate point of intersection of the hyperbolas as center and the appropriate radius draw the required circle that touches all three given circles.
[**Proof**: Proof follows from the analysis.]

For the actual drawing of the hyperbolas Van Roomen referred to standard works on conics:

> Various methods of drawing conic sections are given by various writers, so the method to draw them has to be taken from authors who have written on conic sections. For us it is enough to have solved the problem.[41]

Van Roomen studied the Apollonian tangency problem in response to a challenge by Viète,[42] and Viète was not impressed by the solution, for he knew that the problem could be solved by plane means, so that Van Roomen could be charged with what Pappus had called a "sin" among geometers, namely to solve a problem with inappropriate means. In his reaction to Van Roomen Viète did not spare words in pointing that out (cf. Section 10.4).

5.6 Algebraic analysis of solid problems

Two procedures Similarly to the classical analysis of solid problems, there were two distinct ways of analyzing a solid problem by means of algebra. Each started with deriving the equation (in one unknown) of the problem. The one used this equation to reduce the construction to a standard construction, the other (which was not practiced until Fermat) used it to find the equations (in two unknowns) of the constructing conics. The next two examples illustrate both approaches.

An algebraic analysis by Anderson The first example concerns the construction of the normal to a given parabola through a given point outside the parabola. I have discussed the significance of that problem and Anderson's construction of it above (Construction 4.22). Here is the analysis in Vietean style by which Anderson found the construction:

Analysis 5.9 (Perpendicular to a parabola — Anderson)[43]
Given: a point A (see Figure 5.7), a parabola, not through A, with vertex O,

[41] [Roomen 1596] p. 18: "Conicas porrò sectiones ducendi ratio varia à variis traditur, ideo ratio earundem ducendarum ex authoribus qui de Conicis sectionibus egerunt, petenda est. Nobis namque sufficit problema solvisse."

[42] Three years earlier Van Roomen had challenged mathematicians with the problem of solving a certain equation of degree 45 ([Roomen 1593]); Viète solved this problem in [Viète 1595] and proposed the Apollonian tangency problem as a counter challenge. Van Roomen published the solution discussed here in [Roomen 1596]; Viète criticized it in [Viète 1600].

[43] [Anderson 1619] pp. 25–27.

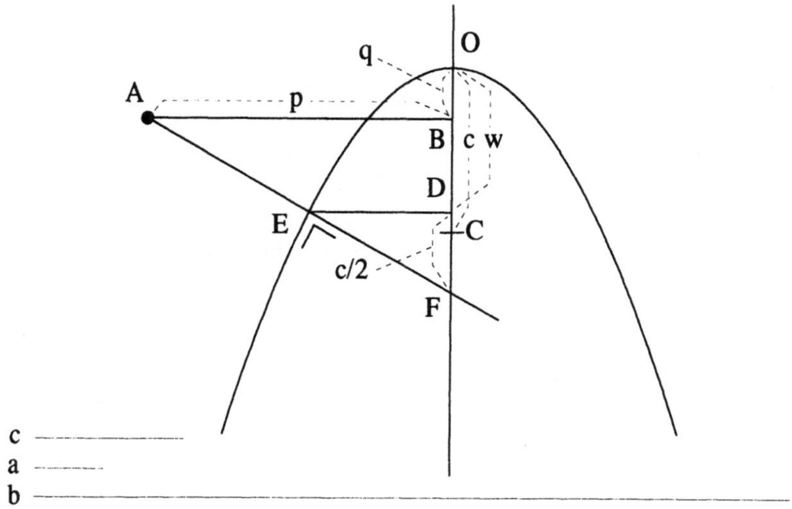

Figure 5.7: The perpendicular to a parabola — Anderson

vertical axis and latus rectum $c = OC$; it is required to draw a line through A intersecting the parabola perpendicularly. [I give the case that A is below O and the vertical distance of A to O is smaller than $\frac{1}{2}c$.]

Analysis:
1. "Factum jam sit" — assume the problem solved; Let AEF be the required perpendicular intersecting the parabola in E and its axis in F; ED is the ordinate of E; call $ED = x$ and $OD = w$; let AB be drawn horizontally, with B on the axis; call $AB = p$ and $OB = q$.
2. By the Apollonian theory of the parabola and its tangents one finds that $x^2 = cw$ and $DF = \frac{1}{2}c$.
3. The triangles EDF and ABF are similar, hence $x : p = \frac{1}{2}c : (w + \frac{1}{2}c - q)$; inserting $w = \frac{x^2}{c}$ yields

$$x : p = \frac{1}{2}c : (\frac{x^2}{c} + \frac{1}{2}c - q) = \frac{1}{2}c^2 : (x^2 + c(\frac{1}{2}c - q)) ;$$

a proportionality in which x is the only unknown.
4. The proportionality is equivalent to one of Viète's standard solid problems (cf. Section 4.2). Anderson here referred to an earlier publication[44] in which he had discussed the Vietean standard equations and problems. The proportionality can be rewritten as a cubic equa-

[44] [Anderson 1615].

tion:

$$x^3 + c(\frac{1}{2}c - q)x = \frac{1}{2}c^2 p\,.$$

5. This corresponds (cf. Equations 4.7 (1) and 4.9 (1)) to the standard equation $x^3 + a^2 x = a^2 b$ and the standard problem: given a and b, to find x, y and z such that $a : x = x : y = y : z$, and $x + z = b$.
6. In order to achieve the construction, it remains to determine line segments a and b such that $a^2 = c(\frac{1}{2}c - q)$ and $a^2 b = \frac{1}{2}c^2 p$. The first relation implies that a is the mean proportional of c, and $\frac{1}{2}c - q$. From the two relations it follows that $b = \frac{\frac{1}{2}cp}{\frac{1}{2}c - q}$, hence b is the fourth proportional of $\frac{1}{2}c - q$, $\frac{1}{2}c$ and p. So both a and b can be determined by standard Euclidean constructions.

Construction 5.10 (Perpendicular to a parabola — Anderson)
Given etc.: see Analysis 5.9.

Construction: (I repeat Construction 4.22, indicating the steps in the analysis corresponding to it; my numbering of the items does not match the one in the analysis.)
1. Draw AB horizontally, with B on the axis; call $AB = p$ and $OB = q$.
2. Construct the mean proportional a of c and $\frac{1}{2}c - q$ (i.e., $c : a = a : (\frac{1}{2}c - q)$).
3. Construct the fourth proportional b of $\frac{1}{2}c - q$, $\frac{1}{2}c$ and p (i.e., $(\frac{1}{2}c - q) : \frac{1}{2}c = p : b$) (steps **2** and **3** implement the relations of step **6** in the analysis).
4. Now solve the following (Vietean) standard problem (cf. Problem 4.5 above): Find three line segments x, y, and z such that a, x, y, and z are in continued proportion and $x + z = b$ (cf. step **5** of the analysis).
5. The segment x is the ordinate DE of the point E in which the required line intersects the parabola; determine E and draw the perpendicular AE to the parabola.
[**Proof**: The proof follows the analysis.]

An algebraic analysis by Fermat, involving loci

Viète himself did not use his algebraic techniques for finding conics by whose intersection a solid problem could be solved, nor did his immediate followers do so. A reason for this comparative indifference toward construction by conics may have been that Viète chose neusis rather than the intersection of curves as basic means for solving solid problems. Fermat seems to have been the first to use Vietean algebra for finding the loci by whose intersection problems were to be solved (cf. Section 13.1). Probably in 1636 he wrote the tract *Introduction to plane and solid loci* (following common practice, I refer to it by the first

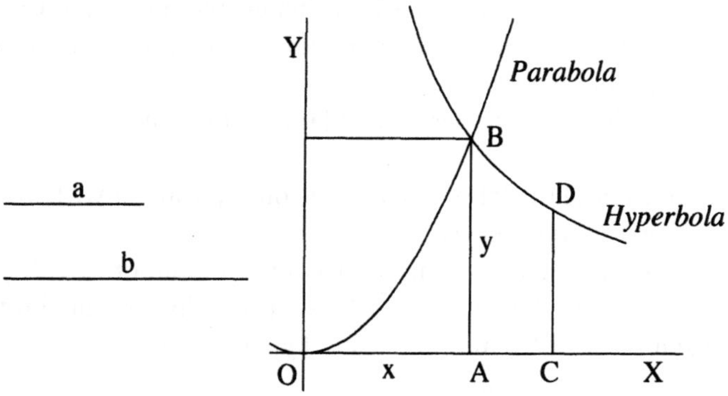

Figure 5.8: Two mean proportionals — Fermat

word of its title, *Isagoge*)[45] to which he added an *Appendix containing the solution of solid problems by loci*.[46] In the *Appendix* he explained how the equations (in two unknowns) of the conic sections needed in a construction could be derived from the cubic or fourth-degree equation (in one unknown) to which the problem first was reduced in the manner taught by Viète. Fermat did so by introducing a second unknown and deriving two quadratic relations that the two unknowns satisfy. The *Appendix* gave only one fully worked example (the others only explain the algebraic part of the process). It is the construction of two mean proportionals — not the most original example but it does make Fermat's approach clear:

Analysis 5.11 (Two mean proportionals — Fermat)[47]
Given: two line segments a and b, $a < b$ (see Figure 5.8); it is required to find their two mean proportionals x and y ($x < y$).

Analysis:
1. Assume the problem solved; the smaller mean proportional x satisfies the equation

$$x^3 = a^2 b\,.$$

Putting each term equal to axy yields

$$(1)\quad x^2 = ay\,,$$

[45] [Fermat Isagoge]; on the dating of the *Isagoge* cf. [Mahoney 1994] p. 405.
[46] [Fermat Appendix], from 1636 or 1637, cf. [Mahoney 1994] pp. 405–406.
[47] [Fermat Appendix] pp. 105–107.

and

$$(2)\quad xy = ab\ ,$$

"hence the problem will be done by a hyperbola and a parabola." (Fermat here referred to the *Introduction* in which equations for conic sections were explained.)
2. Let then, with respect to perpendicular axes OX and OY, $x = OA$ and $y = AB$.
3. From equation (1) it follows that B is on a parabola with vertex O, *latus rectum* a, and axis OY.
4. As to equation (2): take C arbitrarily on OX and draw CD perpendicularly with $CD \times OC = ab$. Now draw a hyperbola through D with asymptotes OX and OY; this hyperbola will be "given in position" and pass through B.
5. But the parabola described earlier is also "given in position" and passes through B, therefore, the point B is "given in position" and if from B we draw the perpendicular BA, A will be given in position, and therefore the required $x = OA$ as well.
6. "Therefore the two mean proportionals are found by the intersection of a parabola and a hyperbola."
[The construction (draw the conics, draw the ordinate BA of their point of intersection, $BA = y$, $OA = x$) and its proof were now obvious to Fermat.]

Fermat added a variant argument, starting from the equivalent fourth-degree equation

$$x^4 = a^2bx\ ,$$

which leads, by setting both terms equal to a^2y^2, to

$$x^2 = ay \quad \text{and} \quad bx = y^2\ ,$$

each defining a parabola. He noted that the two constructions thus found were precisely those by conics in Eutocius' list of constructions of mean proportionals (the ones by Menaechmus, cf. the first of these, Construction 3.1, which indeed is the same as the construction derived above). At this point Fermat explained the advantage of his method over Viète's approach to geometrical problems leading to a fourth-degree equation. Viète's reduced this equation to a third-degree one and then constructed the roots of this equation by neusis or by the reduction to one of the Vietean standard problems. This reduction (basically Ferrari's method[48]) was algebraically rather cumbersome (and was not made easier by the flourish of neologisms with which Viète had adorned it). But if one decided to construct by the intersection of conics, this procedure was superfluous; one could derive the equations of the conics directly from the fourth-degree equation.[49] Fermat proceeded to describe a general procedure for

[48] See Note 18 of Chapter 10.
[49] See also the quotation in Section 13.1.

deriving the equations of the constructing conics for any third- or fourth-degree equation. He also explained how this procedure could be adapted to yield a construction by the intersection of a parabola and a circle. I return to these constructions in Section 13.1.

The basic assumptions of Fermat's approach were that quadratic loci are given by their equations, and that once two such loci were given, their point of intersection was given too. In the *Introduction* Fermat had shown by examples how the parameters of the conics could be determined from their equations, so that, by the Apollonian theory of conics, they could be constructed[50] once their equations were known. On the second assumption Fermat remained silent; apparently he accepted this classical way of determining points by intersection without question.

5.7 Conclusion

Diversity and obstacles

The examples in the present chapter may serve to illustrate the diversity of the early modern practice of geometrical analysis. They should also show that the development of algebraic methods of analysis — the principal dynamics within the early modern tradition of geometrical problem solving — was not a straightforward process. There was a quite powerful alternative analytical method, the classical one, which featured the effective use of a fundamental notion in the geometry of constructions, namely the concept of "given." Moreover, in several ways the algebraic techniques did not immediately correspond to the requirements of geometrical problem solving. In particular, as Ghetaldi's treatment of the special neusis between a circle and a line (Analysis 5.3) makes clear, the geometrical demarcation between plane and solid problems did not correspond to a natural or easily recognizable demarcation between the pertaining equations. Nor was the translation of the results of algebraic analysis in terms of geometrical constructions uncomplicated.

Apart from these technical difficulties of using algebra in the analytical practice, there were more fundamental obstacles. These concerned the traditional dividing lines between arithmetic, geometry, and algebra. They form the subject of the next two chapters.

[50] Cf. Chapter 3, Note 32.

Chapter 6

Arithmetic, geometry, algebra, and analysis

6.1 Introduction

Matters of fusion

The adoption of algebraic methods of analysis induced a gradual fusion of arithmetic, algebra, and geometry, which in turn gave rise to a number of technical, terminological, and conceptual questions. The fusion also spurred issues of legitimation, in particular the question whether it was allowed to use numbers in geometry. The present chapter deals with the technical, terminological, and conceptual questions; the issues of legitimation are the subject of Chapter 7. We find the first successful merging of the fields of algebra, arithmetic, and geometry in the work of Viète; Chapter 8 explains how Viète overcame the obstacles and the legitimation issues involved.

The study of the questions mentioned above is complicated by the fact that the terms "arithmetic," "algebra," "geometry," and "analysis" have modern meanings that are downright misleading if used in describing sixteenth- and seventeenth-century developments. It is therefore useful to fix terminology in these matters, and to do so I have to specify other terms as well, such as "magnitude," "indeterminate," and "operation." With these terminological matters settled it is possible to identify the main obstacles that stood in the way of a smooth merging of arithmetic, geometry, algebra, and analysis.

6.2 Terminology

Terms and their meaning

During the early modern period mathematical terminology was in flux; new terms were introduced and old ones changed their meaning. Yet, in describing the conceptual developments that form the subject matter of my study, I need a reasonable constancy of the meanings of my terms. Moreover, for some mathematical subjects that I want to consider, there were no terms at all. Thus I cannot restrict myself to the terminology of the period, nor can I claim to use that terminology in the meaning it had for its early modern users. On the other hand, I cannot use mathematical terms in their present-day meanings either, because, as will become clear below, these meanings often are far removed from the conceptions that the terms evoked in the early modern period.

In choosing the mathematical terminology for the present study I have tried to remain close to early modern usage. However, I use mathematical terms primarily as analytical tools for describing and explaining developments in earlier mathematical thought. I explain below the senses in which I want the main terms to be understood — in these explanations I do appeal to present-day mathematical understanding. In each case I indicate separately in how far my use of a term corresponds to its use (if such occurred) during the period c. 1580 – c. 1650.

Number I begin with the principal mathematical entities number, geometrical magnitude, magnitude in general, and ratio. I use the term *number* to refer to natural or rational positive numbers, and irrational positive numbers in as far as they occurred at the time in numerical context. This meaning of the term corresponds roughly to its use in the period. The term number was primarily understood in the classical sense of a multitude of units; this concept concerned discrete entities and was thereby essentially different from the concept of geometrical magnitudes such as line segments. Although rational fractions were not multitudes of units, they were usually accepted as numbers, and so were the relatively few irrational numbers which occurred in the writings of the period. Even fewer were the occurrences of imaginaries such as $\sqrt{-6}$; their status (number or not) was left undecided. In the later seventeenth and eighteenth centuries the discrete concept of number fused gradually with the continuous concept of magnitude, but in the period before 1650 such a fusion of the two concepts was very rare.

Magnitude With the term *geometrical magnitude* I refer to line segments, plane figures, and solid figures, in as far as they were considered as to their size. This use of the term corresponds to the classical Greek conception of geometrical magnitude, which was generally known and widely accepted in the sixteenth and seventeenth centuries. Geometrical magnitudes of the same kind (the same dimension) could be added and subtracted in some sense; they coud be compared as to size, and pairs of such magnitudes had a ratio. This adding, subtracting, comparing, and forming of ratios did not presuppose a numerical value of the size (length, area, volume); adding was joining, subtracting was cutting off, or taking away the smaller from the larger, comparing meant deciding which of two line segments (areas, solids) was the longer (larger), equal meant equally long (or large). The size of a line segment was not a number expressing the measure of its length with respect to some unit, but its magnitude, which it shared with congruent line segments. There were no negative magnitudes, nor a magnitude zero.

Abstract magnitude

With the term *abstract magnitude* I refer to mathematical entities that, like geometrical magnitudes, could be joined, separated, and compared, but whose further nature was left unspecified. The concept of such magnitudes was classical, but around 1600 it was something of a novelty, especially if combined with the notion that the operations on such magnitudes could be treated abstractly. The exploration of these ideas was called "universal mathematics" ("mathesis universalis" or similar expressions).

Since opinions differed on whether numbers were magnitudes, I indicate explicitly those cases in which numbers were indeed considered as magnitudes.

Ratio

In its classical sense the term *ratio* referred neither to a number nor to a magnitude but to a relation with respect to size between two magnitudes of the same kind or between two numbers. This interpretation of the term (due to Eudoxus and codified in Def. V-3 of the *Elements*) was the common one in the early modern period. I use the term in the same sense. It should be noted, however, that in the late Medieval and Renaissance periods an alternative approach to ratios had been explored, which understood them in terms of "denominations."[1] The denomination of a ratio of two numbers was its expression in simplest form as a rational number. Thus the denomination of the ratio of 44 to 14 was $3\frac{1}{7}$. Some mathematicians assumed that these denominations could somehow be extended so as to apply for all ratios, including the irrational ones (which arose in particular when ratios of geometrical magnitudes were considered). Thus Campanus, inspired by Jordanus, interpreted the theories of ratio in *Elements* V (for magnitudes) and VII (for numbers) as essentially the same. During the sixteenth century, with the appearance of improved editions of the *Elements*, it became clear that Campanus' interpretation was based on distorted texts, and that the Euclidean theory of ratios of numbers was essentially different from the theory of ratios of geometrical or abstract magnitudes. However, the attempts to understand ratios in terms of their denomination, together with a habituation to the use of irrational roots in algebra, nourished the idea that by considering ratios (irrational or not) as numbers, the complicated definition of equality of ratios in *Elements* V and the intricacies of the theory of irrationals in *Elements* X would become superfluous. Ramus strongly advocated this idea; it was taken over in various degree by several mathematicians. The issue remained alive until well into the seventeenth century as is witnessed by a debate on the matter between Wallis and Barrow in the 1660s.[2]

Notwithstanding these attempts to change the classical conception of ratio, the early modern mathematicians, including those willing to take over Ramus' ideas, knew the classical concept of ratio and were aware of the controversial nature of the alternative.

Operations on numbers

In the following I use the term *operation* in an extended sense comprising

[1]On the denomination theory of ratios and proportions see [Murdoch 1963], [Sylla 1984] and [Sasaki 1985].

[2]Cf. [Sasaki 1985].

addition, multiplication, etc., but also geometrical procedures such as constructions. During the period under discussion there was no general term used in this sense, nor, it seems, was there a general concept corresponding to it.[3]

Numbers, magnitudes, and ratios were submitted to *operations*. The principal operations for *numbers*, to which I refer as "arithmetical operations" (cf. Table 6.1), were addition, subtraction, multiplication, division, square root extraction, and the extraction of higher-order roots. Root extraction is a kind of equation solving (namely the equation $x^n = a$); extending this notion I consider the solution of polynomial equations in one unknown with numerical coefficients as an arithmetical operation as well; this will be useful in studying the transfer of the arithmetical operations to geometry. Certain operations could be performed (i.e., calculated) exactly, others, in general, only approximatively. The first group comprised addition, subtraction, multiplication, and division; I will refer to these four as the "primary arithmetical operations". The second group comprised root extraction and the solution of equations in general; apart from special cases, these operations essentially involved approximative calculations. The nature of the operations from the second group was algebraic, in the sense that (cf. Section 6.3) an unknown was determined from a given relation between one or more of its powers and some given numbers. The primary arithmetical operations, together with square root extraction, correspond to the "plane" constructions, i.e., those that can be performed by straight lines and circles; I will refer to this group as the "quadratic algebraic operations".

The above operations did not involve a change of kind or dimension of the objects, that is, they operated on numbers (or equations with numerical coefficients) and their results were numbers as well. There was one operation that *did* involve a change of kind, namely, taking the ratio of two numbers.

Geometrical operations

In *geometry* the primary operations were the constructions, codified in the first three postulates of the *Elements* (constructions by circles and straight lines), or introduced for special purposes (in particular for problems that could not be constructed by circles and straight lines). These operations were performed upon line segments. Two line segments could be joined (analogous to adding), and a subsegment of a segment could be cut off (analogous to subtracting); these operations could be executed by means of plane constructions.

Two line segments a and b could be combined to form a rectangle rect(a, b) with sides a and b. This operation was in some respects analogous to multiplying numbers in arithmetic; however, it differed from multiplication in that the result, a rectangle, was of a different dimension than the original factors, whereas the product of two numbers was again a number. Consequently, there was no unit element with respect to the formation of rectangles, in contrast to the multiplication of numbers for which 1 was the unit element. One operation resembled, in some respects, the division of two numbers. This was the "application" of a rectangle (or, in general, a plane rectilinear figure) A to a line segment

[3] In particular the modern connotation of the term operation as referring to functions of one or more variables is inapplicable in the early modern period.

NUMBERS and the OPERATIONS acting on them

Numbers: natural or rational positive numbers, and irrational positive numbers in as far as they were used at the time in numerical context.
All operations were executed (if they could be) by calculation, i.e., by numerical algorithms.
There was a unit element, namely, the number 1.

Operation	Notation in this study	Exact	Change of kind or dimension	Analogous operation(s) on geometrical magnitudes
Adding two numbers	$+$	Yes	No	Joining
Subtracting a (smaller) number from a number	$-$	Yes	No	Cutting off
Multiplying two numbers	$\times$	Yes	No	Making a rectangle
Dividing two numbers	$/$ or $\div$	Yes	No	Forming a ratio; applying a rectangle
Extracting the square root of a number	$\sqrt{\cdot}$	No	No	Taking the mean proportional
Solving a quadratic equation with numerical coefficients		No		Plane constructions
Extracting a higher-order root of a number	$\sqrt[k]{\cdot}$	No	No	Taking several $(k-1)$ mean proportionals
Solving cubic and higher-order equations with numerical coefficients		No		Solid or higher-order constructions
Forming the ratio of two numbers	:		Yes	Forming a ratio

Table 6.1: Numbers

GEOMETRICAL MAGNITUDES and the OPERATIONS acting on them

Geometrical magnitudes: line segments, plane figures, and solid figures considered as to their size.
The magnitudes were supposed "given in magnitude" (Section 5.2), i.e., they (or elements equal to them) could be located at will with respect to other given elements.
All operations were executed (if they could be) by geometrical constructions.
Geometrical magnitudes were *not* numbers; there was *no* unit element.

Operation	Exact (i.e., constructible by straight lines and circles)	Change of kind or dimension	Analogous operation(s) on numbers
Joining two line segments, or two plane figures, or two solid figures	Yes for line segments and rectilinear plane figures, otherwise no (cf. Construction 4.15)	No	Adding
Cutting off a subsegment from a line segment, or a plane subfigure from a plane figure, or a solid subfigure from a solid figure	Yes for line segments and rectilinear plane figures, otherwise no	No	Subtracting
Making a rectangle with sides equal to two line segments (notation in this study: rect.$(\cdot,\cdot)$, sq.$(\cdot)$)	Yes	Yes	Multiplying
Applying a rectangle (or rectilinear figure) along a line segment	Yes	Yes	Dividing
Forming a ratio of two line segments or of two plane figures or of two solids (notation: ":")		Yes	Forming a ratio
Plane constructions	Yes		Algorithms involving $+$, $-$, $\times$, $/$, and $\sqrt{\cdot}$ only
Non-plane constructions	No		Algorithms involving the solution of third- or higher-order equations

Table 6.2: Geometrical magnitudes

a. It meant the construction of a line segment b such that $\text{rect}(a, b) = A$; that is, A was placed along a in the form of a rectangle with equal area. The application of rectilinear areas could be performed by plane means. The operation was similar to dividing numbers in the sense that to given figure A and line segment a it supplied a line segment b such that $\text{rect}(a, b) = A$. However, unlike the division of numbers, this operation involved magnitudes of different dimensions, namely, a two-dimensional figure and a line segment.

The application of areas made it possible to join (add) and cut off (subtract) rectilinear plane figures by plane constructions. Indeed, by *Elements* I-45 (cf. Problem 4.12), two rectilinear plane figures A and B could be applied to the same line segment e, yielding line segments a and b such that $\text{rect}(e, a) = A$ and $\text{rect}(e, b) = B$. Having the same side e, these rectangles could be joined or cut off by forming $\text{rect}(e, a + b)$ and $\text{rect}(e, a - b)$, respectively, where $a \pm b$ denotes the results of joining a and b or cutting b from a. Joining or cutting off curvilinear plane figures could in general not be performed by plane means.

Like pairs of numbers, pairs of geometrical magnitudes of the same dimension had ratios.

Operations on abstract magnitudes

As mentioned above, the early modern period witnessed an interest in a "universal mathematics" in which operations such as the arithmetical or geometrical ones were applied to magnitudes independently of their nature. Viète was the first to elaborate such a theory; I discuss it in more detail in Chapter 8, where I also give a table for the Vietean operations on abstract magnitudes (Table 8.1).

Operations on ratios

The most important operation for *ratios* was compounding. Compounding the ratios $a : b$ and $b : c$ produced the ratio $a : c$. In general, two ratios $a : b$ and $P : Q$, in which a and b were not necessarily of the same kind of magnitude as P and Q, could be compounded by determining a magnitude c of the same kind as a and b and such that $P : Q = b : c$; the compounded ratio of $a : b$ and $P : Q$ was then equal to the compounded ratio of $a : b$ and $b : c$, and therefore equal to $a : c$. The operation was defined independently of the arithmetical, algebraic, or geometrical operations. However, the compounding of ratios presupposed the possibility of finding a magnitude c as above, that is, it presupposed the existence (for the definition) and the constructibility (for the execution of the operation) of the fourth proportional of three magnitudes P, Q, and b of which the first two were of the same kind. Neither the existence nor the constructibility of the fourth proportional are obvious for geometrical magnitudes other than line segments or rectilinear plane figures.[4]

If ratios are numerically expressed as fractions, compounding them corresponds to multiplying the fractions. Yet in the early modern period compounding ratios was not seen as a kind of multiplication. From the terminology it appears that the operation was rather seen as akin to adding. For instance, if

[4]For further information about the question of the existence of the fourth proportional see [Becker 1933], [Mueller 1981] pp. 127, 138–139.

RATIOS and the OPERATIONS acting on them

Ratios: relations with respect to size between two magnitudes of the same kind or between two numbers.
A ratio was considered known if its two terms were known.
All operations in this table acted on (one or two) ratios and produced one ratio.
Notation in this study: ":".

Operation	Analogous operation(s) on numbers	Analogous operation(s) on geometrical magnitudes
Compounding two ratios	Multiplying two fractions	Adding two magnitudes
Dividing a ratio in equal parts; general section of a ratio	Partition of a number	Dividing a line segment (plane figure, solid) in equal parts; or in two parts with given ratio and Determining mean proportionals (cf. Section 4.4)
Adding two ratios via their denomination	Adding two fractions	

Table 6.3: Ratios

$a : b$ was equal to $b : c$, the compounded ratio $a : c$ was called the "double" of the ratio $a : b$.[5]

Within the tradition of studying (primarily rational) ratios in terms of their "denomination" (cf. Section 6.2) another concept of adding ratios occurred, whereby the sum of two ratios was equal to the ratio represented by the sum of their denominations. In terms of numerical fractions this operation corresponds to addition.

Arithmetical and geometrical operations compared

The similarity of adding or subtracting numbers and joining or taking away geometrical magnitudes was recognized and put to good use in practical geometry, where lengths and areas were routinely expressed as numbers with respect to certain unit measures. Multiplication and division of numbers were akin to the formation of rectangles from line segments and the application of areas along a line segment. However, because of the changes in dimension involved, the analogy was less obvious. In the employment of numbers in practical geometry, the change of dimension was dealt with by the introduction of corresponding units of lengths and area (as, for example, the yard and the square yard); with respect to such units the area of a rectangle was equal to the product of its sides. However, as I discuss in more detail in Section 6.4, there were strong conceptual obstacles against accepting this numerical interpretation of multiplication for geometrical magnitudes. On the other hand, in his algebra for abstract magnitudes Viète used the analogies in such a way that the arithmetical and the geometrical operations could be interpreted as instances of the same abstract operations in different contexts.

[5] Cf. Section 4.4.

The tradition of treating ratios in terms of their "denomination" (cf. Section 6.2) did conceive ratios to some extent as (fractional) numbers. However, most mathematicians adhered to the conception that a ratio was not a quantity but a relation, so that forming a ratio between magnitudes was essentially different from a division.

Thus we see that, apart from the cases of addition and subtraction, there was no straightforward analogy between the operations on numbers and those on magnitudes. Mathematicians were strongly aware of the differences and the orthodox view at the beginning of the early modern period was that geometrical operations applied to geometrical magnitudes only and arithmetical operations only to numbers. We will see in Chapter 8 how Viète created an approach to the algebraic operations that fully accepted these classical views.

Unknowns, indeterminates, and variables

It will be useful explicitly to distinguish between unknowns, indeterminates, and variables. I use the term *unknowns* to refer to those mathematical objects that, in an arithmetical or geometrical problem, were required to be determined, by calculation or by construction. With the term *indeterminates* I refer to mathematical objects involved in a theorem, a problem, or an algorithm, whose particular values or sizes were inessential for the proof of the theorem, the construction and proof of the problem, or the execution of the algorithm. That is, the proof was valid and the construction or the algorithm could be performed for any (appropriate) values or sizes of the indeterminates. Thus in the classical problem of determining two mean proportionals (cf. Problem 2.1) the two given lengths were the indeterminates, the two mean proportionals were the unknowns. In general solution procedures for equations, as $x^4 + ax^3 + bx^2 + cx + d = 0$, x was the unknown and the coefficients a, b, etc., were the indeterminates.

I use the term *variable* only in connection with a mathematical object involving one or more degrees of freedom. The archetypal object of that kind was the *curve*. Any point on the curve had its ordinate, abscissa, subtangent, etc.; these were variables; they varied (as to magnitude and as to position) according to the point on the curve to which they belonged. A variable, it should be stressed, was not a function in the modern sense, because it did not involve an "independent variable" of which it would be a function.[6] Variables were not unknowns because they were not required to be determined, nor were they indeterminates. In a strict formal sense one might consider indeterminates as variables, considering, for instance, the sides a, b, and c of a generic triangle as variables involved in the totality of triangles seen as a three-dimensional object; but such a conception was alien to early modern mathematics.

The terms unknown, indeterminate, and variable are of recent date and it is not possible to locate early modern terms near to the meanings I specified. Yet the corresponding concepts were implicitly present in early modern mathematics, and I trust that my use of the terms is clarifying.

"Known," "given," "determined," "constant"

I also use to the opposites of the terms "unknown," "indeterminate" and

[6]Cf. [Bos 1974b] pp. 5–12.

"variable" discussed above, namely: *known* or *given*, *determined*, and *constant*. The terms "known" and "given" were actually used in the early modern period, especially in the practice of analysis. "Given" were the entities that, in a problem or a theorem, were assumed given from the outset in such a way that operations and comparisons could be performed on them, cf. Section 5.2. Usually the theorem or problem was studied in sufficient generality for the actual values of the given elements to be nonessential, hence the "given" elements often had the role of indeterminates. In analysis procedures, "known" was used as synonymous with "given:" in the process of solving a problem the unknown was, in a sense, made "known."

With respect to "determinate," the opposite of "indeterminate," it should be remarked that in arithmetic numbers could be determinate independently, whereas geometrical magnitudes like lengths could only be determined in their relation to other magnitudes assumed known or given.[7]

"Constants" or "invariables," which I mention only for completeness, occurred with respect to objects involving one or more degrees of freedom; they were variables that, with respect to the special object under investigation (a curve for instance), happened to be constant (in magnitude, not in position). In the parabola, for instance, the "subnormal," that is the line segment along the axis between ordinate and normal, was constant; in all other curves it was variable.

6.3 Arithmetic, geometry, algebra, analysis

Use of the terms defined

With terminology fixed as above I can now articulate the meaning of the terms arithmetic, geometry, algebra, and analysis (and arithmetical, geometrical, algebraical, analytical) as used in the present study to describe and characterize mathematical subfields, studies, arguments, and procedures from the early modern period.

Arithmetic refers to the mathematical theory and practice that dealt with numbers.

Geometry refers to the mathematical theory and practice that dealt with geometrical magnitudes. Unless expressly indicated (for instance by referring to *practical* geometry), the theory and practice involved no unit measures and the geometrical magnitudes were *not* represented by numbers expressing their measures with respect to these units.

[7] Natural numbers have an individuality that line segments or other geometrical magnitudes lack. The number 21, for instance, is uniquely determined by its name, by its position in the sequence of numbers or by its decomposition in prime factors; by these properties it is distinct from all other natural numbers. Line segments lack properties by which single ones can be uniquely determined; they acquire their individuality only in relation to others. It is tempting to speculate that this difference in individuality between numbers and line segments made (and perhaps still makes) it more difficult to imagine an indeterminate number than an indeterminate line segment. I have not been able to translate this speculation into explanatory argument concerning early modern mathematics, so I restrict myself to mentioning its appeal.

Algebra refers to those mathematical theories and practices that involved unknowns and/or indeterminates, employed the algebraic operations, involved equations, and dealt either with numbers or with geometrical magnitudes or with magnitudes in an abstract more general sense. In as far as it dealt with numbers, algebra was part of arithmetic. Algebra dealing with (geometrical or abstract) magnitudes presupposed (tacitly or explicitly) a redefinition of the algebraic operations so as to apply to such magnitudes.

Analysis comprises mathematical methods for finding the solutions (in geometry: the constructions) of problems or the proofs of theorems,[8] doing so by introducing unknowns. In geometry (cf. Chapter 5) analysis could involve the use of algebra; if no algebra was used, I speak about "classical analysis."

It should be noted that, used in this way, the terms do not induce a subdivision of mathematical activity in mutually disjoint classes. Geometry and arithmetic are disjoint because the former is about geometrical magnitude and the latter about number. Algebra and arithmetic overlap in the theory and solution of numerical equations; that field also constitutes the overlap of arithmetic and analysis. The overlap of geometry and algebra consists of those parts of geometry where algebraic operations (interpreted as applying to geometrical magnitude) and equations are used. The overlap of analysis and geometry consists of the overlap of algebra and geometry, together with classical geometrical analysis.

Early modern meaning of the terms

In fixing my terminology as above I have tried to remain as near to the early modern usage of the terms arithmetic, algebra, geometry, and analysis as is compatible with the requirement of keeping their meaning constant throughout my study. The main difference between my use and the early modern one was that during the seventeenth century analysis gradually became practically synonymous with algebra (later the meanings of the two terms diverged again).

Symbols and notation

In the distinctions above I have not considered the presence or absence of symbols and notation because that is an aspect of communication rather than of conception. Until c. 1590 algebra was almost exclusively performed in the context of numbers and featured symbolic notation only for the primary arithmetical operations, for root extraction, for equality, and for the unknown and its powers.[9] In prose statements indeterminate numbers (such as coefficients of equations) were indicated by special terms (e.g., "the number of the things" for the coefficient of the linear term of an equation[10]), but there were no symbols for them and in calculations they were replaced by chosen numbers ("example numbers" one might say); the reader was supposed to understand that the procedures were valid also for other values of these numbers. Although the

[8]The latter case seldom occurred — cf. Chapter 5 Note 6.

[9]As one of the terms for the unknown was "cosa," the signs for the unknown and its powers were called "cossic" symbols, and by extension algebra was called the cossic art, cf. [Tropfke 1980] pp. 374–378.

[10]"Numerus rerum," cf., e.g., [Cardano 1545] Ch XI, Rule, p. 98 (Witmer translates this as "the coefficient of x").

use of single letter symbols for numerical indeterminates was pioneered by Jordanus in the twelfth century,[11] it was hardly ever, if at all, practiced in later Middle Ages and Renaissance. Effectively, the use of letter symbols for both unknowns and indeterminates was introduced in algebra by Viète in the 1590s; as we will see he did so in an abstract, non-numerical context. In geometry, in contrast, indeterminates had been designated by letters (or pairs of letters) since antiquity.

Difference with modern terminology

In view of the intense developments in mathematics since the early modern period, it is not surprising that the meanings discussed above are markedly different from the modern meanings of the terms arithmetic, algebra, analysis, and geometry. I note in particular that the "structural" connotation (groups, rings, fields, etc.) of the term algebra is entirely modern, and that the restriction of the term analysis to parts of mathematics involving infinitesimal or limit processes dates from the late eighteenth century. Also, the modern mathematician is used to connect the term geometry with linear, affine, or projective spaces over the real numbers, the complex numbers, or some other field. These concepts of space are recent; early modern geometry studied space as the obvious abstract mathematical object corresponding to the physical space known from direct experience, and considered within that space geometrical configurations and magnitudes as described above.

6.4 Obstacles to the merging of arithmetic, geometry, algebra, and analysis

Status of irrational numbers

The gradual fusion of arithmetic, algebra, and geometry that accompanied the adoption of algebraic methods of analysis was not an easy process; there were obstacles, related to the fundamental differences between numbers and aritmetical calculation, on the one hand, and geometrical magnitudes and construction, on the other hand.

The difference between numbers and geometrical magnitudes appeared most pointedly in the matter of irrationality and incommensurability. Classical Greek mathematicians had realized that the system of natural numbers was insufficient for adequately representing geometrical magnitude, even if one were allowed to divide the unit — which in effect is equivalent to introducing rational numbers. Any attempt to use numbers for representing geometrical magnitude exactly (as opposed to approximatively) required the introduction of new, "irrational" numbers. Such numbers (in particular irrational roots) had been introduced by medieval Arabic writers in order to extend the applicability of the algebraic operations and the rules for solving numerical equations. They were readily taken over in the European medieval and Renaissance texts on algebra. Yet mathematicians were aware that their status as numbers (in the classical sense) was problematical. Numbers like $\sqrt{2}$ or $\sqrt[3]{3+\sqrt{2}}$ were indeed called irrational

[11] [Nemore NumDat].

or "surd," because they had no expressible (i.e., rational) ratio to the numerical unit. The insufficiency of rational numbers and the awareness of the doubtful status of irrational numbers formed a considerable obstacle against accepting numbers as adequate medium to deal with or to represent geometrical magnitudes.

Multiplication and division

Another obstacle against the merging of arithmetical and geometrical methods was the dimensional interpretation of the operations in geometry. Unlike multiplication and division in arithmetic, which started and ended with numbers, the analogous geometrical operations, applied to line segments, yielded rectangles and ratios. This circumstance lent conviction to the orthodox opinion that the realms of arithmetic and geometry were fundamentally different and that therefore the use of numbers (other than for practical purposes) was inappropriate in geometry.

Calculational and constructional complexity

The difference between arithmetic and geometry as to the interpretation of the operations is even more marked if we compare the arithmetical operations, that is, calculations, with geometrical operations in general, that is, with constructions. Analogous computations and constructions differed strongly, both in secondary aspects, such as complexity, and in more fundamental aspects, such as exactness.

An example may illustrate these differences. Consider the determination of the height h of a triangle ABC whose basis c and sides a, b are given in magnitude. The geometrical construction of h is straightforward (see Figure 6.1): Draw a line segment AB equal to c, draw circles around A and B with radii b and a, respectively; these circles intersect in C and C'; draw CC', it intersects AB (prolonged if necessary) in D; $h = CD$ is the required height. The calculation of h, if a, b, and c are given as numbers, proceeds as follows: Compute $s = \frac{a+b+c}{2}$; then compute h by the formula

$$h = 2\frac{\sqrt{s(s-a)(s-b)(s-c)}}{c} . \qquad (6.1)$$

Clearly the calculation is more complex than the construction; it involves a square root extraction, several multiplications, and a division. Moreover, the construction hardly requires explanation or proof whereas the formula underlying the calculation is not self-evident at all.

Numerical and geometrical exactness

The example of the height of the triangle also illustrates a fundamental disparity of calculation and construction: there is no direct analogy between exact calculations and exact constructions. The construction of h employs no other means of construction than circles and straight lines and is therefore exact according to the strictest interpretation of geometrical exactness. The calculation of h, however, involves a square root extraction and is therefore in general not exact but approximate. The geometrical procedures consisting of combinations

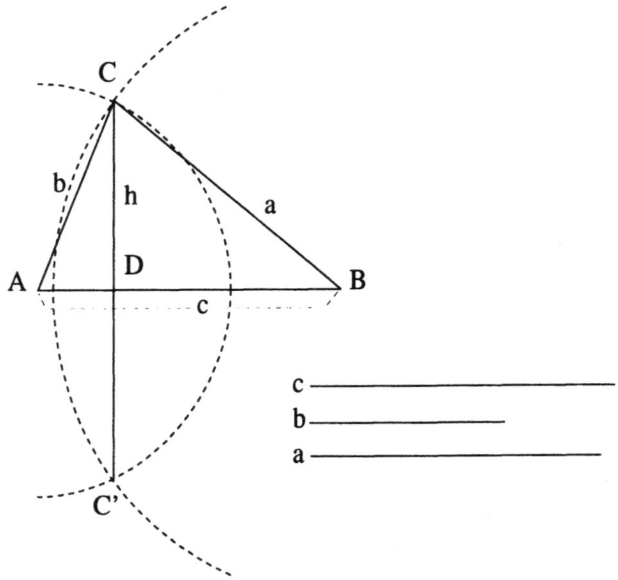

Figure 6.1: Construction of the height of a triangle with given sides

of constructions by straight lines and circles correspond to arithmetical procedures involving the primary arithmetical operations and square root extraction; but in arithmetic the latter operation is not exact, it leads to irrational numbers, which can only be approximated. On the other hand, the group of the exact operations in arithmetic ($+$, $-$, $\times$, $\div$) does not correspond to any naturally distinguishable subcollection of the geometrically exact constructions.

6.5 Conclusion

What could algebra offer to geometry?

The discussion above has indicated the various obstacles to the merging of algebra and geometry. But even if these were considered surmountable, what could algebra, in its late sixteenth-century stage of development, offer to geometry? It could deal with unknowns, it could write down in sufficiently symbolized form equations representing relations between an unknown, its powers, and numbers, the latter serving as "example numbers" for illustrating general procedures of calculation. There was no symbolic technique for dealing with indeterminate numbers.[12] The cossic symbols for the unknown and its powers were slowly being replaced by notations featuring numbers indicating the degree.[13]

Algebra did offer rules for the solution of equations. In the case of quadratic equations these rules could in principle be interpreted in terms of constructions

[12]Cf. note 11.

[13]Notably by Bombelli and Stevin.

by straight lines and circles. The rules for third- and fourth-degree equations were too complicated for geometrical interpretation. Moreover, in special cases (the *"casus irreducibiles"*[14]) they gave rise to "imaginary" numbers.[15] It was known that in such cases roots did exist, but the extant rules for calculating them led to imaginaries, whose interpretation was a major conceptual difficulty, not solved, in fact, until the nineteenth century.[16] In contrast, the geometrical problems corresponding to the *casus irreducibiles* were reducible to the trisection of angles and did not involve conceptual difficulties, nor did they suggest the necessity of introducing anything "imaginary."

Geometry more adequate for general arguments

An algebra beset with such restrictions and uncertainties held little promise for geometry, in which the use of indeterminate lengths, adequately represented symbolically by letters or pairs of letters, was standard, and in which classical analysis dealt competently with unknowns. Indeed with respect to indeterminates geometry was in the advantage, as is illustrated by the fact that it was called in to help algebra rather than the other way around. Thus Cardano used the geometrical configuration of a cube divided by three planes into two smaller cubes and six rectangular prisms, to prove the binomial theorem for third powers. Lacking symbols for indeterminate numbers, he could not write the rule as $(a+b)^3 = a^3 + 3a^2b + 3ab^2 + b^3$ and prove it by twice multiplying $(a+b)$ in itself. Apparently he considered an example worked out for particular numbers insufficient proof. So for generality he turned to geometry, which did provide indeterminates, in this case the edge $a+b$ of the cube and its two parts a and b.[17]

In introducing his "new algebra" for general (including geometrical) problem solving, Viète was probably inspired by the example of Diophantus rather than by the results of sixteenthth century algebra. Descartes' well-known description (in the *Discourse*) of the algebra he learned at school as

> so confined to certain rules and symbols that the end result is a confused and obscure art which encumbers the mind, rather than a science which cultivates it[18]

is another indication that algebra was seen as an unlikely source of help for geometry.

An improbable phenomenon

Thus the adoption of algebraic methods of analysis in geometry was in several

[14]Namely, third-degree equation with three real roots; for these equations the Cardano formulas (see Note 91 of Chapter 4) involve square roots of negative quantities.

[15]Imaginaries also occurred in applying the rules for quadratic equations, but that was not conceptually puzzling, they were signs that no solution existed. Thus Ghetaldi explained ([Ghetaldi 1630], Book V, Ch. II pp. 314 sqq), "how impossible problems are recognized" ("Quomodo Problemata impossibilia cognoscantur"); the examples he gave were problems that led to quadratic equations with negative or imaginary roots.

[16]Cf. [Kline 1972] pp. 265–266.

[17][Cardano 1545] Ch. VI; [Cardano 1966] vol. 4 pp. 235–236, [Cardano 1968] pp. 52–53.

[18][Descartes 1985–1991] vol. 1 pp. 119–120.

respects an improbable phenomenon. The step, usually seen as concentrated in Viète's new algebra and Descartes' innovation of geometry, is generally, and with good reason, recognized as a very important event in the development of mathematics. To us it may also seem a natural and an obvious one. But such an impression would be based too much on the hindsight of successful algebraic geometry and on a historiography that commonly treats the emergence of analytic geometry in the seventeenth century as merely laudable, rather than as enigmatic. Viewed from the perspective of the late sixteenth century, it was not at all obvious that algebra should be of use as a tool for geometrical analysis.

Chapter 7

Using numbers in geometry — Regiomontanus and Stevin

7.1 Introduction

Numbers in geometry

The obstacles to the merging of arithmetic, geometry, algebra, and analysis, discussed in the previous chapter, were primarily of conceptual nature. I now turn to a question on a different level that also played a role in the fusion process, namely whether it was *legitimate* to use numbers in geometry.

Because it concerned legitimation, this question was akin to the issue of legitimate construction, which is the main theme of my study. However, it will become clear (cf. Section 7.5) that in the early modern period the connection between the two issues was less close than might at first be expected. Therefore, a detailed survey of the contemporary arguments about the legitimacy of using numbers in geometry is not necessary for my inquiry. The aim of the present chapter is to explain the issue, to provide some instances of arguments about it voiced by early modern mathematicians, and to assess its relation to the interpretation of exactness of geometrical constructions.

Approximation or extension

Throughout the sixteenth century algebraic techniques were developed and used primarily for solving problems about numbers. As a result the legitimacy of the geometrical use of algebra was bound up with the question of whether it was allowed to use numbers in geometry. Although in practical geometry numbers were in general use, mathematicians concerned with the conceptual foundations of pure geometry felt that there were considerable barriers to this use; number was classically considered to be integer or rational only, and it was well known that the rational numbers alone were insufficient to deal with geometrical magnitudes such as line segments.

The rational numbers are insufficient for dealing with geometrical magnitude; they cannot exactly represent situations involving incommensurable pairs of line segments. On the other hand, rational numbers are an excellent means for representing geometrical situations approximately to any required degree of precision. Using rational numbers in geometry implies that either one accepts approximate rather than exact representation, or one somehow extends the rational number system with sufficient new numbers capable of representing incommensurable magnitudes. Both alternatives for using numbers in pure geometry have fundamental impediments: either the loss of exactness or the introduction of undefined entities.

Sixteenth-century mathematical literature provides examples of either alternative. Two authors in particular held and spiritedly defended archetypical opinions on the use of numbers in geometry: Regiomontanus, who was willing to sacrifice exactness, and Stevin, who was convinced that the number concept was potent enough to deal with continuous geometrical magnitude.

7.2 Regiomontanus

Practical geometry

Numbers and arithmetic were used and applied as a matter of course in such practical disciplines as trigonometry and surveying. One assumed a unit length (Regiomontanus used the term "mensura famosa"[1]); lengths, areas, and volumes were expressed as numbers with respect to the unit length or the corresponding unit square and unit cube. Because the measurement of line segments with respect to the unit involved instrumental and observational errors, the results of the numerical calculations of practical geometry were not exact. In his influential textbook of trigonometry *On triangles*,[2] posthumously published in 1533, Regiomontanus felt called to explain this matter and, despite their approximate nature, to defend the use of numbers as equally, if not more, effective and valid in geometry than abiding to the restrictive requirement of pure geometrical exactness.

"On triangles"

Regiomontanus wrote *On triangles* in the terminology and structure of Euclid's *Data* (cf. Section 5.2), using the concept "given" or "known" in enunciating his propositions. This arrangement required an explanation of these concepts in the context of numbers; Regiomontanus provided it in the first definition of the book:

> A quantity will be called known if the principal measure, or any arbitrarily assumed measure, measures it according to a known number.[3]

The subsequent definitions made clear that Regiomontanus meant the "known" numbers to be integers or rational fractions. Thus a known quantity had a

---

[1]Cf. Note 3.

[2][Regiomontanus 1533].

[3][Regiomontanus 1533] p. 30 (facs. p. 7): "Cognita vocabitur quantitas, quam mensura famosa, aut pro libito sumpta secundum numerum metitur notum." I have modified Hughes' translation of the passage.

rational ratio to the principal measure, and to show that a quantity was known meant to show that this ratio could be found, i.e., calculated.

Irrationals

Predictably, with this definition the irrationals presented a difficulty. Regiomontanus confronted it in Proposition 2 of book I:

> The side of a known square will not remain unknown.[4]

The number expressing the length of the side should be the mean proportional number between 1 and the number expressing the area. But if the latter number was not a square, the mean proportional number did not exist. Regiomontanus resolved this difficulty by adjusting the definition of "known quantities":

> However, it often happens that the numbers by which we measure our squares are not squares; therefore, in order that we will not remain ignorant of the near truth (as are all things knowable by man), we will henceforth use the term "known quantity" in a looser sense than we defined it at the beginning. So we will by the same term call any quantity known if it is either precisely known or almost equal to a known quantity; for in my opinion it is more beautiful to know what is near the truth than to ignore the truth itself completely. There is virtue not only in hitting the mark but also in coming close to it.[5]

Anticipating the question why he did not define known quantities in this way from the beginning, he explained that this would have made the reader confused and suspicious:

> Above I did not want to define a known quantity in this manner by both precision and closeness, fearing that my definition, using an ambiguous term as closeness, would become suspect to the reader. For we are used to take the precise as true and we would hardly satisfy the reader by taking closeness to truth as a definition.[6]

Regiomontanus then explained what to do when the number expressing the area was non-square: approximate the number by a square (integer or fraction) to any desired degree of precision, then accept the root of that square as representing the side:

---

[4] [Regiomontanus 1533] p. 34 (facs. p. 9): "Quadrati noti costa non ignorabitur."

[5] [Regiomontanus 1533] p.34 (facs. p. 9): "Cum autem saepenumero accidat numeros secundum quos quadrata nostra metimur esse non quadratos ne prorsus ignoremus propinquum veritati (ut sunt scibilia humana) laxius posthac utemur vocabulo quantitatis notae, quam initio diffinierimus. Quantitatem igitur omnem quae aut nota praecise fuerit, aut notae quantitati fermè aequalis, univoce notam appellabimus. Pulchrius equidem arbitror scire propinquum veritati, quam veritatem ipsam penitus negligere: non modo enim contingere metam, verumetiam propinque accedere virtuti dabitur."

[6] [Regiomontanus 1533] p. 34 (facs. p. 9): "Non libuit autem hoc pacto superius diffinire quantitatem notam per praecisum et propinquum, ne suspecta lectori diffinitio nostra redderetur, fluctuante vocabulo propinqui id agente: nam etsi praecisum pro vero ponere soleamus, propinquum tamen veritati vix diffinitionem lectori satis facturam accipiet."

> Just as our non-square number is close to the square number we have assumed, so also is the side of our given square close to the precisely known side of the assumed square, and therefore the side of our square will be considered known.[7]

Later on Regiomontanus followed the same line of argument with respect to the trigonometric functions; thus Proposition I-27[8] stated that, if two sides of a rectangular triangle are known, the two acute angles are known as well, because they can be determined with sufficient precision by means of a table of sines.

Significance Regiomontanus' textbook was very influential,[9] so we may assume that his defense of numbers in geometry was generally known and that it supported those who considered numbers as proper and effective means for geometrical investigations. Still the frank admission of the approximate character of numerical procedures and the avoidance of the question concerning the nature of irrationals made it clear that Regiomontanus' approach, although consistent and elegant in its own way, was not compatible with the usual requirements of exactness of pure geometry.

7.3 Stevin

Elements X By the end of the sixteenth century calculations involving square roots of non-square numbers had become so common that, at least in arithmetical and algebraic practice, these irrationals were in effect considered and treated as numbers. The indifference to the more subtle questions concerning irrationality implied in this practice agreed well with the view, put forth by Ramus and taken over in various degrees by several mathematicians, that geometry ought to be the art of measuring well,[10] rather than the pure and strict science exemplified in the theoretical books of Euclid's *Elements*. From that point of view it was especially book X of the *Elements*, with its classification of irrational ratios, which appeared over-theoretical and pointless. Ramus had disqualified the book on these grounds, calling it a "cross for torturing noble minds,"[11] and since then the uselessness of *Elements* X became a kind of partisan slogan of those who favored the use of irrational numbers to simplify matters in geometry.

Simon Stevin provided one of the most articulate statements of this point of view, including an explicit reformulation of *Elements* X in terms of irrational

[7] [Regiomontanus 1533] p. 34 (facs. p. 9): "Quemadmodum autem numerus non quadratus noster, numero quadrato assumpto propinquus est, ita et costa quadrati nostri costae alterius quadrati praecise cognitae propinqua, et ideo nota habebitur."

[8] [Regiomontanus 1533] pp. 64–66 (facs. pp. 24–25).

[9] Cf. [Zinner 1968], p. 320 for sixteenth-century editions of the book.

[10] Cf. for instance [Verdonk 1969].

[11] [Ramus 1599] p. 252 (beginning of book 21): "Equidem toto decimo libro studiosé et accuraté considerato nihil aliud judicare potui quam crucem in eo fixam esse, qua generosae mentes cruciarentur. Quare omni studio diligentiaque connitendum nobis est, ut ista clarissimé evolvantur, miseraque et funesta crux evertatur et prosternatur, atque in perpetuum affligatur."

numbers, serving to show that in those terms the substance of the book was simple and easy.

Irrationals are numbers

Stevin did so in his *Arithmetic*[12] of 1585. The book contained a vigorous vindication of numbers as valid and appropriate means to deal with continuous magnitude, in particular geometrical magnitude. At the outset Stevin claimed that number was not discontinuous[13] and was therefore adequate for dealing with continuous quantities such as length or area. He claimed that all square, cubic, and higher roots of all numbers existed as numbers, and he went on to refute some standard arguments to the contrary. His reasoning[14] may be summarized as follows:

The square root of 8 is a part of 8, and parts are of the same material ("matière") as the whole; the material of 8 is number, hence $\sqrt{8}$ is a number as well as 8. Those who deny that $\sqrt{8}$ and other such roots are numbers call them "absurd, irrational, irregular, inexplicable, surd, etc."[15] because $\sqrt{8}$ is incommensurable with respect to "arithmetical numbers" (i.e., rational numbers) such as 2, 3, etc. But incommensurability concerns the relation of two things, not the things themselves; the incommensurability is not caused by $\sqrt{8}$, so there is no reason to call $\sqrt{8}$ absurd any more than to call 2 or 3 absurd. Indeed a sphere and a cube are dissimilar, but it is absurd to say that this is the fault either of the sphere or of the cube. Some refer to Euclid, who in *Elements* X[16] defined some lines as rational and others as irrational, but that is an empty appeal to authority, because Euclid defined earlier that any line segment could be chosen as rational, so one might as well chose the diameter of the square with side 2 to be rational, which would make $\sqrt{8}$ rational and 2 or 3 irrational. If challenged to explain what is $\sqrt{8}$ ells, Stevin would ask what $\frac{3}{4}$ ell is, and if his adversary would refer to *Elements* VI-9 (construction of aliquot parts of a line segment), Stevin would refer to *Elements* VI-13 and construct $\sqrt{8}$ ells as the mean proportional between line segments of length 1 and 8 ells, respectively. However, there are no constructions corresponding to cube roots (or two mean proportionals) so would Stevin accept that cubic roots of non-cube numbers are surd, irrational, etc.? No, the lack of a legitimate geometrical construction for cubic roots is not the fault of number but of geometry:

> What we proved for $\sqrt{8}$ will be understood as well for $\sqrt[3]{}$ and whatever other roots; it is true that we cannot legitimately cut off from any line its cubic root (because the two mean proportionals between two given lines are not yet geometrically found) whereas we can cut off its square root, but that is not the fault of the numbers;

---

[12] [Stevin 1585]; the reformulation of *Elements* X occurs as an appendix (pp. 189–201) of the third part of the book, entitled "traicte des incommensurables grandeurs avec une appendice de l'explication du dixiesme livre d'Euclide:" the appendix does not occur in the much abridged edition of the *Arithmetique* in [Stevin 1955–1966].

[13] [Stevin 1585] p. 4: "Que nombre n'est poinct quantite discontinue."

[14] [Stevin 1585] pp. 30–37.

[15] [Stevin 1585] p. 33: "absurds, irrationels, irreguliers, inexplicables, sourds, etc."

[16] Stevin here refered to the definitions now numbered 3 and 4.

> for in numbers we easily achieve that which we know not how to do in lines.[17]

Stevin did not explain here which numerical methods of root extraction he had in mind, thus smoothing over the difficulty that such methods are approximate rather than exact.

Irrationals and Elements X

Turning to the use of irrational numbers, Stevin stated that the "absurd" idea that they would be absurd had obscured the doctrine of incommensurable magnitudes so that

> ... many have developed such a horror of the difficulty of the tenth book of Euclid (which treats of this matter) that they call it the cross of mathematicians, too hard a matter for digestion, and they perceive no use at all in it.[18]

But by a reformulation of the book's content in terms of irrational numbers Stevin would make the matter easy and clear. Thus Stevin went further than those who valued useful geometry with numbers more than theoretical geometry as exemplified in book X; he affirmed that numbers could indeed take over that part of pure geometry and make it easy.

Exactness

Later on in the *Arithmetic* Stevin argued against those who were willing to avoid the issue of irrationals and restrict themselves to rational numbers, because ultimate precision was pointless in the practice of measuring. We may discern here an echo of Regiomontanus' argument. In answer Stevin stated (be it without other argument than authority) his adherence to exactness in mathematics:

> one might ask as well why the operations of geometry, as the *Elements* of Euclid, are made to the utmost perfection; but as that seems unworthy of answering, because of the absurdities following from its converse (for those perfect operations provide perfect understandings and these are the source of the perfect and admirable effects that mathematics produces) so much for that one.[19]

---

[17] [Stevin 1585] p. 36: "Ce que nous avons demonstré de $\sqrt{8}$, sera aussi entendu de $\sqrt{}(3)$ [Stevin's sign for a cubic root], et autres racines quelquonques: car combien que de toute ligne ne pouvons legitimement couper racine cubique (à cause que les deux lignes moiennes proportionelles entre deux lignes donnees, ne sont encore geometriquement inventees) comme faisons racine carrée, cela n'est pas la coulpe des nombres; car ce qu'en lignes ne sçavons faire, nous l'achevons par nombres facilement."

[18] [Stevin 1585] pp. 36–37: "...la difficulté du dixiesme livre d'Euclide (qui traicte de ceste matière) est à plusieurs devenu en horreur, voire iusques a l'appeller la croix des mathematiciens, matiere trop dure à digerer, et en laquelle n'apperçoivent aucune utilité."

[19] [Stevin 1585] pp. 169–170: "on pourroit dire pareillement, pourquoi les operations de Geometrie, comme les elemens d'Euclide, sont faistes par l'extreme perfection; Mais comme cela ne semble pas digne de responce, à cause des absurditez suivantes de son contraire (car telles parfaictes operations, donnent parfaictes intelligences, qui sont causes des parfaicts et admirables effects que produict la Mathematique) ainsi de cestui ci."

Significance

It is easy for the modern reader to spot the weakness of Stevin's argument: the absence of a definition of number with respect to which the existence of such numbers as $\sqrt{8}$ could be proven. However, such precision came to mathematics only in the late nineteenth century, so Stevin's defence of numbers may well have seemed, although perhaps not ultimately convincing, yet strong and legitimate enough. The issue of the legitimacy of using numbers in geometry, however, remained undecided as appears from a later polemic on the matter.

7.4 A later discussion

Van Ceulen

Ludolph Van Ceulen, not surprisingly for his love of numerical calculation culminating in a value of π of record precision, considered geometry no forbidden territory for numbers. He died in 1610, leaving several mathematical treatises in manuscript. With the help of Willebrord Snellius his widow Adriana Symons published a compilation of six of these in a Dutch as well as a Latin version. For the latter Snellius provided the translation and added introductions and commentaries.

Snellius on the status of numbers

Snellius favored the free use of numbers but knew that it was a sensitive issue; he used the dedication of the third part of the book to defend Van Ceulen's approach. The passage starts thus:

> You see here, most illustrious Sir, these books on various problems, in which we deal with some geometrical problems in such a way that occasionally we also allow numbers to be associated with this subtle subject. For number is the accurate mediator of all measure, ratio and proportion.[20]

Snellius explained that magnitudes are best expressed by numbers because number is easily divisible *ad infinitum*, whereas it is difficult to actually divide geometrical magnitude in small parts. He even suggested that this might have been Aristotle's opinion expressed in an often quoted passage from the *Metaphysics* in which arithmetic was claimed to be more exact than geometry.[21] Hence the usefulness of numbers, including irrational ones; if alone because — here the slogan duly recurs — they make *Elements* X superfluous:

> For that reason we ought not to deny the philomaths the use of numbers, and especially irrational and surd numbers, such as illustrated in these books; the more so as it should be clear to anyone how useless in practice is that Pythagorean distinction of irrationals in thirteen species to which Euclid devotes the whole of the tenth

[20] [Ceulen 1615b] p. 84: "En itaque tibi vir Amplissime hosce problematum variorum libros, in quibus quorundam Geometricorum problematum tractationem ita exhibemus, ut quandoque numeros quoque in hujus subtilitatis societatem admiserimus. Est enim numerus omnis commensus, rationis et proportionis accuratus interpres."

[21] Snellius' interpretation seems unfounded; cf. Heath's discussion of the passage (*Metaph.* Book I Ch. 2 (982a)) and related ones in [Heath 1949] pp. 4–5.

> book of the *Elements*, because the general laws of that numeration have no value at all if we want to determine the species of this or that number.[22]

Snellius elaborated on the theme:

> Therefore I relax and stop wondering about the usefulness of the tenth book: for there is no element at all among those species that is cited or used anywhere in Archimedes, Apollonius, Serenus, Theodosius, Menelaus, Ptolemy, Theon, Eutocius, Diophantus and even in Euclid himself outside the Elements. . . . and although these things may be preserved as subtleties in the mathematical library, yet, as they are of little use, they have to be separated from the elementary books.[23]

Kepler's reaction

Snellius' words — we do not know what Van Ceulen himself thought of the matter — show that the issue of the status of numbers was alive, as appears also from a reaction published four years later by Kepler, who was enraged by Snellius' statements. He saw the infiltration of number into geometry as an attack, destabilizing the structure and disdaining the harmony of pure geometry. I deal with Kepler's views in more detail in Chapter 11; here one passage from his reaction may suffice. Referring to the "cross for tormenting noble minds," which Ramus had detected in book X,[24] he wrote:

> "Is it only a cross fastened to our talents?" I say, to those who molest the inexpressibles with numbers, that is by expressing them. But I deal with those kinds not with numbers, not by algebra, but by mental processes of reasoning, because of course I do not need them in order to draw up accounts of merchandise, but to explain the causes of things.[25]

---

[22][Ceulen 1615b] p. 84: "Eam ob causam numerorum, maximé irrationalium et surdorum usum istus [sic] libris illustratum philomatis invidere non debuimus: idque adeó tanto magis, ut clarum cuilibet sit, quantopere ad usum inutilis sit Pythagorea illa ⟨alogias⟩ in tredecim species distributio, in qua Euclides, totum 10 Elementorum librum occupavit, cum generales istae numerationis leges nihil pensi habeant ad quamnam speciem hic vel ille numerus sit referendus."

[23][Ceulen 1615b] p. 84: "ideoque de utilitate libri decimi minus sollicitum mirari desino: nullum enim inter eas species elementum extat quod usquam in Archimede, Apollonio, Sereno, Theodosio, Menelao, Ptolomaeo Theone, Eutocio, Diophanto, ipsoque adeo Euclide extra elementa vel citetur vel usum ullum habeat. . . . [crux igitur quaedam istic tantum defixa est, quae solo calculo in abaco facillime tollatur:] et quamvis ista tanquam subtilia in Mathematica bibliotheca conservari possint: attamen ut minus utilia á ⟨stoicheios⟩ segregari debent [nam si ista usum habeant, totum hoc genus, cujus ille liber particulam duntaxat aliquam explicandam sibi sumit haud dubie plus longe reconditae eruditionis et scientiae complectetur.]"

[24]Cf. Note 11.

[25][Kepler 1937–1975] pp. 18–19: "*Crux tantùm defixa est ingeniis?* Equidem iis, qui numeris, hoc est effando vexant Ineffabilia. At ego has species tracto non numeris, non per Algebram, sed ratiocinatione Mentis; sané quia iis mihi non est opus ad subducendas Rationes mercatuum, sed ad explicandas rerum causas." Translation quoted from [Kepler 1997] p. 13; the first sentence is a quotation from [Ramus 1569] p. 258.

7.5 Conclusion

Construction not at issue

The discussions on the acceptability of numbers in geometry mentioned so far concerned primarily the compatibility of number and geometrical magnitude; the main alternatives involved were: approximation, introduction of irrational numbers, or avoiding the use of numbers in geometry altogether. These alternatives did not relate to questions of the exactness of geometrical constructions. On that issue the protagonists of the debate did not disagree strongly. Regiomontanus showed little interest in geometrical construction in his book on triangles, Ramus and Stevin accepted a restriction of legitimate geometrical construction to those by straight lines and circles in so far as they considered the known constructions of two mean proportionals ungeometrical;[26] Snellius agreed with Pappus' classification of constructions;[27] Kepler, as we will see (Chapter 11), rejected all non-plane constructions as ungeometrical.

Algebra dissociated from numbers

During most of the sixteenth century algebraic methods were used only in solving numerical problems and were considered as directly linked to numbers. Therefore, the doubts about the geometrical legitimacy of numbers constituted an additional obstacle against the merging of algebra and geometry. Nevertheless, algebraic methods of analysis were created and adopted within the early modern tradition of geometrical problem solving; they even provided what I have called the principal dynamics in this field. This, however, did not imply that the resistance to the use of numbers in pure geometry was overcome; rather, algebraic methods were dissociated from numbers and redefined to fit geometry. This approach was first developed by Viète.

---

[26] Cf. Note 29.

[27] Cf. [Snellius 1621] pp. *1^v sqq.

Chapter 8

Using algebra — Viète's analysis

8.1 Introduction

Defining the algebraic operations

By 1600 Viète had worked out a consistent and effective apparatus for using algebra in both numerical and geometrical contexts. He had circumvented the issue of numbers in geometry by defining the algebraic operations independently of whether they applied to numbers or to non-numerical quantities (such as line segments or other geometrical magnitudes), thus legitimizing the use of algebra in geometry while retaining the conviction that geometrical magnitude and number were essentially different. Viète called this system "new algebra" or "specious logistics."

Descartes was to address the same issue — the definition of the algebraic operations as applied to geometrical magnitude — some 40 years later, solving it in a manner essentially different from Viète's. His approach will be discussed in Part II (cf. in particular Chapter 21). In the mean time Van Ceulen proposed an interesting but incomplete approach to the same question, which featured some elements of the approach Descartes adopted later. I discuss it at the end of the present chapter.

8.2 Viète's "New Algebra"

Restoring analysis

Several sixteenth-century mathematicians knew how to interpret algebraic equations geometrically as theorems or problems. They used this interpretation primarily in showing the generality and correctness of algebraic relations and procedures for solving equations.[1] However, it was Viète who first introduced

[1] The best known example is Cardano's use of a cube to prove the equivalent of the algebraic relation $(a+b)^3 = a^3 + b^3 + 3ab(a+b)$, cf. Note 17 of Chapter 6. In book II of his *Algebra* ([Bombelli 1572]) Bombelli gave geometrical equivalents of several cubic equations (e.g., $x^3 +$

and promoted the idea that algebra was the proper method for the analysis of problems both in geometry and in the theory of numbers.

Viète's[2] inspiration was Diophantus,[3] whose use of unknowns in the *Arithmetic*[4] he saw as the key to a general method of analysis, which, he thought, had been known to classical mathematicians and was lost, but could be restored. He expounded his reconstruction of this analysis in a series of treatises, together forming what he called the *Book of the restored mathematical analysis or the new algebra* and whose outline he sketched in the first of these treatises, the *Introduction to the analytic art* (*In artem analyticen isagoge*) of 1591 (to which I refer as the *Isagoge*).[5] The *Isagoge* closed with the following proud statement:

> Finally the analytic art, endowed, at last, with its three forms of zetetics, poristics and exegetics, claims for itself the greatest problem of all, which is TO LEAVE NO PROBLEM UNSOLVED.[6]

Zetetics, poristics, exegetics

Thus Viète's analysis — as Descartes' later — was a universal method for solving problems. Its main tool was algebra. Viète introduced a tripartite division of the analytical process, creating neologisms, based on existing Greek terms, to distinguish them: *zetetics*, *poristics*, and *exegetics* or *rhetics*.[7] Viète saw these parts of analysis as related to the consecutive phases in the interplay between the problems themselves, the algebraic methods, and the actual solution of the problems. Zetetics was the art of translating a problem, be it an arithmetical or a geometrical one, into one or more algebraic equations in one

---

$6x = 20$ on pp. 286–288) explaining a geometrical construction of the roots by means of shifting gnomon procedures similar to the construction of two mean proportionals attributed to Plato (Construction 2.3).

[2]Some of the ideas on Viète presented here have been published before in [Bos & Reich 1990] and [Bos 1993]; relevant secondary sources on Viète are: [Klein 1968], the introduction in [Viète 1973], [Egmond 1985], [Brigaglia & Nastasi 1986], [Freguglia 1988] pp. 67–104, [Freguglia 1989], [Giusti 1992], [Garibaldi 1992], [Stefano 1992].

[3]Cf. [Viète 1591] p. 10 ([Viète 1983] p. 27).

[4][Diophantus Arithmetica]; first printed edition [Diophantus 1575]. In his [Viète 1593–1600] Viète incorporated a large number of propositions from the *Arithmetica*.

[5]Viète announced the list of the treatises constituting the planned "Opus restitutae mathematicae analyseos seu algebra nova" in [Viète 1591] (cf. [Egmond 1985] pp. 367–368, the list is not incorporated in the edition of the *Isagoge* in [Viète 1646] nor in the translation in [Viète 1983]). He gave the list in its logical order, but he did not keep to that order in publishing; he postponed the more technical items. By 1593 five treatises had appeared, one more came out in 1600, the remaining three were published posthumously. The list is as follows: (1) *Isagoge* [Viète 1591]; (2) *Ad logisticen speciosam notae priores* [Viète 1631b]; (3) *Zeteticorum libri quinque* [Viète 1593–1600]; (4) *De aequationum recognitione et emendatione tractatus duo* [Viète 1615]; (5) *De numerosa potestatum ad exegesin resolutione* [Viète 1600b]; (6) *Effectionum geometricarum canonica recensio* [Viète 1593b]; (7) *Supplementum geometriae* [Viète 1593]; (8) *Ad angularium sectionum analyticen theoremata* [Viète 1615b]; (9) *Variorum de rebus mathematicis responsorum liber VIII* [Viète 1593b].

[6][Viète 1591] p. 12: "Denique fastuosum problema problematum ars Analytice, triplicem Zetetices, Poristices et Exegetices formam tandem induta, jure sibi adrogat, Quod est, NULLUM NON PROBLEMA SOLVERE" (translation based on the one in [Viète 1983] p. 32).

[7][Viète 1591] p. 1 ([Viète 1983] pp. 11–12): "zetetice," "poristice," "exegetice," "rhetice:" Viète retained the Greek declension of his neologisms.

or more unknowns. The interpretation of "poristics" is difficult. The relevant sections in the *Isagoge* admit various translation and interpretations. Viète did not use or discuss the term in later treatises, apart from two passages in the *The supplement to geometry* and one in *Book VIII of various replies on mathematical matters* where he mentioned results derived "in poristics."[8] However, it seems safe to assume on the basis of these passages that Viète's poristics, as the middle phase of the analytical process, concerned techniques of transforming algebraic proportionalities and equalities, and that Viète envisaged to present these techniques independently of whether some of the quantities involved were unknown and others not. Finally exegetics, also called rhetics, was the art of deriving the arithmetical or geometrical solutions from the equations supplied by zetetics and, if necessary, transformed to amenable forms by poristics.

Abstract, "specious" algebra

Viète did not see the algebra, which was to be the essential tool in his analysis, as a technique concerning numbers, but as a method of symbolic calculation concerning abstract magnitudes. In elaborating this conception he created symbolic procedures of calculation applying to magnitudes irrespectively of their nature (number, geometrical magnitude, or otherwise — note that he considered number to be a kind of magnitude). For this purpose he introduced letter symbols for indeterminates as well as unknowns. Although letter symbols for indeterminates were common in geometry and had occasionally been used in arithmetic (notably by Jordanus[9]), Viète was the first to employ letter symbols for general indeterminates. This was not a self-evident step because it raised the question of the status and nature of these general indeterminate magnitudes and of the operations performed on them. It is clear from his writings that this question was a serious one for Viète. His answer is difficult to reconstruct precisely,[10] but in outline it may be summarized as follows. In his "new algebra" mathematical entities such as numbers, line segments, figures etc., whether known, unknown, or indeterminate, were considered only in their aspect of being a magnitude, abstracted from their actual nature. Viète himself spoke of "in species," "in form" or "in kind," calling his new algebra a "calculation regarding forms," or "regarding species:" he also used the term "specious logistics"[11] — which I take over in the sequel. Thus his specious logistics dealt with abstract magnitudes symbolically represented by letters.

[8][Viète 1593] pp. 244 and 245 ([Viète 1983] pp. 395 and 396), [Viète 1593b] p. 353. See also Reich and Gericke's discussion of the terms zetetics, poristics exegetics and rhetics in [Viète 1973] pp. 22–23 and 31, Witmer's note 6 on p. 12 of [Viète 1983], and [Garibaldi 1992] p. 170. Witmer cites translations of Viète's definition of poristics by Vaulézard, Vasset, Durret, Ritter, and Smith, which differ considerably. Both Witmer and Reich and Gericke suggest that the words "in poristics" in *The supplement to geometry* might refer to a treatise *Ad logisticem speciosam notae posteriores*, a no longer extant sequel to the *Notae priores* ([Viète 1631] and [Viète 1631b]).

[9]Cf. Chapter 6 Note 11.

[10]Cf. [Viète 1973] pp. 26–27.

[11]Cf. [Viète 1591]: "Logisticen sub specie" (p. 1), "Logistice numerosa est quae per numeros, Speciosa quae per species seu rerum formas exhibitur, utpote per alphabetica elementa" (p. 4); cf. [Viète 1983] pp. 13 and 17.

Multiplication and the law of homogeneity

An algebra for abstract magnitudes required an appropriate reinterpretation of the arithmetical and algebraic operations. As we have seen in Section 6.2, there are important dissimilarities between the multiplication of numbers and the multiplication of line segments, so in order to arrive at a consistent interpretation of multiplication for abstract magnitudes, a choice had to be made. Viète chose to be inspired by geometry. Geometrical multiplication involved a change of dimension; the product of two line segments was a rectangle, that is, a magnitude of a different kind than the original line segments. Viète accepted this dimensional aspect of multiplication. However, in geometry the highest dimension for magnitudes was the dimension of space itself, so products of more than three line segments defied interpretation. Viète did not want to take over this aspect, and his abstract conception of magnitude enabled him to avoid it. In his conception any species of magnitude was accompanied by a scale of successive higher-dimensional species of magnitudes, constituted in analogy to the first three geometrical dimensional magnitudes — line, area, solid — but continued ad infinitum. Viète used the term "degree"[12] for these higher dimensional species and refered to their sequence with words derived from "scalae" (stairs, ladder),[13] I refer to such a sequence of species of magnitudes as a "scale." Within one scale only magnitudes of the same degree could be compared, added, or subtracted. This constituted the "law of homogeneity,"[14] which Viète considered fundamental to his new algebra. Multiplication linked the degrees within one scale in the way suggested by geometry: the product of two magnitudes of the first degree was of the second degree, the successive powers of a first-degree magnitude ran through all the degrees, etc. Viète described the degrees with terms inspired by geometry.

8.3 The "New Algebra" as a formal system

Assumptions and axioms

While considering abstract magnitudes Viète could obviously not specify how a multiplication (or any other operation) was actually performed but only how it was symbolically represented. Thereby the "specious" part of his new algebra was indeed a fully abstract formal system implicitly defined by basic assumptions about magnitudes, dimensions, and scales as explained above, and by axioms concerning the operations. The operations were addition, subtraction, multiplication, division, root extraction, and the formation of ratios.[15] It will be useful to list these assumptions and axioms explicitly (cf. Table 8.1). They

[12] [Viète 1591] p. 2: "gradus:" Witmer translates "grade" [Viète 1983] p. 15.

[13] Thus the successive powers of one magnitude of the first dimension (cf. the second of the assumptions and axioms mentioned below) were called "magnitudines scalares" [Viète 1591] p. 3; Witmer translates "scalar magnitudes" [Viète 1983] p. 16.

[14] [Viète 1591] p. 2: "lex homogeneorum," Witmer translates more literally: "law of homogeous terms" [Viète 1983] p. 15.

[15] Viète's specious algebra may indeed be considered as the first occurrence in mathematics of a fully abstract formal system with some complexity. In modern structural terms it could be characterized as a graded ring in which the modules are continuous semigroups and in which only the homogeneous elements are considered.

are as follows:[16]

> **1.** The *law of homogeneity*: Only magnitudes of the same kind (which implies that they have the same dimension) can be compared, added, and subtracted (the smaller from the larger).
> **2.** Within one scale of magnitudes there is an element, called "side" or "root"[17] that by multiplication in itself successively generates magnitudes of all dimensions within the scale; this induces a numbering of the dimensions as successive degrees.
> **3.** Within one scale of magnitudes any two magnitudes can be multiplied; the dimension of the product is the sum of the dimensions of the factors. Division is the inverse operation to multiplication; any magnitude can be divided by a magnitude of smaller dimension, the dimension of the quotient is the difference of the two dimensions.
> **4.** Multiplication is commutative; addition and multiplication interact distributively.
> **5.** Any two magnitudes of the same dimension have a ratio satisfying the usual rules for ratios and proportionalities. Proportionalities can be transformed into equivalent equalities by the equivalence $a : b = c : d \Leftrightarrow ad = bc$.

With respect to the comparison of Viète's system with that of Descartes it is important to note that Viète introduced no unit element with respect to multiplication. Viète did not consider ratios as magnitudes but as relations. A ratio was not the result of a division of two magnitudes of the same dimension, but a relation that two such magnitudes had with respect to their size. In adopting this classical conception of ratio Viète rejected the idea that ratios (including irrational ones) could be understood and treated as numbers. As we have seen, this idea had been explored in connection with the study of ratios in terms of their "denomination," and some mathematicians, notably Ramus, regarded it as the means to avoid the intricacies of treatment of irrational ratios in *Elements* V and X (cf. Sections 6.2 and 7.3).

Terminology and notation

Viète indicated the successive powers of the "side" (cf. item **2** above) by the terms "square," "cube," "square-square," "square-cube," etc.[18] Indeterminate magnitudes of the same dimensions as the side, the square etc. were called "length" or "width," "plane," "solid," "plane-plane," "plane-solid," etc.[19] Viète then proceeded to introduce rules of symbolic notation and rules for the manipulation of equations involving abstract magnitudes. He used (capital) letters to represent the magnitudes, together with a terminological system to indicate the dimension. If relevant, he distinguished between unknown and indeterminate (or

[16] The list summarizes Chapters 2–4 (pp. 1–8) of [Viète 1591] ([Viète 1983] pp. 14–23).

[17] "Latus," "radix."

[18] [Viète 1591] p. 3: "latus, quadratum, cubus, quadrato-quadratum, quadrato-cubum, . . . "— Viète continued to the 9th dimension.

[19] [Viète 1591] p. 3: "1. Longitudo latitudóve. 2. Planum. 3. Solidum. 4. Plano-planum. 5. Plano-solidum . . .."

ABSTRACT MAGNITUDES and the OPERATIONS acting on them — According to Viète

Abstract magnitudes: magnitudes that could be joined, separated, and compared, but whose further nature was left unspecified; they were represented by letters.
All operations were performed abstractly, that is, they were represented symbolically.
There was no unit element.

Operation	Vietean notation	Change of dimension	Corresponding operation(s) on numbers	Corresponding operation(s) on geometrical magnitudes
Adding two magnitudes of the same dimension	+	No	Adding	Joining
Subtracting two magnitudes of the same dimension	−	No	Subtracting	Cutting off
Multiplying two magnitudes of equal or different dimensions	in	dim(A in B) = dimA + dimB	Multiplying	Making a rectangle
Dividing a magnitude by another of lower dimension	$\div$	$\dim(\frac{A}{B}) = \dim A - \dim B$	Dividing	Applying a rectangle
Forming a ratio of two magnitudes of the same dimension	No special symbol	The ratio was a relation, not a magnitude	Forming a ratio	Forming a ratio
Extracting square or higher-order (k-th) roots of magnitudes whose dimension is a multiple of k	"Latus" or "Radix:" order indicated by 'q', 'c' etc.	dim(Latus [order k] M) = (dim M)/k	Root extraction	Determining the side of a square, a cube, etc. (**NB** not determining mean proportionals because they have the same dimension)
Solving equations			Solving equations	Constructing problems

Table 8.1: Abstract magnitudes — Viète

given) magnitudes by using vowels for the former and consonants for the latter. He usually took an unknown, A, as the "side" generating, by multiplication in itself, the sequence of magnitudes mentioned in item **2** above; these he denoted as "A, A quad., A cub., A quad.quad.," etc. Indeterminate magnitudes were in principle indicated by single letters with an indication (sometimes abbreviated) of the dimension, for instance: "B planum, C solidum, D plano-planum," etc..

As the magnitudes were unspecified, the execution of the operations could not be specified either; the magnitudes were merely represented by letters and the operations were executed abstractly, namely by expressing their result in words or in notation. The operations were denoted by "plus" or "+," "minus" or "−," "in" (for multiplication), and the horizontal bar for division. In general Viète used few symbols apart from these, whereby his equations are often nearer to sentences than to formulas. A characteristic Vietean equation, in translation, reads as:

> X squared times thrice E minus E cubed will be equal to X squared times B.[20]

This is the equation for the trisection of an angle; in a circle of radius X, E is the unknown chord subtending a third of the given angle whose chord is B. In modernized notation the equation is:

$$3X^2E - E^3 = X^2B \,, \tag{8.1}$$

where E is the unknown.

Existence questions

It may be noted that Viète assumed division to be possible within one scale of magnitudes (cf. item **3** above). Thus implicitly he presupposed that for any magnitudes a and b with $\dim b > \dim a$ there is a magnitude x such that $ax = b$. In modern terms this presupposition implies (as is easily seen) that within one scale the sets M_n consisting of the magnitudes of dimension n are isomorphic — which most likely conforms to Viète's idea of magnitude.

In Viète's dimensional interpretation, root extraction can only be performed on magnitudes with appropriate dimension. The cubic root of a three-dimensional magnitude D sol. was the one-dimensional magnitude A, satisfying A cub. = D sol. Viète probably assumed that, provided the magnitudes were continuous, such an A did exist, but he made no general statements on this issue. His general symbolic calculus was independent of whether radicals or roots of equations existed. The actual effectuation of root extraction or equation solving did not belong to specious logistics; these procedures depended on the special nature of the magnitudes. If the magnitudes were numbers, extracting a cubic root was a numerical, in general merely approximate, procedure; if they were geometrical, cube root extraction required some construction procedure, involving two mean proportionals; for abstract magnitudes the actual effectuation of the extraction of cubic roots was meaningless.[21]

[20] [Viète 1615b] pp. 301: "X quadratum in E ter, minus E cubo, aequetur X quadrato in B." Cf. [Viète 1983] p. 445.

[21] Thus in a passage on mean proportionals in [Viète 1631b] (Prop. 5 p. 15, cf. [Viète 1983]

8.4 The "New Algebra" and the operations of arithmetic and geometry

The effectuation of operations

Viète usually reserved the term specious logistics for that part of his new algebra that dealt with abstract magnitude and in which therefore no assumptions could be made about the actual effectuation of the algebraic operations. But the new algebra was developed to serve in the solution of problems in either arithmetic or geometry and for that purpose the operations had actually to be performed. The double link between the actual operations and specious logistics was provided by zetetics and exegetics or rhetics. Zetetics translated the arithmetical or geometrical problems under consideration in the general terms of specious logistics. In exegetics or rhetics the results gained by applying the general methods of specious logistics were retranslated back in the context of the original problems, arithmetical or geometrical as the case might be. This division of tasks among the parts of analysis implied that there was no "specious" exegetics or rhetics. In arithmetical context the exegetics interpreted the final equation provided by specious logistics as a numerical one and calculated its solution, whereas in geometrical context exegetics derived from this equation a geometrical construction of the problem.

Three of Viète's treatises concern exegetics, two in geometrical and one in numerical context. The numerical treatise was *On the numerical resolution of powers for exegetics*,[22] which explained the algorithms for root extraction and the solution of numerical third-degree equations.

Geometrical exegetics

The first of the geometrical exegetic treatises, *A canonical survey of geometrical constructions*[23] concerned the geometrical effectuation of the algebraic operations as far as they could be achieved by the Euclidean means of construction, circles and straight lines. Line segments a and b could be added and subtracted; their product, a rectangle with the sides a and b (rect.(a, b)), could be formed. To any rectangle with sides a and b an equal square could be constructed by taking the mean proportional c of a and b; if rect.(a, b) was to be divided by some line segment d, the third proportional e of d and c (i.e., e satisfying $d : c = c : e$) was the quotient.[24] Viète then proceeded to the solution of quadratic equations, that is, the geometrical construction of their roots; one of his constructions was discussed above (Construction 4.4).

Viète's Euclidean effectuations of the quadratic operations were neither new nor subtle, but his treatise on the subject testified to the thoroughness with which he pursued his program of a general logistics with various exegetics depending on the nature of the magnitudes.

---

pp. 36–37) Viète explained that a series of, say, four mean proportionals between two one-dimensional magnitudes A and B could be "exhibited" by: A, $\sqrt[5]{A^4B}$, $\sqrt[5]{A^3B^2}$, $\sqrt[5]{A^2B^3}$, $\sqrt[5]{AB^4}$, B (his sign for $\sqrt[5]{\ }$ was "Latus qc." for "Latus quadrato cubum"). He did not at that point discuss whether these radicals existed and how they were to be achieved.

[22] [Viète 1600b].

[23] [Viète 1592].

[24] [Viète 1592] Props 1–6, pp. 229–231, [Viète 1983] pp. 371–374.

In the second treatise on geometrical exegetics, *The supplement to geometry*,[25] Viète turned to the construction of geometrical problems whose analysis, by zetetics and poristics, led to third- or fourth-degree equations. He proved that all such problems could be reduced either to the trisection of an angle or to the determination of two mean proportionals between two line segments; he gave constructions of these problems by neusis. I return to these Vietean constructions in Chapter 10.

A new urgency about constructions

Viète's new algebra constituted the first major development in the principal dynamics within early modern geometrical problem solving: the introduction of algebraic methods of analysis. This development necessitated a rethinking of the methods of construction. Algebraic analysis made clear the extent and some of the structure of the class of non-plane problems. No longer could these problems be seen as exceptional cases to be dealt with when and if they occurred; for advanced geometrical problem solving they were the rule rather than the exception.

On the other hand, the new analysis gave no guidance in choosing constructions and interpretating their exactness. For instance, the property of one of the two mean proportionals between two line segments a and b, was adequately expressed by the equation

$$x^3 = a^2 b \,, \tag{8.2}$$

as well as by the explicit algebraic expression of the root of this equation:

$$x = \sqrt[3]{a^2 b} \,. \tag{8.3}$$

But when it came to constructing this mean proportional these equations gave no direct help. Even when, as here, an explicit algebraic expression could be given for the root, the question remained how x was to be constructed; the translation of the algebraic symbol of a cubic root into a geometrical procedure was not a self-evident matter. I explain Viète's approach to these questions in Chapter 10.

8.5 The significance of Viète's "New Algebra"

Symbolic representation underestimated

Viète's work constituted a most important step in the development of symbolic algebra; the great abstractness of his system and his introduction of letters to represent indeterminates combine to justify such an assessment. Yet Viète himself appears to have underestimated the value of symbolic representation; his love was in words and sentences, not in abbreviations and symbols. He was a prolific creator of technical terms, preferably of Greek origin. He did not invent new symbols and used only a few (mainly $+$, $-$, and the quotient bar), relying rather on abbreviations in those cases where the typography did not allow him to write out the technical terms in full (with their correct declension). As a

[25] [Viète 1593].

result Viète's algebra was yet rather involved as regards mathematical formulation and thereby he and his followers did not harvest the full versatility, clarity, and power that algebra could have provided. However, this was caused by incomplete symbolization rather than by any obstacle intrinsic in the system; in particular, the dimensional interpretation of the magnitudes and the resulting necessity of homogeneity did not produce the complexity.

No analytic geometry Also in another sense Viète and his early followers did not gather all the benefits of his symbolic algebra — they did not apply it to the study of curves. The first to do so were Fermat and Descartes, more than forty years after Viète began publishing his new methods. Descartes' first motivation to do so concerned locus problems (cf. Chapter 19). Fermat (cf. Section 13.1) stated the correspondence between equations in two unknowns and curves in the context of constructing solid problems by the intersection of conics. He used equations to describe these conics; apart from that, he only employed techniques that were available already to Viète. Had Viète opted, in the classical manner, for construction by the intersection of curves, he might well have been led to the relation between curves and equations. Thus we may consider Viète's adoption of the neusis as postulate to supplement geometry and his interest in special problems and standard forms of equations in one unknown, as reasons for a delay in the development of analytic geometry.

Modernity A proliferation of terminological subtleties, an underestimation of symbolic notation, and an interest deflected from curves, were obstacles for a full deployment of the powers of algebra within the Vietean school. Yet this should not detract from the value of Viète's achievement. By introducing the use of letter symbols for general indeterminate and unknown magnitudes, Viète provided mathematics with what was to become the most essential means of communicating mathematical argument. The importance of this event can hardly be overestimated. Moreover, his "new algebra" was an abstract and daring mathematical creation, totally untypical of the period, elaborated with great consistency and thoroughness. It strikes the late-twentieth-century reader as remarkably modern in its abstract axiomatic approach, and it testifies to Viète's deep awareness of the foundational issues of mathematics. Indeed, it seems that in his case, as in Descartes' later, few if any of the epigones appreciated the philosophical motivation and subtleties of the system their master had created.

8.6 Another approach: Van Ceulen

Homogeneity and the unit In his *Geometry* of 1637 Descartes gave an alternative approach to introducing algebra into geometry. Like Viète he did not use the number concept as vehicle for the merging of algebra and geometry, but redefined the algebraic operations so as to be applicable outside the domain of numbers. Like Viète, he also realized and elaborated the consequences of his innovation with respect to the concept of construction. However, Descartes avoided the requirement of homogeneity and the resulting scales of successive dimensions. He did so — as we will see (Chapter 21) — by introducing a unit segment.

In view of the two different approaches, it is of interest here to discuss a third, less elaborate attempt to use algebraic operations in geometry, which was chronologically intermediate between the two others. The author of the attempt was Van Ceulen, and it was published in the work, discussed above (Section 7.4), which Snellius edited after his death. The title of the volume well indicated the interest of its author in the numerical, the algebraical, as well as the geometrical approach to problems:

> *Arithmetical and geometrical elements, with their uses in solving various geometrical problems, partly by the tracing of lines only, partly by irrational numbers, sine tables and algebra.*[26]

The third book of this work contained an attempt[27] to explore the correspondence between the arithmetic of irrational numbers of the form $a + \sqrt{b}$ (with a and b rational numbers) and the geometry of line segments. Van Ceulen introduced a unit length ("famosa mensura," cf. above Section 7.2) to relate lengths to numbers, but he realized that the algebraic operations still had to be translated into geometrical ones, that is, into constructions. Thus he showed that, given the unit, any length $a + \sqrt{b}$ could be constructed (by straight lines and circles), and that, conversely, when any such length was given, the unit could be constructed. He then proceeded to explain the geometrical equivalents of the arithmetical operations. Addition and subtraction were obvious. Multiplication involved the unit.

Multiplication

Van Ceulen's treatment of multiplication is interesting, if alone because Descartes later adopted another one. He formulated it as a problem:

> Problem. The rectangle, formed by two lines whose lengths are defined with respect to an assumed unit, has to be applied along that unit.[28]

Thus the unit played a role in Van Ceulen's interpretation of multiplication; the product of two line segments a and b was not the rectangle with sides a and b but a rectangle with the same area and one side equal to the unit. The other side of this rectangle was, as geometers well knew, the fourth proportional between the unit, a and b. Van Ceulen's solution of the problem is a construction of that fourth proportional, but not the usual Euclidean one of *Elements* VI-12. Van Ceulen did not use symbols for indeterminate numbers and therefore had to explain the operation by an example, for which he chose the numbers 3 and $\sqrt{19}$. He gave the following construction.

---

[26] [Ceulen 1615b].

[27] [Ceulen 1615b] pp. 105 sqq.

[28] [Ceulen 1615b] p. 112, Prop. 33: "Rectangulum a duabus lineis in assumpta mensura longitudine definitis comprehensum ad ejusdem unitatem applicare."

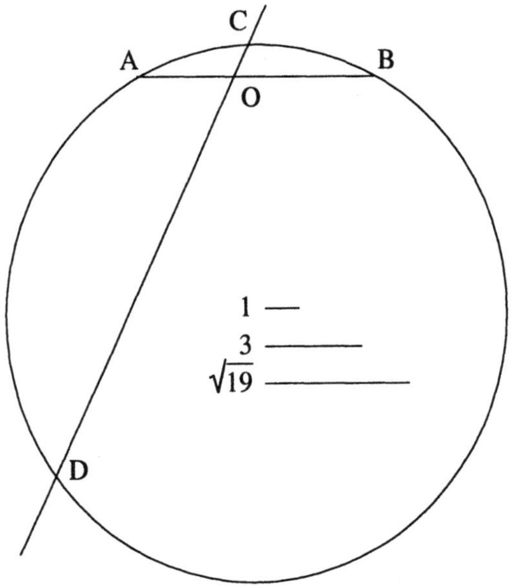

Figure 8.1: Multiplication of line segments — Van Ceulen

Construction 8.1 (Multiplication of line segments — Van Ceulen)[29]
Given: a unit length (see Figure 8.1) and two segments of lengths 3 *and* $\sqrt{19}$*, respectively. It is required to construct a line segment with length equal to the product of* 3 *and* $\sqrt{19}$.

Construction:
1. Mark $AO = 3$ and $OB = \sqrt{19}$ along a line.
2. Along any other line through O mark OC equal to the unit.
3. Draw a circle through A, B, and C; it intersects CO prolonged in D.
4. OD is the required line (that is, the rectangle with length OD and unit width is equal to the rectangle with sides of length 3 and $\sqrt{19}$).
[**Proof**: By *Elements* III-35 rect(OA, OB) is equal to rect(OC, OD).]

In the example the length of the resulting segment is $3\sqrt{19}$. Yet Van Ceulen did not take the step of stating that therefore the product of the two segments OA and OB would be equal to the segment OD. And Snellius added a rather confusing note[30] to the effect that the product of two line segments certainly was an area and not a line.[31]

[29] [Ceulen 1615b] p. 112–113.
[30] [Ceulen 1615b] p. 113.
[31] Remarkably, Bombelli had used a similar construction based on *Elements* III-35 and on the use of a unit length for a geometrical interpretation of solving the equation $ax = b$. He did

Significance

Proceeding in this way Van Ceulen achieved a complete correspondence between quadratic irrational numbers and line segments in geometrical figures generated by Euclidean constructions starting from the unit length. He applied it to various problems, and used it in his fourth book for calculating the quadratic irrationals occurring in various geometrical configurations. Snellius provided this fourth book with a title suggesting that numbers could replace classical geometrical analysis: "On the numerical solution of geometrical data."[32] This title may have contributed to Kepler's anger about the book (cf. above Section 7.4).

Van Ceulen's treatise shows that the issue of correlating the algebraic and the geometrical operations remained alive in the period between Viète and Descartes. Apparently Viète's consistent separation between numbers and geometry was too high a price for Van Ceulen to pay for the use of algebra; he favored a more direct connection by the introduction of a unit length. On that basis he worked out the geometrical constructions corresponding to the quadratic algebraic ones, but he did not pursue the matter to higher-order equations and thus was not confronted with questions of construction beyond circles and straight lines.

8.7 Conclusion: numbers, algebra, and geometrical construction

Number and construction

We have seen that, rather than introducing numbers into geometry, mathematicians prefered redefining the algebraic operations so as to apply for geometrical magnitude. Consequently the issue of construction remained alive — introduction of numbers would have made construction superfluous by reducing geometrical problem solving to calculation. Yet on a longer time scale the transfer of algebra into geometrical context was part of the process whereby number infiltrated and finally took over geometry, in the sense that geometry came to be the study of spaces over number fields or fields in general. However, this process, which did entail a considerable loss of interest in geometrical constructions and their exactness, was hardly, if at all, in evidence in the early modern period, and the debates on construction were virtually independent of the discussions on the introduction of number into geometry.

Reconsidering construction procedures

Not only did the the issue of construction remain alive while algebraic methods of analysis were introduced in geometry, this introduction — the principal dynamics within the field of geometrical problem solving — actually induced a new concern about constructional procedures. In conclusion of this chapter I summarize the principal aspects of this development.

First, from an algebraic point of view there was no reason for a restriction to quadratic operations, that is, to operations that could be performed geomet-

so in a projected fourth book of his *Algebra*; this fourth book, however, was not published at the time. [Bombelli 1572] only contained books 1–3, books 4 and 5 remained in manuscript until 1929 (cf. [Bombelli 1929]). For details see [Giusti 1992], pp. 309–311.

[32] [Ceulen 1615b] pp. 137–183: "De ⟨dedomenoon⟩ geometricorum per numeros solutione."

rically by means of straight lines and circles. Geometrical equivalents had to be found for cube root extraction and in general the solution of higher-degree equations. Thus geometers were forced to choose methods for constructing beyond straight lines and circles and to legitimate their choices.

Second, the algebraic translation of geometrical problems into equations introduced a new method of surveying and classifying these problems. One aspect of this was the tendency within the Vietean school to search for matching sets of standard geometrical problems and standard equations. In general such classifications of problems and equations made clear what the extent should be of a complete theory of geometrical construction: To solve "all problems" entailed dealing with all polynomial equations in one unknown. Geometrical problem solving in its full generality meant finding geometrical procedures to construct the roots of any equation. We will see that Descartes fully realized this extent of the program of geometry.

Third, there was an interaction between the choice of methods of construction and the algebraic study of curves. Pappus had advocated construction by intersection of curves. But his book also paid considerable attention to alternative constructions, such as by neusis or reduction to standard problems, and these alternatives were more popular in the decades after the publication of the *Collection*. As we have seen, Viète's lack of interest in construction by conics may well have delayed the development of analytic geometry.

Fourth, it should be noted that although the adoption of algebraic methods made questions about construction acute and pressed for generality in their investigation, it did not provide clues for the interpretation of exactness with respect to geometrical construction. Here the principal dynamics in their field forced geometers to forge new answers but it provided no guidance. In the next chapters I review the reactions of various mathematicians to that challenge.

Chapter 9

Clavius

9.1 Introduction

Principal methodological questions about construction

Having reviewed in the previous five chapters the early modern tradition of geometrical problem solving, the emergence of algebraic analysis as the principal dynamics in that field, and the questions of legitimation that were raised in relation to this process, I now turn to the actual debates on the interpretation of geometrical exactness with respect to constructions.

After the appearance of Commandino's edition of Pappus' *Collection* in 1588, the principal methodological questions with respect to geometrical construction were clear (cf. also Section 3.7). Constructions should be performed by geometrically acceptable means and they should be as simple as possible. Thus the questions were:

> **A.** Which means of construction are acceptable in geometry for solving problems that cannot be solved by straight lines and circles?
> **B.** How should the constructions by acceptable means be classified according to simplicity?

Pappus' classification, his precept, and the examples of constructions in his work had articulated these questions and had provided answers, be it that Pappus' practices of construction were not always consistent with his general statements. Nevertheless, by the 1590s the discussions on geometrical exactness had a definite object and a clear structure. The object was geometrical construction, and the questions of acceptability and simplicity provided the structure.

The next chapters sketch the pre-Descartes debates on the exactness of constructions by presenting the opinions of five mathematicians that represent archetypal positions with respect to the interpretation of exactness. The present chapter introduces the first of these mathematicians, Clavius; the others are Viète, Kepler, Molther, and Fermat. I discuss the debates in a more general way in Chapters 14 and 28.

Clavi
const

Christoph Clavius (1537–1612) was one of the most influential mathematicians around 1600. Several of his constructions have been discussed in earlier chapters (Constructions 4.3, 4.8, 4.9, 4.18). His strength was in editing and commenting on classical texts and in composing textbooks. The textbooks, although not very innovative, were solid and comprehensive. His works were widely used throughout the seventeenth century; Descartes, for instance, was taught mathematics from Clavius' works.[1]

At a number of places in these books Clavius dealt with geometrical constructions and the criteria for accepting them as genuinely geometrical. In his treatment of the matter we recognize the first effects of the *Collection*, be it that he was inspired by Pappus' constructions themselves rather than by his classification and precept. Clavius' arguments on geometrical exactness had much more precision than earlier ones such as those discussed in Chapter 2, and he was the first to take up a theme that was to become crucial later on, namely, the legitimacy of various methods of generating curves, in particular tracing by motion and pointwise construction.

The most informative item in Clavius' work with respect to construction was a small treatise on the quadratrix, composed in 1588–1589. His textbook on practical geometry of 1604 also contained relevant remarks. In the present chapter I confine myself to these two sources and use them to assess Clavius' interpretation of the exactness of geometrical constructions.

9.2 The treatise on the quadratrix

Publication

In 1574 Clavius published an edition of Euclid's *Elements* with notes and additions.[2] It was a successful work and a second, much enlarged edition appeared in 1589.[3] A year before,[4] Clavius had read the passages on the quadratrix in book IV of Pappus' *Collection* (cf. Section 3.2). These inspired him to compose a small separate treatise on this curve, which he inserted after the sixth book of the *Elements*.[5] The treatise was also incorporated in his *Practical Geometry* of 1604.[6] Both the Euclid edition and the *Practical Geometry* were reprinted in Clavius' *Mathematical Works* of 1611–1612, so that the treatise even recurs twice in that work.[7]

Clavius gave his treatise the title:

> *On the amazing nature of a certain curved line, by help of which a figure of arbitrarily many equal sides can be inscribed in a circle, and*

[1] Cf. [Milhaud 1921], p. 235.

[2] [Euclid 1574].

[3] [Euclid 1589].

[4] Cf. [Euclid 1589], vol. 1, p. 894: "Forte superiori anno incidi in librum 4 Pappi Alexandrini . . . ." Clavius may have read the passage in the printed edition [Pappus 1588] or in one of the then extant manuscripts, cf. Chapter 3, Note 1.

[5] [Euclid 1589], vol. 1, pp. 894–918.

[6] [Clavius 1604] pp. 359–370.

[7] [Clavius 1611–1612] vol. 1, pp. 296–304; vol. 2, pp. 188–194.

the circle can be squared, and many other things can be performed, very pleasing to know.[8]

The quadratrix

As the title indicated, Clavius claimed no less than having found the genuinely geometrical constructions of all regular polygons, and the quadrature of the circle. In the treatise he also gave the construction of the general angular section. His starting point was Pappus' treatment of the quadratrix in Book IV of the *Collection*; the constructions he gave were basically the same as Pappus' (cf. Definition 3.3, Construction 3.4 and Equation 3.2). What Clavius added was an argument why these constructions should be considered as genuinely geometrical. As the constructions presupposed the quadratrix to be given, he had to explain how that curve could be traced, or otherwise given, in a geometrically acceptable manner. Pappus had defined the curve by specifying a procedure for tracing it by motion, but his remarks about this procedure suggested that its geometrical status was dubious. Clavius described Pappus' construction and explained the objections to it reported by Pappus, notably the alleged *petitio principii* in the definition (cf. Section 3.2 and Note 15 of Chapter 3).

Clavius' construction of the quadratrix

Clavius accepted the objections and concluded that if a better, truly geometrical[9] construction of the quadratrix could be given, its use in solving problems would be legitimized. He proceeded to provide a construction of the curve that he deemed geometrical. It was a pointwise construction (and not a very surprising one, given the definition of the curve):

Construction 9.1 (Quadratrix — Clavius)[10]
Given: a square $OACB$ (see Figure 9.1); it is required to construct the quadratrix within the square.

Construction:
1. Draw the quarter arc BA.
2. Bisect BO and CA in D and E, respectively; draw DE; bisect arc BA in F, draw OF, OF intersects DE in G; G is on the quadratrix.
3. Bisect BD and CE in D' and E', respectively; draw $D'E'$; bisect arc BF in F'; draw OF'; OF' intersects $D'E'$ in G'; G' is on the quadratrix.
4. Repeat this procedure with other segments and corresponding arcs until sufficiently many points on the quadratrix are determined.
5. "The quadratrix line should be drawn through these points in an appropriate way, so as not to be sinuous, but it should at all times

[8] [Euclid 1589] vol. 1, p. 894: "De mirabilia natura lineae cuiusdam inflexae, per quam et in circulo figura quotlibet laterum aequalium inscribitur, et circulus quadratur, et plura alia scitu iucundissima perficiuntur." Compare the remarks on Clavius' treatise in [Gäbe 1972] pp. 120–128 and [Mancosu 1996] pp. 74–77.
[9] [Euclid 1589] vol. 1, p. 895: "geometricè."
[10] [Euclid 1589] vol. 1, pp. 895–896.

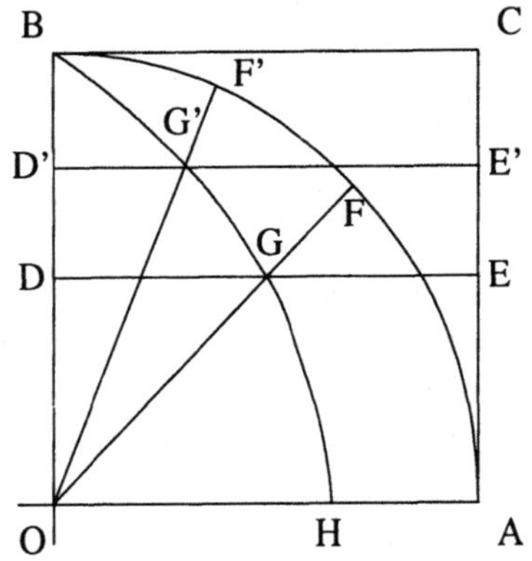

Figure 9.1: Pointwise construction of the quadratrix — Clavius

continue smoothly, making no bents or angles anywhere."[11]
6. To construct point H, the intersection of the quadratrix with the axis OA (which cannot be among the points G constructed in **5**) mark points below OA laying symmetrically with points on the quadratrix above the axis near H; draw a curved line through these points on both sides; its intersection with OA gives H "without noticeable error, that is, an error which can be detected by the senses."[12]
[**Proof**: The construction employs the property of the quadratrix expressed in Equation 3.1.]

Like the construction of the cissoid and the proportionatrix (Definitions 4.10, 3.5) discussed above, this construction of the quadratrix is a pointwise construction, that is, a procedure whereby arbitrarily many points, evenly spaced along the curve, can be found.

9.3 The geometrical status of the construction

Practical precision

Such, then, was the construction that in 1589 seemed so convincingly geomet-

[11][Euclid 1589] vol. 1, p. 896: "per ea puncta quadratrix linea congruenter ducenda est, ita ut non sit sinuosa, sed aequabiliter semper progrediatur nullum efficiens gibbum, aut angulum alicubi."

[12][Euclid 1589] vol. 1, p. 896: "sine notabili errore, qui scilicet sub sensum cadat."

rical to its author that he dared advance the claim to have solved the classical problems of squaring the circle, dividing the angle, and constructing regular polygons. The claim was far-reaching, it implied an extension of the arsenal of legitimately geometrical means of construction and thereby a new interpretation of geometrical exactness. Not surprisingly, Clavius felt the need to justify this step.

Clavius' principal argument was that his construction was more accurate than Pappus' procedure of tracing the curve by motion. However, the terms he used in describing how the curve should actually be drawn, make clear that he did not claim absolute precision for his own construction. Moreover, he later added a variant of the construction from which it appears once more that he aimed at great but not at ultimate precision. In this variant construction[13] Clavius skilfully avoided intersections of straight lines and circles under small angles, thus steering clear from situations in which the use of a ruler and a compass gave notably imprecise results.

Clavius further claimed that, as to precision, his construction compared favorably with the usual pointwise constructions of conic sections. These constructions were performed by repeatedly determining a mean proportional,[14] and Clavius considered the construction of a mean proportional (cf. Construction 4.2) as less precise than the bisections he used in finding points on the quadratrix. He concluded:

> Hence unless someone wants to reject as useless and ungeometrical the whole doctrine of conic sections which Apollonius of Perga has pursued with such acuity of mind that because of that he has been called a great geometer . . . one is forced to accept our present description of the line as entirely geometrical.[15]

And, he added, not only would Apollonius' work be repudiated by rejecting pointwise constructions,[16] but also the achievements of Archimedes and

[13]The addition of the variant construction occurs in the re-edition of the *Elements* in 1603 ([Euclid 1603]) and in the *Practical Geometry* [Clavius 1604] p. 361; it also occurs in both versions of the quadratrix treatise in [Clavius 1611–1612] (vol. 1, p. 297 and vol. 2, p. 190).

[14]cf. Note 16.

[15][Euclid 1589] vol. 1, pp. 897–898, the full text is: "Haec igitur est descriptio quadratricis, quae geometrica appellari potest, quemadmodum et conicarum sectionum descriptiones, quae per puncta etiam fiunt, ab Apollonio traditur, geometricae dicuntur, cum tamen magis errori sint obnoxiae, quam nostra descriptio, propter inventionem plurimarum linearum mediarum proportionalium, quae ad earum descriptiones sunt necessariae, quibus in quadratricis descriptione opus non est. Quare nisi quis totam sectionum conicarum doctrinam quam tanto ingenii acumine Apollonius Pergaeus persecutus est, ut propterea magnus geometra apellatus sit, reiiecere velit tanquam inutilem et non geometricam, . . . admittere omnino cogetur descriptionem hanc nostram quadratricis lineae, ut geometricam."

[16]In fact, Clavius was wrong in attributing pointwise constructions to Apollonius; in the *Conics* the construction of the conic sections is performed by the intersection of a cone and a plane. However, Apollonius gave properties of the conics that could be readily translated into pointwise constructions, and in his commentary Eutocius indeed provided such pointwise constructions ([Apollonius Conics] I-20–21; for Eutocius' commentaries see [Apollonius 1891–1893] vol. 2, pp. 232–235); the commentary was available in Commandino's edition [Apollonius 1566]. Eutocius' pointwise procedure indeed involved repeated construc-

Menaechmus, who both used conics, not to mention the great usefulness of these curves in gnomonics; moreover, the conchoid of Nicomedes, useful in finding two mean proportionals, was also constructed pointwise.[17] Remarkably, Clavius did not refer to Pappus' classification or his precept in his defense of the pointwise constructed quadratrix.

Clavius' later views

In 1589 Clavius presented his case with conviction and confidence; he announced his construction of the quadratrix as "geometrice"[18] ("geometrical"), and he stated without qualifications that by means of the curve it was possible to square the circle,

> a matter which until the present day has kept the minds of mathematicians in suspense.[19]

Later Clavius was more cautious about the geometrical status of the pointwise construction of the quadratrix. When he reprinted the treatise in his *Practical Geometry* of 1604, he wrote that the pointwise construction of the quadratrix was not fully geometrical

> but it is more accurate than all others which I could find until now, so by using it in practice we could hardly miss our goal. [20]

And where in 1589 he had written:

> This then is a description of the quadratrix that can be called geometrical in the same way as the descriptions of conic sections,[21]

he now added a cautious "quodammodo":

> This then is a description of the quadratrix which is in a certain way [quodammodo] geometrical...[22]

In the *Mathematical Works* of 1611–1612 the "quodammodo" was inserted in both versions of the treatise.

tion of geometric means.

[17]In the relevant passages of Eutocius ([Eutocius CommSphrCyl] pp. 615–620) and Pappus ([Pappus Collection] IV-22 (§§ 26–27) pp. 185–187) the conchoid is not constructed pointwise, but by a tracing procedure (cf. Section 2.4). Clavius himself gave a pointwise construction of the curve in his [Clavius 1604] pp. 301–304.

[18][Euclid 1589] vol. 1, p. 494.

[19][Euclid 1589] vol. 1, p 494: "Quae res ad hunc usque diem animos mathematicorum tenuit suspensos."

[20][Clavius 1604] p. 359: "accuratior tamen est omnibus aliis quas hactenus videri potui, ita ut practicè a scopo aberrare non possimus."

[21][Euclid 1589] p. 897: "Haec igitur est descriptio quadratricis, quae geometrica appellari potest, quemadmodum et conicarum sectionum descriptiones."

[22][Clavius 1604] p. 362: "Haec igitur est descriptio lineae quadratricis geometrica quodammodo, quemadmodum et conicarum descriptiones." Elsewhere (p. 362) he made a similar change: "admittere omnino cogetur hanc descriptionem nostram quadratricis lineae esse quodammodo geometricam" — the "quodammodo" was not there in the 1589 text.

It seems, then, that Clavius' belief in the geometrical legitimacy of his construction of the quadratrix lasted for only a short period; by 1604 he had returned to the usual sixteenth-century view on construction: truly geometrical methods for trisecting the angle, finding two mean proportionals, and squaring the circle had not (yet) been found. In the section on two mean proportionals in the *Practical Geometry*, for instance, he discussed the classical solutions from Eutocius' list "although till the present day nobody has truly and geometrically found the two mean proportionals between two given straight lines."[23] True to his appreciation for practical precision he presented the constructions attributed to Hero, Diocles, and Nicomedes in some detail because he considered them to be "the handier, easier ones, less liable to error."[24] Clavius conceded that the other constructions from the list were "most ingenious and subtle"[25] but he did not explain them. The constructions by Diocles and Nicomedes employed the cissoid and the conchoid, respectively; Clavius added pointwise constructions for these curves.[26]

Why Clavius' change of mind?

We do not know why precisely Clavius withdrew the claim that by his pointwise construction of the quadratrix the angular sections and the circle quadrature could geometrically be performed. The treatise generated at least some discussion; in 1592 Van Roomen, for instance, wrote to Clavius, politely but with some disappointment, that the construction was attractive but of no help in calculations. But Van Roomen did not comment on the geometrical status of Clavius' procedure.[27] In a book on cyclometry published in 1616[28] Lansbergen referred to Clavius' attempt to overcome the classical objections against the quadratrix by a pointwise construction. He suggested that the great Greek mathematicians would have been aware of the possibility to construct the curve in such a way and that Clavius himself would have to admit that the very point of intersection of the quadratrix and the axis could not be constructed in this way.

9.4 Clavius' interpretation of geometrical exactness

Idealization of practice

As we have seen, in most of his works Clavius displayed a rather traditional

[23][Clavius 1604] p. 297: "...quamvis nemo ad hunc usque diem, vere ac geometricè duas medias proportionales inter duas rectas datas invenerit." Elsewhere in the *Practical Geometry* (p. 399) the trisection received a similar comment; it was much studied but "until the present day not solved geometrically by anybody" ("Neque ab ullo ad hunc usque diem geometrice est solutum").

[24][Clavius 1604] p. 297: "quos commodiores, facilioresque et errori minus abnoxios iudicavimus."

[25][Clavius 1604] p. 297: "quamvis acutissimis subtilissimisque."

[26][Clavius 1604] pp. 297–304.

[27][Bockstaele 1976] p. 94.

[28][Lansbergen 1616] pp. 37–38.

attitude to geometrical exactness: he gave no positive criteria for geometrical legitimacy and he maintained that the classical problems had not yet been geometrically solved. During a brief period, however, inspired by Pappus' treatment of the quadratrix (but not, apparently, by his classification and precept) Clavius held a more exposed (and more interesting) view, defending a new interpretation of the exactness of geometrical constructions. Although he did not formulate his position explicitly, we may characterize it as the assumption that criteria of exactness in pure geometry should be parallel to criteria of precision in geometrical practice.

Thus Clavius took practical precision as guideline for deciding on geometrical exactness and therefore his approach belongs to the category that in Section 1.6 I have called "idealization of practical methods." Yet he did not explain why the criteria of pure and practical geometry should be parallel. In fact, the distinction between practical and pure geometry remained opaque in his arguments. We will see in Chapter 12 a more explicit justification of the passage from practical precision to pure geometrical exactness, provided by Molther in 1619.

Influence Interesting though Clavius' venture was, its direct influence on later mathematicians was presumably limited, especially because he soon mitigated his statements. Yet his treatise probably helped to generate an interest in the pointwise construction of those curves that themselves served the construction of particular geometrical problems. Indeed Clavius' attempt to legitimate constructions involving curves — in his case the quadratrix — showed that if one did not accept authority or mere postulate as a basis for such a legitimation, the legitimatory arguments should concern the process of generating the constructing curves. Descartes was the first to pursue this line of argument in great depth. He was aware of Clavius' treatise and, as we will see in Section 24.3, probably several of his arguments arose in direct critical reflection on Clavius' pointwise construction of the quadratrix.

Chapter 10

Viète

10.1 A new postulate

Treatises on construction

In Chapter 8 I have discussed Viète's crucial role in the creation and understanding of algebraic methods of analysis. The present chapter is devoted to his ideas about construction. As we have seen, Viète considered problem solving by means of algebra as consisting of three parts, zetetics, poristics, and exegetics. Construction belonged to the exegetical part of geometrical problem solving. Two of Viète's treatises were specially devoted to geometrical exegetics, namely, *A canonical survey of geometrical constructions*[1] and *The supplement of geometry*.[2] I therefore first deal with Viète's opinion on construction as expressed in these works.

A new postulate

The *Canonical survey* concerned quadratic equations only; the constructions in that case were classical Euclidean. The *Supplement of geometry* dealt with geometrical problems that led (by zetetics and poristics) to third- or fourth-degree equations. Here the Euclidean constructions no longer sufficed; Viète confronted the methodological question of construction beyond straight lines and circles. He presented an answer and the importance he attached to this answer is reflected in the title of the treatise: Geometry was in need of a "supplement:" the reason was precisely the absence of means of construction beyond straight lines and circles. Earlier, at the end of his programmatic introductory treatise, the *Isagoge* of 1591, Viète had announced how that defect should be amended:

> In order that, so to say, geometry itself supplies a deficiency of geometry in the case of cubic and biquadratic equations, he [the learned analyst] assumes, when dealing with cubes and squared squares, that it is possible

[1] [Viète 1592].
[2] [Viète 1593].

> *to draw, from any given point, a straight line intercepting any two given lines, the segment included between the two lines being prescribed beforehand, and possible.*

> This being conceded (it is, moreover, not a difficult assumption) famous problems that have heretofore been called irrational can be solved artfully: the mesographic problem, that of the trisection of an angle, finding the side of a heptagon, and all others that fall within those formulae for equations in which cubes, either pure or affected, are compared with solids and fourth powers with plano-planes.[3]

The assumption was not new: it was the classical "neusis": Given two lines, a point O and a segment a, to draw a straight line through O intersecting the two lines in points A and B such that $AB = a$ (cf. Problem 2.4 and 3.7). New was that Viète emphatically gave it the position of a postulate and decided to use it as the preferred construction beyond straight lines and circles. By accepting it as a postulate Viète circumvented the question of how the neusis was to be effectuated; it was, he implied, as obviously possible a procedure as drawing straight lines and circles. In particular the postulate status made reference to construction by conics or higher-order curves superfluous. Thus Viète's choice constituted a significant deviation from Pappus' precept which prescribed that solid problems, including neusis, should be reduced to "solid" constructions, that is, construction by the intersection of conic sections (cf. Section 3.6).

10.2 *A supplement to geometry*

The neusis postulate

The opening sentence of *A supplement to geometry* took up the line indicated at the end of the *Isagoge*:

> To supply the defect of geometry, let it be conceded
>
> *To draw a straight line from any point to any two given lines, the intercept between these being any possible predefined distance.*[4]

[3] [Viète 1591] p. 12: "Ad Cubos et Quadrato-quadrata postulat [sc. Analysta edoctus], ut quasi Geometria suppleatur Geometriae defectus, *A quovis puncto ad duas quasvis lineas rectam ducere interceptam ab iis praefinito possibili quocumque intersegmento.* Hoc concesso (est autem ⟨aitema⟩ non ⟨dusmechanon⟩) famosiora, quae hactenus ⟨aloga⟩ dicta fuere, problemata solvit ⟨entechnoos⟩, mesographicum, sectionis anguli in tres partes aequales, inventionem lateris Heptagoni, ac alia quotcumque in eas aequationum formulas incidunt, quibus Cubi solidis, Quadrato-quadrata Plano-planis, sive pure sive cum adfectione, comparantur." (The translation is a modified version of [Viète 1983] p. 32.) On the interpretation of the term "possible" see Note 4.

[4] [Viète 1593] p. 240: "Ad supplendum Geometriae defectum, concedatur *A quovis puncto ad duas quasvis lineas rectam ducere, interceptam ab iis praefinito possibili quocumque intersegmento.*" (Cf. the translation in [Viète 1983] p. 388.) By adding the word "possible" Viète evidently wanted to exclude cases in which no solution of the neusis existed. Such absence of solutions might occur if the segment had to be located in the same quadrant as the pole of the neusis or if a segment had to be inserted between a circle and a line outside the circle. In these cases the segment has to be larger than a certain minimum value. In the cases oc-

Viète explained that the given lines could be either two straight lines or one circle and a straight line.

Viète added a few short paragraphs with further comments.[5] He mentioned that Nicomedes' conchoids were probably devised in order to perform the neusis construction and that Archimedes accepted the postulate without question. But, he wrote, Archimedes had also accepted the tracing of parabolas and spirals, and he went on to criticize Archimedes' use of the spiral in rectifying the circle. He then added that it would be better to postpone further discussion of these matters until after the explanation of angular sections — but I have not been able to locate such a discussion in Viète's published work. The comments did not explicitly deal with the status of the postulate, indeed Viète provided hardly any justification of this status. Yet his readers could easily interpret the critical tone toward Archimedes' use of the spiral as casting doubt as well on his and Nicomedes' use of curves like the parabola and the conchoid.

Main result

The main aim of *A supplement to geometry* was to show the power of the neusis postulate. Viète did so by proving an elegant and striking result; he showed that any geometrical problem leading to a third- or fourth-degree equation could be reduced to either finding two mean proportionals between two given lines, or to trisecting a given angle. The result established the true centrality of the two classical problems within a large class of non-plane problems. Through Pappus' *Collection* it was known that both these classical problems could be constructed by a neusis (cf. Constructions 2.6 and 3.9), and thus by supplementing geometry with the neusis postulate all problems leading to equations of degree less than five were duly brought within the power of legitimate geometry.[6] We will see (cf. Section 16.4) that the central position of the two classical problems, constructing two mean proportionals and trisecting an angle, was an important theme in Descartes' early studies on construction.

Trisection and mean proportionals by neusis

As Viète's neusis constructions for two mean proportionals and trisection were

curring in *A supplement to geometry* solutions do exist. — Note that the terminology here provides evidence that construction in early modern mathematics did not concern existence (cf. Section 1.1); existence is here presupposed and still constructibility is postulated.

[5] [Viète 1593] p. 240, [Viète 1983] pp. 388–389.

[6] It is of interest to restate Viète's result in terms of real extensions of the field Q of rational numbers. We may identify the set of numbers constructible by straight lines and circles as the smallest real extension field K of Q such that $a \in K$ and $a > 0$ implies $\sqrt{a} \in K$. The set of numbers constructible by straight lines, circles, and the construction of two mean proportionals is the smallest extension field M of K such that for all $a \in M$ also $\sqrt[3]{a} \in M$. Similarly, the set of numbers constructible by straight lines, circles, and trisection is the smallest extension field T of K such that for all $a \in T$ with $-1 < a < 1$ the trisection equation $4x^3 - 3x = a$ (cf. the proof of Construction 4.6) has three distinct roots in T. In these terms Viète's result may be rephrased as stating that any real root of a third- or fourth-degree equation is either in M or in T. It is easily seen that the intersection of M and T is equal to K; so Viète's division of solid problems into those that are constructible by two mean proportionals and those that are constructible by trisection is in fact a division into mutually exclusive classes. — I assume that the results mentioned in this note, which are not difficult to derive, can be found in recent algebraic literature, but I have been unable to locate a discussion of them.

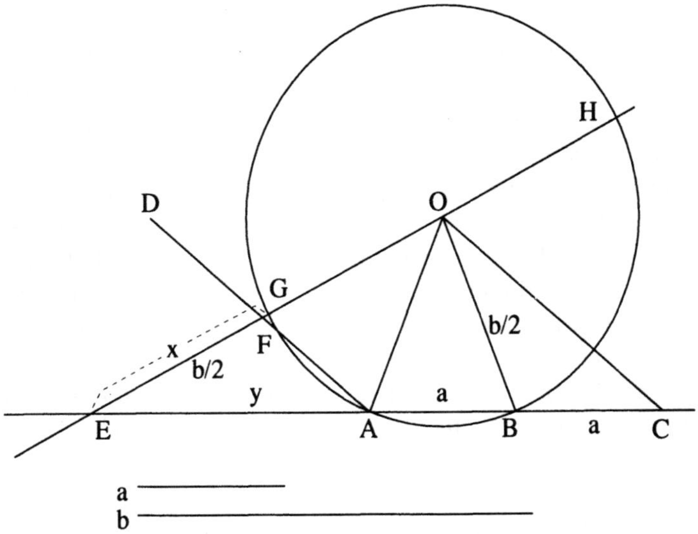

Figure 10.1: Construction of two mean proportionals by neusis — Viète

crucial in his theory of geometrical exegetics, I discuss them both. Viète did not take over the neusis constructions from Pappus' *Collection* but presented others, both closely related to classical procedures. His construction of two mean proportionals was as follows:

Construction 10.1 (Two mean proportionals — Viète)[7]
Given: two line segments a and b ($a < b$) (see Figure 10.1); it is required to find their two mean proportionals x and y.

Construction:
1. Draw a circle with center O and diameter b; draw a cord $AB = a$; prolong AB to both sides; take $BC = a$ on AB prolonged; draw CO; draw $AD \parallel CO$.
2. By neusis, draw EFO through O, intersecting BA prolonged and AD in E and F, respectively, such that $EF = \frac{1}{2}b$.
3. EFO intersects the circle in G and H.
4. $x = EG$ and $y = EA$ are the two required mean proportionals.
[**Proof**: In Proposition 4[8] of the *Supplement* Viète had considered (see Figure 10.2) two straight lines through a point E intersecting a circle in points A, B, H, and G, respectively. Call $AB = a$, $GH = b$, $EG = x$, and $EA = y$. The proposition asserted that if $xy = ab$ (or equivalently $a : x = y : b$), then $a : x = x : y = y : b$, that is,

[7][Viète 1593] Prop. 5 (p. 243) (Tr. [Viète 1983] pp. 392–394).
[8][Viète 1593] p. 242, [Viète 1983] p. 392.

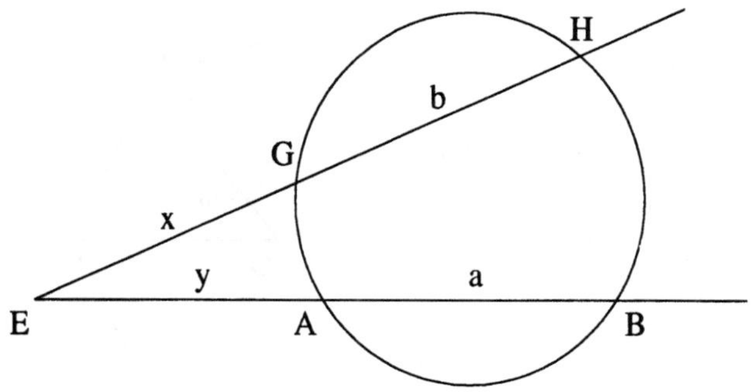

Figure 10.2: Proposition used in the construction

x and y are the two mean proportionals of a and b. He proved Prop. 4 straighforwardly by means of a corollary to *Elements* III-36 from which it follows that $x(x+b) = y(y+a)$. Now in the constructed figure (Figure 10.1) the neusis implies $EF = b/2 = GO$, hence $x = EG = FO$. By similarity $EF : EA = FO : AC$, so $b/2 : y = x : 2a$, whence $xy = ab$. By Prop. 4, then, x and y are the two mean proportionals of a and b.]

A comparison with Nicomedes' construction of two mean proportionals (Construction 2.6, Figure 2.6) reveals that Viète's construction is basically the same; if one removes the circle and turns the figure over 180 degrees one arrives at a configuration much similar to the one below line GK in Nicomedes' figure, and the steps in the construction correspond. Viète used the circle in his proof of the construction, which is different from Nicomedes'. Viète did not refer to Nicomedes; he may have found the construction and the proof independently or he may have thought the alternative proof important enough to consider the whole construction as independent of Nicomedes'.

For trisecting angles Viète gave the following neusis construction:

Construction 10.2 (Trisection — Viète)[9]
Given: an angle ψ (see Figure 10.3); it is required to find an angle φ equal to one third of ψ.

[9] [Viète 1593] Prop. 9, pp. 245–246 tr. [Viète 1983] p. 398.

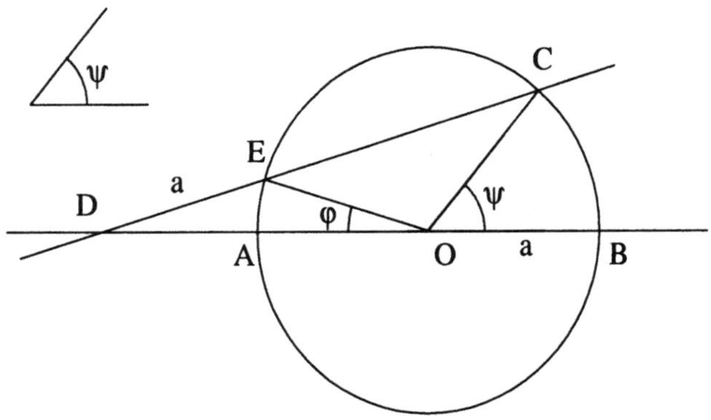

Figure 10.3: Trisection by neusis — Viète

Construction:
1. Draw a circle with center O and radius a, and mark the horizontal diameter AOB; prolong AB to the left; take C on the circle such that $\angle COB$ is equal to the given angle ψ.
2. By neusis, draw DEC through C intersecting AB prolonged and the circle in D and E, respectively, such that $DE = a$.
3. Draw EO.
4. Then $\varphi = \angle EOA$ will be the required angle; that is, $\angle EOA = \frac{1}{3}\angle COB$.
[**Proof**: $DE = EO = OC$; the triangles DEO and EOC are isosceles; $\angle OEC = 2\varphi$, hence $\angle EOC = \pi - 4\varphi$, so $\angle COB = \pi - (\varphi + (\pi - 4\varphi)) = 3\varphi$.]

Note that Viète did not prefer neusis between straight lines over neusis between straight line and circle; his trisection is of the latter type and from Pappus' *Collection* he could have taken a trisection by neusis between straight lines (cf. Construction 3.9).

It is also noteworthy that in the *Book of Lemmas*, which may in part be of Archimedean origin,[10] there is a proposition (nr 8) stating (cf. Figure 10.4) that if a chord CE in a circle with radius a is prolonged to D, with $ED = a$, and $DAOB$ is the diameter through D, then arcCB = 3(arcAE). Viète's construction is based on precisely this theorem. However, the *Book of Lemma's*

[10][Archimedes Lemmas]; cf. [Archimedes nd] p. xxxii.

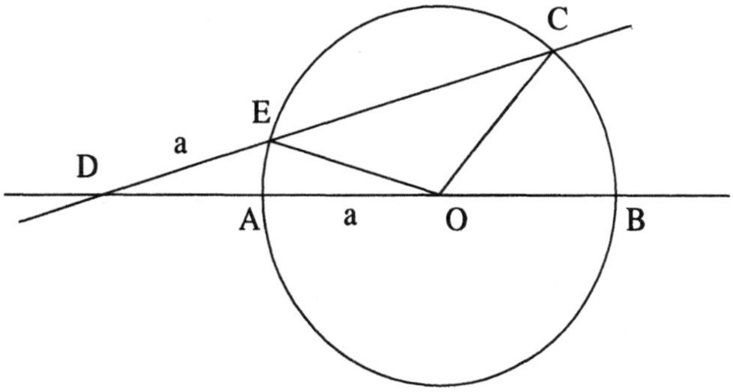

Figure 10.4: Angles in ratio 1 : 3 — Archimedes

arrived late in Western Europe through Arabic translations and the first Latin edition appeared as late as in 1659,[11] so probably Viète found the construction himself.

10.3 The construction of roots of third- and fourth-degree equations

Standard cubic equations

It will be useful to study in somewhat more detail how Viète argued that any problem leading to a third- or fourth-degree equation could be reduced to either a construction of two mean proportionals or a trisection. He did so in the final proposition of the *Supplement to geometry*. His argument is characteristic for his particular mixture of algebraic analysis and geometrical construction.

In Section 4.2 I discussed Viète's standard solid problems corresponding to the following set of standard cubic equations (cf. Equation 4.7):

$$\begin{aligned} x^3 + a^2x &= a^2b\,, \quad (1) \\ x^3 - a^2x &= a^2b\,, \quad (2) \\ a^2x - x^3 &= a^2b\,, \quad (3) \end{aligned} \tag{10.1}$$

(with $a,\ b > 0$). Together with the equation $x^3 = a^2b$, corresponding to the problem of two mean proportionals, they were a complete set in the sense that

[11] Cf. [Archimedes nd] p. xxxii.

any cubic equation could be reduced to one of them by removing the quadratic term; the three different forms in Equation 10.1 arise because Viète only considered positive coefficients. The standard problems corresponding to these equations were formulated in terms of four proportional line segments. Viète did not provide constructions of these three problems, but he suggested that other problems could be reduced to them.

In the *Supplement to geometry* Viète considered variant forms of the last three equations which made it easier to introduce further case distinctions. These were:[12]

$$\begin{aligned} x^3+3a^2x &= 2a^2b\,, & &(1)\\ x^3-3a^2x &= 2a^2b\,,\ b>a>0\,, & &(2.1)\\ x^3-3a^2x &= 2a^2b\,,\ a>b>0\,, & &(2.2)\\ 3a^2x-x^3 &= 2a^2b\,,\ b>a>0\,, & &(3.1)\\ 3a^2x-x^3 &= 2a^2b\,,\ a>b>0\,. & &(3.2) \end{aligned} \tag{10.2}$$

Viète did not consider the cases in which $a=b$; he realized, no doubt, that then one root is a or $-a$, and the equation can be reduced to a quadratic and a linear one. I note that (cf. Note 12) the distinctions $a <$ or $> b$ correspond to the now familiar conditions $(\frac{p}{3})^3 <$ or $> (\frac{q}{2})^2$, which, for a cubic equation $x^3+px+q=0$, distinguish the "casus irreducibilis" (three real roots, complex numbers under the cubic root signs in the Cardano-formula, cases 3.2, 4.2) from the opposite case (one real and two complex roots, no complex numbers under the cubic root signs in the Cardano-formula, cases 2, 3.1, 4.1).[13]

---

[12]Proposition 25 (pp. 256–257) of [Viète 1593] (tr. [Viète 1983] p. 416–417). The argument is given in a very condensed way and it is not quite complete. I have modernized Viète's exposition by translating his prose statements and his formulas into formulas of more modern style. I have retained, however, Viète's use of positive magnitudes only. For easier comparison I have used the expressions $3a^2$ and $2a^2b$ for the coefficients in all equations. Viète himself used these only in cases (2.2) and (3.2) (these are the ones related to trisection); he described the other cases in general terms. Case (2.1), for example, was formulated as follows (p. 257):

> Adfectos vero cubos sub latere negate ita demum reduci ad puros, cum solidum, à quo adficitur cubus, negatur de cubo, & praeterea triens plani coëfficientis cum latere adficiens solidum, cedit quadrato semissis latitudinis oriundae ex adplicatione adfecti cubi ad praedictum trientem.

This may be translated as (I have added between square brackets the interpretation in terms of equations):

> Cubes affected at the sides negatively [$x^3-Px=Q$] can thus be reduced to pure ones [$y^3=R$], if the solid by which the cube is affected [Px] is subtracted from the cube, and moreover if the third of the plane that together with the side makes the affecting solid [$P/3$] cedes to [$<$] the square of the half of the side [$(\frac{1}{2}S)^2$] that would arise from the application of the affected cube to the said third [$S=\frac{Q}{P/3}$].

So the condition is $\frac{P}{3} < (\frac{1}{2}\frac{Q}{P/3})^2$, equivalent to the familiar condition $(\frac{P}{3})^3 < (\frac{Q}{2})^2$. If we rewrite the inequality in the form of 10.2, i.e., taking $P=3a^2$, $Q=2a^2b$, the condition is $(a^2)^3 < (a^2b)^2$, equivalent to $a<b$, so this is case (2.1).

[13]See Note 91 of Chapter 4.

Viète did not make a distinction between cases (3.1) and (3.2), his formulation suggested that the equation $3a^2x - x^3 = 2a^2b$ is reducible to trisection for any values of a and b. The omission may be related to the fact that in case (3.1) the only real root is negative and Viète only considered positive roots.

Viète claimed, with reference to his *Treatise on the understanding of equations*,[14] that cases (1) and (2.1) could be reduced to "pure cubes," that is, to an equation $x^3 = a^2b$, that is solved by constructing the two mean proportionals of a and b.[15]

Viète then showed that in cases (2.2) and (3.2) a root can be found by angle trisection. He did so by providing explicit constructions; I have presented one of these (for case 3.2) above (cf. Construction 4.6) as an example of a construction by trisection.

Referring to the same treatise he had stated at the beginning of his argument that all fourth-degree equations can be reduced to third-degree ones via quadratic equations[16] and that third-degree equations can be reduced to forms lacking the quadratic term, that is, to one of the cases in Equation 10.2 above. The geometrical procedures corresponding to these reductions could be performed by straight lines and circles. As a result the construction of the roots of any fourth- or third-degree equation could be reduced to the construction of the roots of the standard third-degree equations and could therefore be performed by straight lines and circles together with either a trisection or a construction of two mean proportionals, and therefore, by straight lines, circles, and neusis.

The algebraic reductions

Viète's result relied heavily on the algebraic results in *Treatise on the understanding of equations*. This treatise, however, was published in 1615 only. His immediate readers could not consult it.[17] Nevertheless, algebraists would be familiar with the results at that time; the standard solution of third- and fourth-degree equations by the rules of Ferrari[18] and Cardano relied on the reduction of fourth-degree equations to third-degree ones, and of third-degree equations to standard forms without quadratic term. For the latter, Cardano's

[14] Published only much later: [Viète 1615].

[15] He referred to the procedure for solving third-degree equations, which has become known as "Viète's method." In the case of $x^3 + 3a^2x = 2a^2b$ (his first example, here abbreviated and modernized), he solved the auxiliary quadratic equation $Y^2 + 2a^2bY = a^6$, determined y from $y^3 = Y$ (this is the "pure cube") and showed that $x = \frac{a^2-y^2}{y}$ solved the original equation. Cf. [Viète 1615] pp. 149–150 (tr. [Viète 1983] pp. 286–289).

[16] The reference is to Viète's method for solving fourth-degree equations in [Viète 1615] pp. 140-148 (tr. [Viète 1983] pp. 266–286). It consisted in the reduction (I abbreviate and use modern notation) of the equation $x^4 + ax^2 + bx + c = 0$ (i) to the two equations $x^2 + xy + \frac{1}{2}y^2 + \frac{1}{2}a - \frac{b}{2y} = 0$ (ii) and $y^6 + 2ay^4 + (a - 4c)y^2 + b^2 = 0$ (iii). The reduction is achieved by determining an expression $F(y)$ in y such that inserting $x^2 + xy + F = 0$ into (i) eliminates x.

[17] Some mathematicians, such as Ghetaldi and Anderson (who saw [Viète 1615] through the press), had earlier access to Viète's manuscripts; it is unclear whether these manuscripts were known in wider circles.

[18] In his *The great art* of 1545 Cardano published, with due acknowledgment, the rule found by Ferrari for reducing a fourth-degree equation to a third-degree one; [Cardano 1545] Ch. 39 (pp. 235–253 of [Cardano 1968]), cf. the presentations of the rule in [Kline 1972] pp. 267–268 and [Tropfke 1980] pp. 452–453.

rule[19] provided expressions for the roots in terms of square and cubic roots. As I noted above, the cases in which the square roots lead to imaginary quantities were covered precisely by Viète's trisection solution; in the remaining cases, the cubic roots indeed corresponded to "pure cubes".

It should be remarked that Viète did not note the fact that all the necessary transformations can be performed by the usual Euclidean means of construction — an essential step in his argument — let alone that he actually spelled out the geometric equivalents to these transformations. Thereby his proof of the constructibility, by neusis, of roots of third- and fourth-degree equations was highly abstract. In particular the proof does not imply a feasible unified general construction applicable for the whole class of problems leading to third- and fourth-degree equations. We will see (Chapter 17) that Descartes found such a construction in the 1620s.

Solid problems

Viète did not explicitly mention one important consequence of his main result, namely, that all geometrical problems leading to equations of degree less than five were either "plane" or "solid" in terms of Pappus' classification, that is, that they could be constructed by straight lines, circles, and conic sections. The consequence would be obvious to those familiar with the *Collection* because that work contained a "solid" construction of the neusis. Thus it seems likely that by 1600 informed mathematicians would be aware of this consequence.[20]

10.4 Further statements by Viète on constructions

"Book VIII of various replies . . ."

The *Supplement to geometry*, then, presented a clear-cut and definite position in the matter of geometrical construction: non-plane constructions were to be performed by neusis, legitimated by a new geometrical postulate. The *Supplement* appeared in 1593. In the same year Viète also published a volume on various mathematical subjects, the *Book VIII of various replies on mathematical matters*. Here he touched upon several matters concerning construction, such as mean proportionals and the use of the quadratrix and the spiral. He stated that construction by shifting rulers was mechanical and therefore not geometrical; for instance, he referred to one such construction of a regular heptagon by François Foix de Candale as "accurate, but not geometrical."[21] He stressed that the usual postulates of geometry did not allow construction of cubic equations:

> Geometry admits no cubic equations at all within the limits of its

[19]See Note 91 of Chapter 4.

[20]The converse of this consequence, namely, that all solid problems in Pappus' sense lead to equations of degree less than five, applies as well, because the intersection of two conics (implied in Pappus' criterion for solid problems) leads to a fourth-degree equation. However, an argument to this effect could only be formulated when a general theory of the equations of conic sections was available, that is, after Descartes' *Geometry*.

[21][Viète 1593b] p. 359: "accuratam, sed non geometricam."

usual postulates.[22]

Still Viète was interested in constructions outside the traditional restriction to "plane" constructions; the *Book VIII of various replies on mathematical matters* contained substantial sections on the spiral and the quadratrix and their use in squaring the circle and dividing angles. Like Clavius (to whom he did not refer) Viète gave a pointwise construction of the quadratrix, but he did not claim (as Clavius had done, cf. Section 9.2) that such a construction legitimated the use of the curve in solving geometrical problems. However, he did claim that it was possible to derive theorems about the curves. Thus he wrote on the use of spirals in squaring the circle:

> Although the spirals are not described in the way of true knowledge, and neither are their tangents found in that way, still we can reason truly about questions of how large the angles are in the case of tangents, how large the lines are that are subtended by these angles, and thus art helps mechanics and mechanics helps art. This I wanted to show in this chapter, as well as a good method to square the circle as near to the true value as one wishes; it is a not too difficult method and I don't think that a more general and artful method can be proposed.[23]

The fact that mathematicians may derive true theorems about figures they cannot construct in a strictly legitimate fashion was later also remarked and discussed by Kepler (cf. Section 11.3).

"Apollonius from Gaul"

In the treatises on his "new algebra"[24] Viète did not explicitly mention the construction of solid problems by the intersection of conics. He did so in his *Apollonius from Gaul* of 1600,[25] but there he did not refer to neusis constructions. So we have no text in which he directly compared the two methods of construction.

In 1595 Viète had proposed the Apollonian tangency problem — to find the circle that is tangent to three given circles — at the end of his own solution of the famous problems posed by Van Roomen (one of these problems required the solution of a 45-degree equation).[26] Van Roomen took up the challenge and gave a solution of the Apollonian problem by the intersection of conics, not believing that a solution by straight lines and circles could be found.[27] I have given Van

[22] [Viète 1593b] p. 362: "Omnino aequalitates cubicas non adgnoscit Geometria suis contenta solitis postulatis."

[23] [Viète 1593b] p. 393: "Etsi non describantur volutae, neque tangantur ⟨kat epistemonikon logon⟩, attamen quanti sint anguli in volutarum contactu, quantaeve rectae, quae iis angulis subtenduntur, ratiocinamur ⟨epistemonikoos⟩, et ⟨mechaniken⟩ juvat ⟨technike⟩, ⟨techniken mechanike⟩, ut hoc capite placet exemplificari, et quadrandi circulum tam proxime quam placuerit vero, methodum bene paratam, neque ⟨dusmechanon⟩ exhibere, qua haud scio an alia possit proponi generalior et artificiosior."

[24] Cf. Note 5 of Chapter 6.

[25] [Viète 1600].

[26] Van Roomen had proposed his problems in [Roomen 1593] (fol. **iij^v); Viète countered by proposing the Apollonian tangency problem in [Viète 1595] p. 324.

[27] [Roomen 1596].

Roomen's construction in Section 5.5 (Analysis 5.7, Construction 5.8). However, as Viète knew, the problem could in fact be solved by straight lines and circles, and so he took Van Roomen to task in his *Apollonius from Gaul*,[28] published in 1600. He gave the plane construction of the Apollonian problem and in the preface of the book he dwelled upon the status of constructions by intersection of conics, such as the one Van Roomen had given. He stated plainly that in his opinion these constructions could not be accepted as truly geometrical solutions of problems. The argument gave him the occasion for some stylistic flourish:

> When I proposed the philomaths to solve Apollonius' problem about the circle that has to be drawn tangent to three given ones, I meant it to be constructed in the geometrical way (illustrious Adrianus), not mechanically. So when you hit the circle by hyperbolas you miss the mark. For in geometry hyperbolas are not described in the way of true knowledge. Menaechmus doubled the cube by parabolas, Nicomedes by conchoids, but is the cube thereby doubled geometrically? Dinostratus squared the circle by the inordinate winding curve [i.e., the quadratrix] Archimedes by the ordinate one [i.e., the spiral], but is the circle thereby geometrically squared? No geometer would make that proposition. Euclid and all his disciples would raise in protest. Therefore illustrious Adrianus, Belgian Apollonius if you want, your work is in vain; the problem I proposed is plane but you have dealt with it as a solid one, so you have not established the meeting of hyperbolas which you assume in your procedure, nor could you do so except in the case that the asymptotes of these hyperbolas are parallel.[29] And besides, the ancients have always been wary of drawing conic sections in the plane — therefore, do away with your mixed lines and take from an Apollonius reborn at the shores of the Aquitanian Ocean the construction that accords with the art and with true knowledge.[30]

It should be remarked that if Viète had only wanted to rebut Van Roomen's construction, he would not have needed the arguments against conics; he could

[28] [Viète 1600].

[29] In that case the intersection of the hyperbolas can be found by straight lines and circles.

[30] [Viète 1600] p. 325: "Problema Apollonii de describendo circulo, quem tres dati contingant (clarissime Adriane) geometrica ratione construendum proposui ⟨filomatheisi⟩, non mechanica. Dum itaque circulum per hyperbolas tangis, rem acu non tangis. Neque enim hyperbolae describuntur in geometricis ⟨kat' epistemonikon logon⟩. Duplicavit cubum per parabolas Menechmus, per conchoidas Nicomedes, an igitur duplicatus est geometrice cubus? Quadravit circulum per volutam inordinatam Dinostratus, per ordinatam Archimedes, an igitur geometrice quadratus est circulus? Id vero nemo pronunciabit Geometra. Reclameret Euclides et tota Euclideorum schola. Ergo clarissime Adriane, ac si placet Apolloni Belga, quoniam Problema quod proposui planum est, tu vero ceu solidum explicasti, neque ideo occursum hyperbolarum, quem ad factionem tuam adsumis, firmasti, neque etiamnum potes firmare, quoniam revera si asymptoti fuerint parallelae, erit irritus labor, et alioqui conicas sectiones in plano describere semper veriti sunt antiqui, missas fac lineas mixtas, et jam ab Apollonio ad ripas Oceani Aquitanici exsuscitato ultro accipe ⟨techniken kai epistemoniken cheirourgian⟩."

simply have censured him for committing the "sin" of constructing a plane problem by solid means. It is tempting, therefore, to read in these lines the reason why Viète preferred neusis over construction by conic sections: Euclid would protest and the ancients thought conics difficult to trace, perhaps more difficult than performing a neusis, of which Viète had stated in the *Isagoge* that it was "not difficult."[31] However, the text does not entirely justify such an interpretation; it may well be that the arguments were more inspired by Van Roomen's solution than by a definite point of view about constructions in general — after all, Viète had not stated these arguments against curves when he introduced the neusis in the *Isagoge* and *A supplement to geometry*.

Angular sections

Viète did not restrict his geometrical investigations to problems leading to cubic and biquadratic equations. Indeed, as mentioned earlier, one of his claims to fame was that he solved a 45th-degree equation proposed by Van Roomen.[32] This equation related to angular sections (cf. Section 4.3). One of the treatises belonging to Viète's planned *Book of the restored mathematical analysis or the new algebra* was devoted to angular sections.[33] Angular sections, as well as mean proportionals, were types of problems that, if translated into algebra, could lead to equations of any degree. In the early modern tradition of geometrical problem solving they provided the only examples of problems leading to equations of degrees higher than four. In the *Isagoge* Viète had claimed that his new method could solve "the greatest problem of all, which is to leave no problem unsolved."[34] One would therefore expect that Viète expressed himself somewhere, and especially in connection with angular sections, on the question how to construct beyond neusis. As far as I could ascertain, however, Viète did not leave any statements about this matter. It seems that his interest in the angular section problem was predominantly algebraic.

10.5 Viète's interpretation of geometrical exactness

Classifying non-plane problems

Having surveyed Viète's statements on geometrical construction I may now summarize his ideas on what I have called the interpretation of exactness in geometrical problem solving. Viète saw the restriction to constructions by straight lines and circles as a defect of geometry and claimed that this defect should be remedied by introducing a new postulate. In choosing the postulate he opted against the intersection of conics, thus disregarding Pappus' canon and precept; instead he adopted the neusis construction as new postulate. By applying his "new algebra" as analytical tool he was able to characterise and classify a large class of non-plane geometrical problems, namely, those leading to third- and fourth-degree equations. He showed that these problems were reducible

[31] Cf. Note 3.
[32] Cf. Note 26.
[33] [Viète 1615b] cf. Chapter 8, Note 5.
[34] Cf. Chapter 8 Note 6.

either to two mean proportionals or to trisection, and he proved that, within geometry supplemented by the neusis postulate, they were geometrically constructible. The result is indeed beautiful. It showed that algebra, combined with a carefully formulated position on construction, provided a well-structured and interesting extension of the Euclidean geometrical procedures. However, Viète did not address the question of how to proceed beyond the possibilities of neusis.

Context: Viète's program

Viète formulated his ideas about construction in the context of his new analysis. The grand structure of his analytic program of problem solving, with a zetetics introducing algebra, a poristics dealing with proportionalities and equations, and an exegetics requiring the constructional procedures corresponding to equations, forced him to devote attention to the question of construction. Thus Viète's case clearly illustrates how the principal dynamics within the early modern tradition of geometrical problem solving, namely, the adoption of algebraic analysis as a tool for geometry, demanded new approaches to the concept of construction.

Viète's reasons

Viète adduced few if any explicit arguments for assuming the neusis postulate and rejecting other procedures; we can hardly count the qualification "not difficult" for the neusis as substantial support, and the comments on the use of conics in the answer to Van Roomen were written long after Viète had made his choices and may well have been induced primarily by their polemical context. Why, then, did he choose the neusis postulate in his interpretation of exactness of constructions beyond straight lines and circles?

Adopting the neusis postulate meant choosing against the classical alternative of constructing by the intersection of conics and higher-order curves. It may be that Clavius' attempt to legitimate the quadratrix had failed to convince Viète and had made him aware that construction by curves begged the question of how to construct these curves. We may also look for an explanation in connection with the possibilities of algebraic analysis of curves. As I remarked earlier (cf. Section 8.5) the algebraic representation of curves by equations in two unknowns is absent in Viète's work, although all the algebraic prerequisites for such a conception were present. This conception, the key idea of what is now called analytic geometry, is first found with Fermat and Descartes, motivated by the analytic treatment of locus problems and of construction by curves. Viète seems not to have been interested in locus problems. It may be that in the matter of construction he decided against the use of curves because he did not immediately see how his new algebra could be applied to curves. And even if the decision was not induced in this way, it certainly steered the course of his later researches away from the possibilities of representing curves by equations.

Thus we may tentatively link the negative side of his choice — the rejection of construction by curves — to a distrust of curves and to a failure to see the possibilities of algebraic analysis for dealing with curves. As to the positive side — the selection of the neusis as one single powerful postulate — I think the explanation must be sought in the structure of the theory Viète was creating.

Legitimating significant results

Indeed, the Vietean texts analyzed in the present chapter and in Chapter 8 strongly suggest that the main reason for Viète to adopt the neusis postulate lay in a wish to provide the field of non-plane geometrical problems with a relatively simple postulational base, rather than a conviction that neusis was intrinsically the true and obvious next postulate after the Euclidean ones. It appears from the *Isagoge* that by the time he started publishing his series of treatises the structure of his theory of non-plane geometrical problems had taken definite form; he had seen the central position of the two classical problems of trisection and finding two mean proportionals, he had identified a large class of problems that could be reduced to these two, namely, those reducible to third- or fourth-degree equations, and he knew that both classical problems could be reduced to neusis. Choosing the neusis as additional constructive postulate was clearly the simplest way to secure the foundations of these well-structured and significant results. Seen in this light, it is understandable that Viète adduced no explicit arguments for choosing his new postulate; the arguments lay in the results that the postulate legitimated. Thus Viète's approach to the interpretation of exactness of geometrical construction represents the attitude that in the classification of Section 1.6 I have termed the "appreciation of the resulting mathematics."

Chapter 11

Kepler

11.1 Constructibility and creation

"On world harmony"

In 1619 Kepler published his *Five books on world harmony*, the grand synthesis of his cosmic theories based on his vision of a harmonic creation.[1] It contained a profound analysis of the concept of constructibility, as well as a precise and sharp criticism of the constructional practices of contemporary geometers and their use of algebra. For Kepler the issue was of central philosophical importance, and for that reason his analysis of constructional exactness in geometry was both more detailed and more critical than any in his period.

Harmony

The key concept of Kepler's philosophy was harmony.[2] Harmony underlay God's creation; the ability of the mind to recognize harmony was man's key for understanding the creation. Harmony was fundamentally mathematical, and geometry was the field in which harmonies could be recognized and known. Kepler related these harmonies in particular to the five Platonic solids and to the regular triangles, squares, pentagons, and hexagons that form the faces of these solids. The ratios of the sides of these polygons to the diameters of their circumscribed circles, were the crucial elements in Kepler's mathematics of harmony. They were harmonious ratios, and they could be known because these regular polygons could be constructed within a given circle by the Euclidean means of straight lines and circles. The regular heptagon, in contrast, was not knowable because it could not be constructed by straight lines and circles.[3]

Knowable

Harmonious ratios were knowable, unharmonious ones were not. As Kepler

[1] [Kepler 1619]; I quote from the edition of this work in [Kepler 1937–1975] taking the translations from [Kepler 1997].

[2] Cf. [Field 1988].

[3] Kepler's did not actually prove that the heptagon was not constructible by straight lines and circles — the means for such proofs became available only in the nineteenth century. His arguments, [Kepler 1937–1975] pp. 47–55 consisted of criticism of the extant constructions.

equated knowable with constructible, the question of demarcation between constructible and nonconstructible geometrical figures was essential. Moreover, it was necessary for Kepler's argument that this demarcation should be strongly restrictive. If the class of harmonious ratios were too large, the interpretation of their occurrence in nature as signs of God's deliberate choices would be meaningless. To allow any means of construction beyond straight lines and circles would extend the class of ratios and harmonious figures (no longer, for instance, excluding the heptagon); hence, these means had to be rejected.

11.2 Kepler's demarcation of geometry

Harmonious proportions The case for a restrictive demarcation of geometry was clearly stated in the opening sentence of the Prooemium of Book I:

> We must seek the causes of the harmonic proportions in the divisions of a circle into equal aliquot parts, which are made geometrically and knowably, that is, from the constructible regular plane figures. I thus considered that to start with it should be intimated that the features which distinguish geometrical objects to the mind are today, as far as is apparent from published books, totally unknown.[4]

Definitions What was needed, Kepler continued, was a detailed exposition of the concepts of constructibility in accordance with the original intentions of Euclid, as explained by Proclus and in general misunderstood by all other mathematicians. Kepler gave such an explanation in book I,[5] showing which regular polygons were constructible, and how these could be further distinguished into different classes. The relation between knowable and constructible was articulated in the seventh and eighth definitions of Book I. Knowing was measuring by a known measure. A magnitude was measurable if its ratio to the basic measure was rational. Lines that could be measured by a basic measure (in the case of regular polygons, the diameter of the given circle) and areas that could be measured by the square of the basic measure were knowable. All further figures that could be formed "by some definite geometrical connection" from measurable lines and areas were knowable as well:

> VII Definition. In geometrical matters, to know is to measure by a known measure, which known measure in our present concern, the inscription of Figures in a circle, is the diameter of the circle.
> VIII Definition. A quantity is said to be knowable if it is either itself immediately measurable by the diameter, if it is a line; or by its [the

[4] [Kepler 1937–1975] p. 15: "Cum a divisionibus circuli in partes aliquotas aequales, quae fiunt geometricè et scientificè, hoc est, à figuris planis regularibus demonstrabilibus, sint nobis petendae causae proportionum harmonicarum: illud initio significandum duxi, differentias rerum geometricarum mentales, hodie, quantum apparet in libris editis, in solidum ignorari." (Translation quoted from [Kepler 1997] p. 9.)

[5] In particular Props 30–50, [Kepler 1937–1975] pp. 34–64.

> diameter's] square if a surface; or the quantity in question is at least formed from quantities such that by some definite geometrical connection, in some series [of operations] however long, they at last depend upon the diameter or its square. . . .[6]

Thus knowable magnitudes could have an irrational ratio to the basic measure, but only if these magnitudes were constructed in a certain and geometrical way, that is, by Euclidean means.

Within the class of knowable objects Kepler introduced a further subdivision according to a degree of knowability, related to the classification of irrationals in Euclid's *Elements X*. As I am here primarily concerned with the demarcation between knowable and not knowable (constructible and not constructible). I don't discuss this further subdivision.

11.3 Constructibility and existence

The regular heptagon

Inevitably within this line of argument, Kepler was confronted with the question of the status of non-knowable figures. He provided a detailed discussion of the matter in Proposition 45 (book I),[7] which claimed that the regular heptagon could not be constructed by circles and straight lines. The regular heptagon was not knowable; but Kepler did not question its existence. Knowability or constructibility did not coincide with existence. Although nobody could ever construct a regular heptagon, such a figure might exist, formed by accident perhaps:

> So no Regular Heptagon has ever been constructed by anyone knowingly and deliberately, and working as proposed; nor can it be constructed as proposed; but it can well be constructed fortuitously; yet it is, all the same [logically] necessary that it cannot be known whether the figure has been constructed or no.[8]

Properties of non-constructible figures

One could speak about the properties of a regular heptagon, but these were conditional properties. Kepler considered himself here in philosophically and theologically dangerous waters; a friend even advised him to leave the arguments out. He did not do so and stated that there might exist non-knowable entities,

[6] [Kepler 1937–1975] pp. 21–22: "VII. Definitio. Scire in geometricis, est mensurare per notam mensuram; quae mensura nota in hoc negocio inscriptionis figurarum in circulum, est diameter circuli. VIII. Definitio. Scibile dicitur, quod vel ipsum per se immediate est mensurabile per diametrum, si linea; vel per ejus quadratum, si superficies: vel quod formatur ad minimum ex talibus quantitatibus, certâ et geometricâ ratione, quae quantumcunque longâ serie, tandem tamen à diametro, ejusve quadrato dependeant. . . ." (translation quoted from [Kepler 1997] p. 19.)

[7] [Kepler 1937–1975] pp. 47–56.

[8] [Kepler 1937–1975] p. 50: "Itaque nullum unquam regulare septangulum à quoquam constructum est, sciente et volente, et ex proposito agente: nec construi potest ex proposito: sed benè fortuitò construi posset: et tamen ignorari necesse est, sit ne constructum an non." (Translation quoted from [Kepler 1997] p. 66.)

not even knowable to God, which still had knowable properties. Kepler called them conditional entities ("Entia conditionalia"):

> Now it is appropriate to put a word in here for Metaphysicians in connection with this algebraic treatment: let them consider if they can take anything over from it to explain its Axioms, since they say that which does not exist [a Non-entity] has no characteristics and no properties. For here, indeed, we are concerning ourselves with Entities susceptible of knowledge; and we correctly maintain that the side of the Heptagon is among Non-Entities that are not susceptible of knowledge. For a formal description of it is impossible; thus neither can it be known by the human mind, since the possibility of being constructed is prior to the possibility of being known: nor can it be known by the Omniscient Mind by a simple eternal act: because by its nature it is among unknowable things. And yet this which is not a knowable entity has some properties which are susceptible of knowledge; just as if [they were] conditional Entities. For if there were a Heptagon inscribed in a circle, the proportion of its sides [to the semidiameter] would have such properties. Let this indication suffice.[9]

In the margin of this passage we read:

> In case it should be supposed that these comments are blasphemous. One of my friends, a very practiced mathematician, thought they could be left out. But nothing is more habitual among Theologians than to claim that things are impossible if they involve a contradiction: and that God's knowledge does not extend to such impossible things, particularly since these formal ratios of Geometrical entities are nothing else but the Essence of God; because whatever in God is eternal, that thing is one inseparable divine essence: so it would be to know Himself as in some way other than He is if He knew things that are incommunicable as being communicable. And what kind of subservient respect would it be, on account of the inexpert who will not read the book, to defraud the rest.[10]

---

[9] [Kepler 1937–1975] pp. 55: "Illud autem obiter monendi sunt metaphysici occasione hujus cossae; considerent, si quid hinc transsumere possint ad explicationem illius axiomatis, *cùm non entis nullae dicuntur esse conditiones, nullae proprietates*. Nam hic quidem versamur nos in entibus scientialibus; et pronunciamus rectè, quod latus septanguli sit ex *non entibus*; puta scientialibus. Cum enim sit impossibilis ejus formalis descriptio; neque igitur sciri potest à mente humana, cum scientiae possibilitatem praecedat descriptionis possibilitas: neque scitur a Mente Omniscia actu simplici aeterno: quia suâ naturâ ex inscibilibus est. Et tamen hujus non entis scientialis sunt aliquae proprietates scientiales; tanquam entia conditionalia. Si enim e s s e t septangulum descriptum in circulo, laterum ejus proportio tales haberet affectiones. Sufficiat monuisse." (Translation quoted from [Kepler 1997] p. 74; I modified the translation of "conditionalia".)

[10] [Kepler 1937–1975] p. 55: "Haec ne blasphemè dicta putentur, omitti posse censuit amicorum unus, mathematum peritissimus. Atqui nihil est vulgatius apud theologos quàm impossibilia esse, quae contradictionem involvunt: et Dei scientiam ad talia impossibilia se non

11.4 Kepler's criticism of non-plane constructions

Figures studied for their own sake

Kepler was well aware that mathematicians did not keep to the strong demarcation of geometry he advocated and that constructional means other than straight lines and circles were freely used to trisect angles, find mean proportionals, and construct regular heptagons. He referred to Pappus' classification of problems in this connection and emphasized that Pappus accepted plane, solid, and line-like problems all as geometrical, whereas he (Kepler) restricted the qualification "geometrical" to Pappus' class of plane problems.[11] He claimed that the figures dealt with by non-plane means were studied not for their own sake but for certain external purposes. Constructible regular figures, in contrast, were studied for their own sake as Archetypes of harmony.[12]

Legitimacy of Euclidean means

Kepler offered little direct argument to explain why the Euclidean means of construction, straight lines and circles, were legitimately geometrical and certain. He invoked the authority of Euclid and Proclus, and argued that the whole structure of the Euclidean *Elements* was based on the restriction to construction by straight lines and circles; removing this restriction would lead to the disintegration of the structure. He blamed his contemporaries for doing so: Ramus by his disregard for the theory of irrationals and for the Platonic solids in *Elements* X and XIII, and Snellius and others by rejecting *Elements* X as useless (cf. Section 7.4).[13] He wrote:

> Ramus removed the form from Euclid's edifice, and tore down the coping stone, the five solids. By their removal every joint was loosened, the walls stand split, the arches threatening to collapse. Snellius therefore takes away the stonework as well, seeing that there is no application for it except for the stability of the house which was joined together under the five solids.[14]

In fact both Ramus and Snellius acknowledged the special status of constructions with straight lines and circles in geometry.[15] Yet Kepler rightly saw that the willingness of his contemporaries to study non-plane procedures and to introduce

extendere, praesertim cum hae formales rerum geometricarum rationes nihil sint aliud, quam ipsa essentia Dei; quia quicquid in Deo est ab aeterno, id una individua est essentia divina: esset igitur seipsum quodammodo alium scire, quàm est; si quae sunt incommunicabilia, sciret ut communicabilia. Et quae haec adulatio, propter imperitos librum non lecturos, defraudare caeteros." (Translation quoted from [Kepler 1997] p. 74.)

[11] [Kepler 1937–1975] p. 59.

[12] [Kepler 1937–1975] p. 19.

[13] [Kepler 1937–1975] pp. 15–19.

[14] [Kepler 1937–1975] p. 19: "Ramus aedificio Euclideo formam ademit, culmen proruit, quinque corpora; quibus ablatis, compages omnis dissoluta fuit, stant muri fissi, fornices in ruinam minaces: Snellius igitur etiam caementum aufert, ut cujus nisi ad soliditatem domus sub quinque figuris coagmentatae nullus est usus." (Translation quoted from [Kepler 1997] p. 13.)

[15] For Ramus, cf. Section 2.5; For Snellius cf. Section 14.3.

numbers and algebra into geometry undermined the classical authority for a restriction to construction by circles and straight lines.

Criticism of Pappus' trisection

Kepler was more articulate about his reasons for rejecting constructions beyond straight lines and circles. In a critical analysis of various methods of angle trisection[16] he identified, much more explicitly than had been done before, all the assumptions that were implied in the usual constructions of solid problems. He discussed in particular Pappus' trisection. In the *Collection* Pappus had given a trisection by a neusis and a construction of the neusis by the intersection of a circle and a hyperbola, that is, by solid means (Constructions 3.9 and 3.8 respectively). Kepler combined these two constructions and criticized them. He argued as follows.[17]

Construction of curves geometrically unacceptable

The crucial step in the construction as given by Pappus was the following: Given two intersecting straight lines and a point outside them, draw a hyperbola through the point with the given lines as asymptotes (cf. step **3** of Construction 3.8). Kepler questioned the possibility of drawing this hyperbola. If one followed Apollonius' way of constructing conics,[18] one had to find, in the space surrounding the given plane, the positions of the top and the base circle of a cone, and determine the intersection of that cone with the plane. As Kepler formulated it, one had to incline a cone in such a way that it produced the required hyperbola. Kepler rejected this procedure as non-scientific because it remained unclear how this positing of a cone in space could be done. The procedure was "solid" and Kepler considered Pappus' solid and line-like procedures as ungeometrical. He then turned to the alternative: to draw the hyperbola in the plane. To do so one could construct arbitrarily many points of the hyperbola or one could try to trace the curve by some motion. In the latter case the procedure was line-like in Pappus' classification because it was by such motions that curves like the quadratrix were traced; this method should therefore also be rejected. Pointwise construction was no acceptable alternative either because the curve segments between the constructed points would still have to be traced by some line-like procedure.

Thus to Kepler all three standard ways of constructing curves — intersection of solids, tracing by motion, and pointwise construction — were geometrically unacceptable, at least if these curves were to serve as means to produce knowable figures.

Existence and constructibility

In this context Kepler again articulated the distinction between existence and constructibility. He admitted that the required hyperbola existed and was even unique, but that, he claimed, was not at issue. What was required was that the hyperbola be made, be constructed. And for the actual construction, the usual methods ordered him to do things that could not scientifically be done, such

[16] [Kepler 1937–1975] Book I Prop. 46 pp. 56–61.
[17] [Kepler 1937–1975] pp. 59–60.
[18] [Apollonius Conics] I-52–60.

as inclining a cone in space or drawing a curve through points by some tracing motion. He wrote:

> But because we are not investigating what it will be, once the construction is carried out, but rather by what means, in order to give it existence, a thing not yet constructed is to be constructed: accordingly, we get nothing more from the Solid and Line-like Problems of the ancients, as far as obtaining knowledge of the required lines is concerned, than we got before from the Analytical method of the moderns. There is clearly only one line of a Hyperbola [that lies] between the given Asymptotes, passes through the given point, and can be drawn in their plane. But when it is not yet drawn, I am required to adjust the inclination of the Cone over the point of application until it [the hyperbola] comes into being and is drawn: alternatively, not using the Cone, I am required to change the construction lines that plot the Hyperbola by repeatedly finding points, until the curve is long enough: and the parts that lie between the points I have plotted I am required to suppose to have been plotted also: in either case, I am required to pass over by a single act or motion something which potentially involves infinite division; so that by this passage something may be attained which is concealed in that potential infinity, without the light of perfect knowledge, which the problems the ancients dubbed Plane do have.[19]

11.5 Kepler's objections to algebraic methods

Cossic methods

Kepler also dealt with the algebraic approach to geometrical problems, or, as he referred to it, the use of "cossic" methods.[20] Van Roomen, Viète, Bürgi,

[19] [Kepler 1937–1975] p. 60: "At quia non de hoc quaerimus, quid sit, re jam factâ, sed quomodò, ut sit quidque, res nondum facta, sit facienda demum: ideò nihilò plus habemus ex problematibus solidis et linearibus veterum, quod ad quaesitam linearum scientiam faciat; quàm priùs ex doctrina analyticâ modernorum. Est sanè una sola hyperbolae linea, inter asymptotos positas, per punctum propositum, in earum plano ductilis. At eâ nondum ductâ, conum jubeor tantisper inclinare super puncto applicationis, donec existat illa, ductaque sit: vel sine cono, lineas, quae hyperbolam delineant per continuata puncta, jubeor tantisper mutare, donec satis prolongata sit hyperbola: et quae partes inter facta puncta cadunt intermediae, eas jubeor imaginari factas: jubeor utrinque, id quod est potestate divisionis infinitae, actu seu motu uno transire; ut hoc transitu etiam id attingatur, quod latet in illâ infinitate potestativâ, sine perfectae scientiae luce, qualem habent problemata à veteribus plana cognominata." (Translation quoted from [Kepler 1997] p. 87; I changed "Linear" into "Line-like.")

[20] The relevant passages are on pp. 50–55, 57–58 of [Kepler 1937–1975]. Kepler had earlier questioned the use of algebra in geometry: in [Kepler 1615] pp. 111–113, he proposed two geometrical problems and challenged "cossists" such as Van Roomen, to try their art, predicting that they would find equations involving three or even five continuous proportionals (that is third- or fifth degree equations) which would be useless in finding a geometrical solution. Van Roomen actually died in 1615 but Anderson took up the challenge, cf. F. Hammer's notes on pp. 518–523 in de *Werke* edition.

and others had studied algebraic equations satisfied by the sides of regular polygons.[21] Kepler saw the "Cossic art" primarily as a method for solving numerical problems; he did not refer to Viète's more sophisticated use of algebraic methods for general abstract magnitudes. He acknowledged the value of the cossic methods for finding useful inequalities or approximations and for calculating trigonometric tables. But the method had no value in genuinely geometrical study:

> . . . the former [the use of equations] is particularly excellent and noble in this semimechanical Cossa, but base and degraded in geometry which produces knowledge . . ..[22]

Bürgi's heptagon equation

In the case of non-plane problems, scientific knowledge could not be attained by geometrical methods nor by algebra. Kepler argued this point extensively and in doing so he gave one of the first explicit and critical discussions of the relation between algebra and geometry. He related his arguments to certain equations for the side of the regular heptagon derived by Jobst Bürgi.[23] One of these was (in modern notation):[24]

$$7 - 14x^2 + 7x^4 - x^6 = 0 , \tag{11.1}$$

x being the side of a regular heptagon inscribed in a circle with radius 1. Kepler interpreted the equation in terms of a continued proportion: in the series of seven proportionals, the first two of which were the radius and the side respectively of a regular heptagon, (i.e., 1, x, x^2, x^3, x^4, x^5, x^6), 7 times the first together with 7 times the fifth was equal to 14 times the third and once the seventh.

Objections against equations in geometry

Kepler then gave five separate arguments in support of his view that algebra, in particular equations, did not provide the means to solve non-plane problems in geometry. First he remarked that, although the equation implied a property of the side of the heptagon, it did not offer a method to find that side. The proposition stating that the property applied in the case of a regular heptagon was geometrical and provable, but the important question was in how far the knowledge of that property made the side of the heptagon known. As long as the heptagon itself was not yet known, the property was of little value:

---

[21] The famous 45th degree equation which Van Roomen proposed in [Roomen 1593] (cf. Chapter 10 Note 26) originated in such studies; Viète dealt with the subject in [Viète 1615b]; on Bürgi's algebraic studies cf. M. Caspar's notes in [Kepler 1973] pp. 370–373.

[22] [Kepler 1937–1975] p. 58: ". . . illud imprimis excellens et nobile est in hac cossâ semimechanicâ, degener verò et abjectum in Geometriâ scientificâ . . ." (Translation quoted from [Kepler 1997] p. 84.)

[23] Kepler had also corresponded with Van Roomen about these equations, cf. [Bockstaele 1976] pp. 289-291 and [Kepler 1937–1975] vol. 15 p. 244.

[24] [Kepler 1937–1975] p. 52.

> ... but since I do not yet have that proportion described by geometrical means: therefore I waited for someone to explain to me how to set up that proportion first.[25]

And:

> But how am I to represent the relationship, by what Geometrical procedure? No other means of doing it are afforded me save using the proportion I seek; there is a circular argument: and the unhappy Calculator, robbed of all Geometrical defenses, held fast in the thorny thicket of Numbers, looks in vain to his algebra. This is one distinction between Algebraic and Geometrical determinations.[26]

Kepler then noted a second objection:[27] the argument leading to the equation essentially presupposed the quantity to be discrete rather than continuous, and was based on a particular choice of the unit (namely as half the diameter). Geometrical reasoning, did not in this way depend on discreteness and choice of unit, numbers were never used for irrational geometrical quantities. I have noted this objection of Kepler against the use of numbers in geometry above in Section 7.4.

A third objection of Kepler concerned the fact that the equation had more than one solution.[28] Kepler noted that not only the side of a heptagon satisfied the equation but also any of its diagonals. Thus, as far as the equation was concerned, no distinction could be made between side and diagonal, which Kepler considered alien, even repellent, to the geometer.[29]

Kepler's fourth objection was that if, as in this case, the unknown was irrational, the equation, based on calculations with numbers, could never provide more than approximations. Approximations did not provide the kind of knowledge the geometer seeks:

> Fourth, assuming that a single proportion would [suffice to] define what is required; I am not told how to bring the matter to a conclusion but only how to stalk the quarry, from a distance. ... this is not to know the thing itself but only something close to it, either greater or less than it; and some later calculator can always get closer to it [still]; but to none it is ever given to arrive at it exactly. Such

[25] [Kepler 1937–1975] p. 52: "... sed cùm eam proportionem nondum habeam ullo geometrico actu descriptam: illud igitur expectabam, ut quis me doceret priùs illam proportionem constituere." (Translation quoted from [Kepler 1997] p. 70.)

[26] [Kepler 1937–1975] p. 53: "At quomodo repraesentabo affectionem, quo actu geometrico? Nullo alio id doceor facere, quam usurpando proportionem, quam quaero; principium petitur: et miser calculator, destitutus omnibus geometriae praesidiis, haerens inter spineta numerorum, frustrà cossam suam respectat. Hoc unum est discrimen inter cossicas et inter geometricas determinationes." (Translation quoted from [Kepler 1997] p. 71.)

[27] [Kepler 1937–1975] p. 53.

[28] [Kepler 1937–1975] pp. 53–54.

[29] [Kepler 1937–1975] p. 53: "illud maxime mirum est (quanvis geometram praecipuè absterreat)."

> indeed are all quantities which are only to be found in the properties of matter of a definite amount; and they do not have a knowable construction by which in practice they might be accessible to human knowledge.[30]

Kepler added a fifth argument[31] relating the equation to the construction of mean proportionals. The argument, however, is difficult to interpret and I leave it aside here.

Kepler concluded his criticism of the use of algebra in constructing regular polygons as follows:

> So we conclude that these Algebraic Analyses make no contribution to our present concerns; nor do they set up any degree of knowledge that can be compared with what we discussed earlier.[32]

The arguments Surveying Kepler's objections to using algebraic methods in geometry, we may conclude that they were strong and pertinent as far as constructions were concerned. His first argument pointed to the fundamental restriction of using algebra in geometry: equations and algebraic expressions representing unknown line segments do not themselves constitute solutions, i.e., constructions of geometrical problems, nor do they automatically provide such constructions.

However, in other respects Kepler's objections reveal an underestimation of the conceptual power of algebra. His third argument concerned the fact that an equation (usually) has more than one root and does not itself provide means to distinguish between these roots. But the same applies for geometrical problems with multiple solutions. So the objection does not bring more than the first one, namely, that an equation is not a solution but still a problem, and Kepler's wonder about multiple roots seems to indicate some lack of familiarity with equations.

Kepler's second and fourth arguments related to the idea that numbers, being discrete, did not belong in geometry, which concerned continuous quantity. We have seen (cf. Sections 7.2, 7.3, 7.4) that this idea was widely discussed around 1600. However, Viète had shown, well before the publication of *On world harmony*, that the rejection of numbers from geometry did not necessarily obstruct the geometrical use of algebra.

Comparison with Viète Kepler's arguments invite a comparison with Viète's ideas on construction and the use of algebra. Indeed Viète had dealt with the same issues and arrived

[30] [Kepler 1937–1975] p. 54: "Quartò, posito, quod una sola proportio faciat imperatum; illam non doceor absolvere, sed saltem venari eminus. . . . Hoc non est scire rem ipsam, sed saltem aliquid proximè majus vel minus; potestque semper posterior aliquis computator approximare magis; pervenire ad punctum ipsum, nulli unquam datur." (Translation quoted from [Kepler 1997] p. 72.)

[31] [Kepler 1937–1975] pp. 54–55.

[32] [Kepler 1937–1975] p. 55: "Concludimus igitur, analyses istas cossicas, alienas esse à praesenti contemplatione; nec ullum constituere gradum scientiae, cum iis comparabilem, quos explicavimus in superioribus." (Translation quoted from [Kepler 1997] p. 74.)

at a contrary conclusion, namely, that algebra was an appropriate means for use in geometry. Rather than accepting a restriction to numerical subject matter, Viète had realized that algebra could be applied to magnitude in general, and to do so he had elaborated a profound reinterpretation of the basic algebraic concepts. Like Kepler, Viète realized that an equation in itself was no solution. But rather than seeing this as an unanswerable objection, he had elaborated the further "exegetic" procedures for deriving geometrical constructions from algebraic equations.

Kepler did not refer to Viète; it is unclear whether he was familiar with the Vietean treatises. Even if he was, he might have refrained from referring to them because Viète did accept construction beyond straight lines and circles. Moreover, Viète's arguments were formulated with ornate terminology in the context of a grand theoretical structure, which may have prevented the separate ideas to become clear. In contrast, the polemical context of Kepler's argumentation made him formulate the point that equations were not equivalent to solutions in a much more precise and challenging way than Viète ever did.

11.6 Kepler's interpretation of geometrical exactness

A restrictive point of view

Kepler summed up the conclusion of his discussion of contemporary constructional practice beyond straight lines and circles by a metaphor: there was no scientific way to cross the frontier between plane and solid problems, the bridge over the water was broken, the shores of the two territories remained disconnected. The metaphor occurred in connection with the problem of two mean proportionals:

> For if a proportion between solids is not given in a form such as [a ratio between] two cubic numbers: we cannot, as an intellectual procedure, measure the proposed solid in terms of another known one: because two intermediate proportionals cannot be constructed exactly in the plane: though they may be present in the cubes, yet there is no passage from the plane figures to form any of those cubes without the two means: [it is] as if the bridge were broken.
>
> And for finding two mean proportionals some give instructions to use Geometrical motion, thereby ordering one to do something that is useless for achieving certainty through an appropriate Geometrical act: indeed Pappus himself gives instructions that use Conic sections, to be produced with the help of two [mean] proportionals, although the Cone itself is a solid. So we are always assuming what is required to prove; and the bridge lies on the other bank.[33]

[33][Kepler 1937–1975] p. 61: "Nam nisi solidorum proportio fuerit data talis, qualis est inter duos numeros cubicos: mensurare solidum propositum alio solido noto non poterimus, ad mentem informandam: quia duae intermediae proportionales exactè in plano constitui non possunt: in cubis etsi possunt inesse, at à planis ad cubos illos quoscunque formandos, non

The basis for Kepler's restrictive interpretation of geometrical exactness was philosophical; he needed the classification of ratios that resulted from the strict adherence to Euclidean constructions; any extension would explode his theory of harmony. Thus he took an orthodox and very restrictive view in the matter of geometrical constructions: only those by straight lines and circles were genuinely geometrical. Compared with the ideas of his contemporaries, Kepler's interpretation of geometrical exactness was unusually rigid.

Appeal to authority and tradition

Clearly Kepler's strategy in defense of his interpretation of exactness was what in Section 1.6 I have called appeal to authority and tradition. His authorities were classical: Euclid and Proclus. The former had, he claimed, established the proper geometrical approach to construction, namely, the restriction to straight lines and circles, in his *Elements*; the latter had provided the arguments for this approach in his commentary to the *Elements*.

Kepler's strategy was not successful. For all his strong words and images, Kepler convinced few, if any, geometers to refrain from dealing with higher-order problems and from searching for solutions that could be considered exact.

datur transitus sine ipsis duabus mediis, veluti ponte abrupto.

Et duas medias proportionales invenire, docent alii per motum geometricum, imperantes quod est impraestabile, quoad certitudinem actus geometrici adaequati: docet et ipse Pappus, per sectiones conicas, beneficio duarum proportionalium expediendas, cum et conus sit solidum quid. Ita semper principium petitur; et pons jacet in adversâ ripâ." (Translation quoted from [Kepler 1997] p. 88, slightly modified.) The margin title of this passage reads: "Duas medias proportionales scientificè invenire impossibile."

Chapter 12

Molther

12.1 *The Delian Problem*

In the same year as Kepler's *World harmony* a book by a much lesser known mathematician was published in Frankfurt. It was entitled *Molther's book*

> *The Delian problem of doubling the cube, that is, given any solid, to make a similar solid in a given ratio, by means of the second Mesolabum, by which two continuously proportional means can be obtained. Now at last easily and geometrically solved after innumerous attempts of the most eminent mathematicians. The history of the problem is given first and some results are added about the trisection of an angle, the construction of a heptagon, the quadrature of the circle and two very convenient designs of proportional instruments.*[1]

About its author, Johannes Molther, little is known.[2] The *Delian Problem* of 1619 was his only mathematical book. It contained a number of neusis constructions for two mean proportionals. These constructions were not new; they were the same as, or at least very similar to, the ones given by Pappus, Archimedes, and Viète. Molther did not explicitly present the constructions themselves as original; but he did claim that he was the first to prove conclusively that the neusis construction was truly geometrical. In his opinion a convincing argument for the geometrical legitimacy of neusis constructions had not yet been given,

[1] [Molther 1619] titlepage: "Problema Deliacum de cubi duplicatione, hoc est de quorumlibet solidorum, interventu Mesolabii secundi, quo duae capiantur mediae continue proportionales sub data ratione similium fabrica. Nunc tandem post infinitos praestantissimorum mathematicorum conatus expedite et geometrice solutum. Ubi historia problematis praemittitur, et simul nonnulla de anguli trisectione, heptagoni fabrica, circulique quadratura et duabus commodissimis instrumentorum proportionum formis inserentur." Doubling the cube was called the "Delian problem" because of a tradition that people from Delos were once told by an oracle to double their altar.

[2] Molther, born in 1591, was professor of Medicine at Marburg, his date of death is unknown; cf. [Bos 1993b] p. 29.

and so, by presenting such an argument, he could claim that he had for the first time really solved the problems of doubling the cube and constructing two mean proportionals.

Legitimation

As to mathematical techniques the book was unoriginal, occasionally coming near to plagiarism. The historical interest of the book lies in the arguments about the legitimacy of constructions. Molther based these arguments on an analysis of how draftsmen achieved precision in using instruments. His further reasoning may be summarized as saying that exactness in pure geometry is the mental analogue of the practical precision in applying instruments such as rulers and compasses.

We have met this idea in Clavius' legitimation of his construction of the quadratrix. Clavius, however, did not elaborate the argument further, nor did he fully maintain it in later publications. It seems indeed that Molther's book is the only early modern text in which this approach to geometrical legitimation is expounded in great detail. It is, thereby, an extreme example of the interpretation of exactness that I have termed "idealization of practical methods," and for that reason his arguments deserve to be discussed in my present study.

12.2 Arguments

Against earlier constructions

In the first, historical sections of the book Molther critically reviewed earlier attempts to construct two mean proportionals. In Molther's opinion, none of these constructions was truly geometrical. Some of them used curves, which were traced by machines or constructed pointwise; neither method could be accepted as fully geometrical. Other constructions were impractical, merely approximate, or just false. Among all the constructions Molther preferred that of Nicomedes by means of the conchoid (cf. Construction 2.6), because this curve could be traced more easily than the conics or the cissoid, and also because its pointwise construction (here he referred to Clavius, who gave such a construction of the conchoid in his *Practical Geometry*[3]) was simple and practical. Yet, curves as the conchoid were not traced in a truly geometrical way, and it was still an open problem

> how the required positionings of such lines could be performed at one stroke without other instruments than those which the geometer is perfectly allowed to use, with such truth and precision that they would also sustain the scrutiny of reason.[4]

Thus Molther's criteria, as far as he articulated them in his historical survey, concerned precision and accuracy. He explicitly rejected the use of curves that were merely constructed pointwise or traced by instruments.

---

[3]Cf. Chapter 9 Note 17 of Chapter 9.

[4][Molther 1619] p. 25: "(—) quomodo eiusmodi linearum requisitae applicationes Geometrice sine alio quam Geometrae absolute concesso instrumento, mox prima actione praestarentur tanta veritate ac praecisione ut etiam rationis censuram sustinerent."

At the end of the historical introduction Molther announced his own method for finding two mean proportionals in the following terms:

Molther's solution: the neusis postulate

> We, however, have finally realized that the business of this problem which was explored through many centuries and which left the most ingenious of mortals at a loss, is really so smooth, easy, obvious, ready and evident, that it has by right to be counted as the next postulate, because it meets the very terms for a legitimate postulate; hence it does not at all require a belaboured construction and proof, as difficult problems do, but, as a principle clear in itself, it needs only a simple explanation, after which anyone can understand it and give it its due assent.[5]

So Molther's solution was simple and direct indeed. He claimed that the neusis construction could be accepted as a postulate in geometry on the same level as the traditional postulates that canonized the usual Euclidean constructions, and that therefore the duplication problem could indeed be solved geometrically by means of neusis.

Now this was precisely what Viète had done (cf. Section 10.2) in his *Supplement of geometry*. But Viète had given no arguments why the neusis should be given postulate status. For Molther that was the heart of the matter, which may explain how easily he dismissed Viète's work in his survey of earlier constructions:

> With all his subtlety Viète gathered nothing that could withstand criticism.[6]

12.3 Molther's justification of the postulate

The crux of Molther's reasoning was the argument that neusis construction was as obviously possible and acceptable in geometry as the construction of straight lines and circles by a ruler and a compass. The first chapter of his book was devoted to that argument. It opened with the formulation of the postulate:

The postulate

> It is postulated that, given two lines and a point in position in the same plane, a line can be drawn through that point such that the segment intercepted on that line by the two given lines is equal to another straight line given in length.[7]

[5] [Molther 1619] p. 27: "At vero nos rem istam exploratae per plurima secula difficultatis, in qua mortalium ingeniosissimi haesitarunt, ita expeditam, facilem, obviam, parabilem promtamque dudum animadvertimus, ut quia hasce Postulati legitimi conditiones obtinet, Postulatus sit proxima meritoque annumeranda, adeo ut nequaquam ceu Problema contentiosum anxiam constructionem et demonstrationem requirat, sed tanquam Principium per se manifestum, levi contenta sit explicatione, qua adhibita à quolibet capi et assensum meriri possit."

[6] [Molther 1619] p. 26: "Subtillissimus Vieta nihil quod censuram sustineret venatus est."

[7] [Molther 1619] p. 29: "Postuletur, duabus lineis punctoque in eodem plano situ datis, ut è puncto isto linea recta applicetur, cuius portio a lineis illis intersecta alteri rectae longitudine datae sit aequalis."

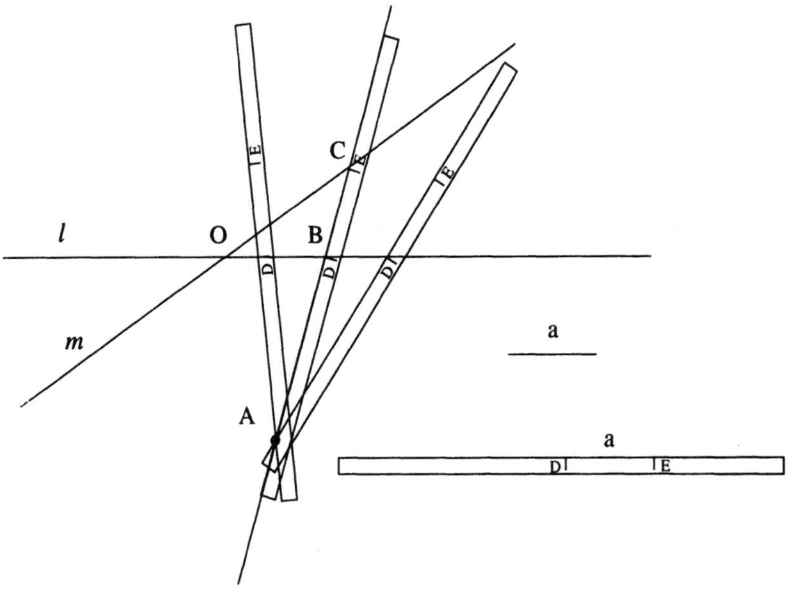

Figure 12.1: Neusis — Molther

Performing a neusis Molther then explained the procedure for performing a neusis. His description of it can be summarized in the form of a construction:

Construction 12.1 (Neusis — Molther)[8]
Given lines l and m intersecting in O, a point A and a length a (see Figure 12.1); it is required to construct the straight line ABC intersecting l and m in B and C, respectively, such that $BC = a$.

Construction:
1. Take a ruler and mark points D and E on it with distance a.
2. Move the ruler over the plane in such a way that it slides along the point A and that the point D on the ruler moves along the given line l.
3. Stop that motion at the moment that the point E is on the line m; then draw a straight line along the ruler; it intersects l and m in B and C, respectively.
4. ABC is the required line.

Legitimation To justify the status of postulate for the neusis, Molther analyzed the procedure at great length, arguing that each of its steps was legitimately geometrical. Since the first Euclidean postulate established the tracing of straight lines, the ruler, by which this was done, was a geometrical instrument. Moreover, such

[8][Molther 1619] pp. 29–30.

rulers could be made very precise, for instance, by making them of metal or by folding a piece of paper. The segment $DE = a$ could be marked on the ruler by a compass, which was also a geometrical instrument. The required movements of the ruler could be performed with great precision. This precision was guaranteed by our senses, which could judge whether the ruler remained along A during the motion and whether the point D moved along the given line l. It was also by the senses that the geometer decided to stop the motion at the moment that the point E lay on the second given line m. Thus these procedures essentially involved motion and the judgment and action of the geometer to stop that motion when the required position is reached. Molther argued that both motion and the testimony of the senses were implicitly assumed in the usual Euclidean postulates:

> For we have to use sense [sc. in performing a neusis] to move the ruler and to assess whether it is placed in the way as postulated: because what can be clear by itself should not be made known through any demonstration. For in the other case [sc. the Euclidean postulates] we assess in no other way than by sense whether a ruler is duly posited along the two points, from one of which to the other a straight line can be drawn according to the postulate; similarly whether the given interval with which a circle is to be described is justly contained in a compass; and again similarly whether the one leg of the compass is rightly placed in the given centre around which the circle is to be drawn. Certainly one sees with the same ease whether the ruler is along the point A and at the same time whether the point C on the ruler is at the line l and the point B on it at the line m.[9]

Molther pointed out that it was precisely the role of the senses that underlay the distinction between axioms and postulates. Axioms regarded the mind; postulates, "principles of construction"[10] as Molther called them, regarded the senses.

Pure geometry

Despite his detailed analysis of the procedure by which geometers actually draw lines, it was not Molther's view that geometry coincided with the physical actions of draftsmen. He left room for a step of idealization from practice to pure geometry. He argued that also in the abstract mental world of pure geometry the usual Euclidean postulates required motion and the testimony of the senses (in this case the inner senses), and that therefore the neusis construction, using

[9][Molther 1619] pp. 33–34: "Sensu enim advertendum et agnoscendum sitne Regula ita ut postulatur applicata: quia nulla demonstratione id innotescat, quod per se liquidum esse potest. Quemadmodum alias haud aliter quam sensu cognoscimus sitne Regula ad duo puncta, è quorum uno ad alterum trahi postulatur Recta, debite applicata: sitne intervallum pro Circulo describendo datum Circini apertura iuste comprehensum: sitne pes Circini alter in centro dato circa quod gyranda est Peripheria, recte positus. Nempe eadem facilitate protinus cernitur, sitne Regula iuxta punctum A simulque Regulae punctum C in linea l, et Regulae punctum B in linea m." (Letters changed to correspond to Construction 12.1.)

[10][Molther 1619] p. 34: "Principia fabricandi."

the same motions and the same testimony of the senses, should be granted the same postulate status in pure geometry.

Motion in pure geometry

Molther stressed that motion was very common within pure geometry; a line was generated by motion of a point; spheres, cones, and cylinders were generated by the motion of circles and straight lines. Even if one considered pure geometry as an action of the mind alone based on postulates, constructions still had to be performed in the mind by an inner sense, and this was done by procedures idealized from the actual physical construction procedures. Indeed, the analysis of the role of motion and the senses in the actual physical procedure served to help the inner sense to perform the neusis abstractly in the mind as easily as it performed the mental operations corresponding to the use of a ruler and a compass. Therefore, neusis should be accepted as a postulate in pure abstract geometry. Molther formulated this argument as follows:

> But if someone would judge that geometry at its most pure should be practiced just by action of the mind and based only on its postulates, then even he would idealize, by mathematical abstraction, the ideas of a material ruler, or a compass, and he would grasp these ideas in his mind so that rulers and compasses would do their work by an interior sense in the imagination. And thus it would be easy to imagine thinkingly the process of which we have shown how it is performed in reality.[11]

Instruments

Although Molther probably considered his arguments about the acceptability of the neusis in pure geometry as his most important contribution, he also presented some practical fruits of his investigation, namely, an instrument and a procedure that he had devised for neusis construction.[12] The instrument was a combination of rulers sliding along each other with points and pins adjustable as required (cf. Figure 12.2); the procedure employed a cord along which the given distance $BC = a$ was marked and which, while being kept straight, was moved along the plane such that it wrapped about a pin in the given pole A and such that one of the given points on it followed one of the given lines; the required position of the line was reached at the moment that the other point on the cord coincided with the second given line.

Solid constructions

Molther concluded the first chapter with two solid constructions. The first[13] was a trisection by neusis. It was a variant of (and probably inspired by) the neusis construction that occurs in Pappus' *Collection* (cf. Construction 3.9);

[11][Molther 1619] p. 36: "Si quis autem existimet oportere puram puram [sic] Geometriam sola mentis actione, etiam secundum Postulata sua, exerceri; is quoque Regulae Circini materialis ideas ⟨afairesei⟩ Mathematica abstrahat et mente complectatur, ut in Phantasia per sensum interiorem Regulae ac Circini opera faciant. Sic enim proclive fuerit cogitando illud fingere, quod quomodo reipsa praestetur monstravimus."

[12][Molther 1619] pp. 38–39.

[13][Molther 1619] pp. 40–45.

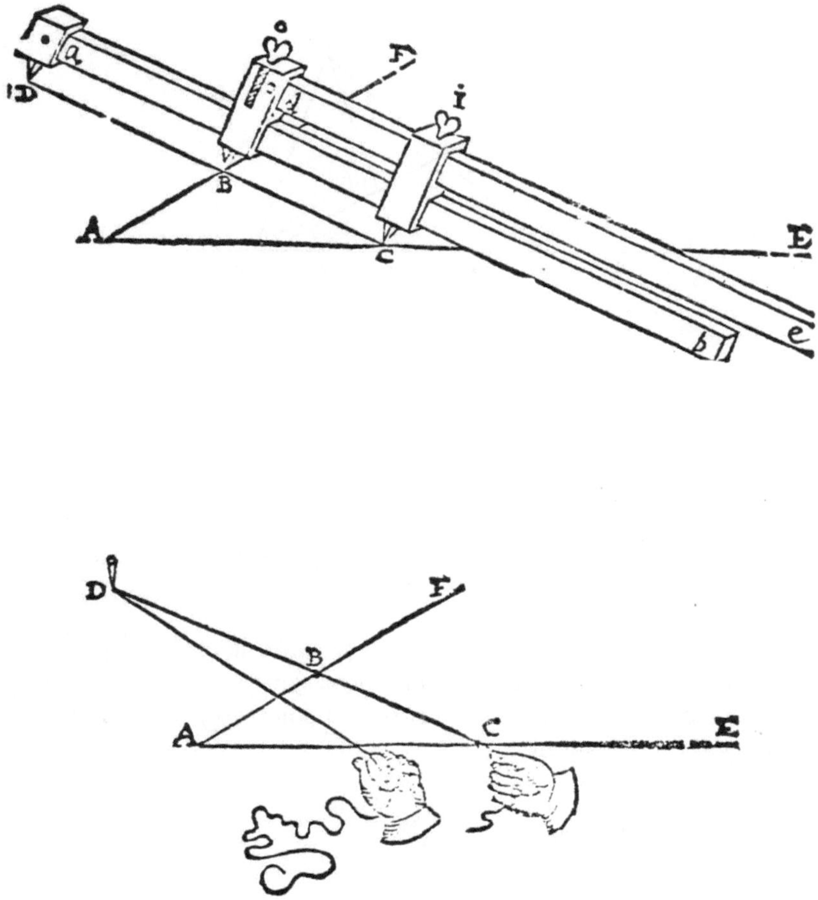

Figure 12.2: Molther's instrument and his procedure for neusis construction, *Delian problem* pp. 38–39

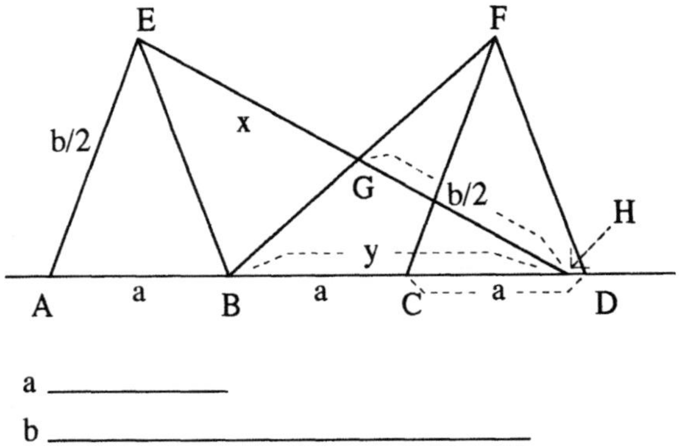

Figure 12.3: Construction of two mean proportionals — Molther

Molther modified the construction slightly to make it easier in practice. The second[14] was a construction of the regular heptagon by means of a curve (introduced for this special purpose and constructed pointwise) which Molther called the "Heptagon tracing spiral".

12.4 The "Mesolabum"

The construction The second chapter of Molther's book was devoted to the *Mesolabum* proper, that is, to the construction of two mean proportionals. The construction and the proof given by Molther were essentially the same as Viète's (cf. Construction 10.1), but Molther omitted any reference. For easier reference I repeat the construction here:

Construction 12.2 (Two mean proportionals — Molther)[15]
Given two line segments a and b $(a < b)$ (see Figure 12.3); it is required to find their two mean proportionals x and y.

Construction:
1. Mark points A, B, C, D along a straight line such that $AB = BC = CD = a$.
2. Construct two congruent isosceles triangles AEB and CFD, with

[14][Molther 1619] pp. 45–48, on the "Helix ⟨heptagoonografousa⟩."
[15][Molther 1619] pp. 51–58.

$AE = EB = CF = FD = \frac{1}{2}b$; draw BF.
3. By neusis (Molther here refers to his postulate) from E between the lines BF and AD, construct the line EGH with intercept $GH = \frac{1}{2}b$.
4. $x = EG$ and $y = BH$ are the required mean proportionals.
[**Proof**: See Construction 10.1.]

Other topics

In the remainder of the book Molther discussed various more or less related topics, such as an approximate quadrature of the circle by means of mean proportionals (equivalent to $\pi = \sqrt[3]{31}$), the duplication of the cube, the various legends around that problem, and the construction of a solid similar to a given solid and with its volume in a given ratio to the given solid, a problem reducible to two mean proportionals.[16]

12.5 Molther's interpretation of geometrical exactness

No demarcation

Molther's legitimation of the neusis construction belongs to the category "idealization of practice." It was elaborated in considerable detail. Yet it was incomplete in that it did not take up the issue of the demarcation between legitimate and illegitimate procedures in geometry. Should all motions and all testimonies of the senses be accepted, in idealized form, in pure geometry? And if not, how should the demarcation be drawn between acceptable and unacceptable procedures? Molther did not address these questions. He seems to have been preoccupied more by the solvability of one famous problem (and by the triumph to be gained by solving it) than by the issue of construction and demarcation in general. He did not discuss line-like problems nor the construction of such curves as the quadratrix or the spiral. Actually, he may not have been aware of problems, other than the quadrature of the circle, which could not be solved by straight lines, circles, and neusis.

Interest

It is unlikely that Molther exerted much influence; there is hardly any documentable interest in his book.[17] Yet he followed a line of argument that, witness Clavius' adoption of it earlier, had a certain appeal, invoking, as it did, an obvious parallel between pure and practical geometry. We may therefore assume that the arguments have occurred, in various degrees of sophistication, to a number of early modern geometers. Thus Molther's attempt at legitimizing geometrical procedures is of interest because it made explicit the tacit arguments and also the weaknesses involved in a rather natural approach with respect to the interpretation of geometrical exactness.

[16] This problem was mentioned in he title of the book. Clavius had dealt with it in his *Practical Geometry*, cf. Note 45 of Chapter 4.
[17] The only 17th century reference I have found is in [Mersenne 1636] p. 68.

We may locate the basic weakness of the approach in the fact that precision in practice is a matter of degrees, whereas the interpretation of exactness requires a strict demarcation between exact or geometrical, on the one hand, and inexact or ungeometrical, on the other hand. It is true that Molther made no attempt at formulating a strict demarcation. But he shifted the earlier, Euclidean demarcation between exact and non-exact, and thereby his reasoning begged the question of where this demarcation would ultimately be located. The interpretation of exactness by idealization of practice involves an attempt to derive a strict demarcation from a gradual qualitative scale; such an attempt necessarily implies an arbitrary and thereby unconvincing choice.

Chapter 13

Fermat

13.1 Geometrical problems and their analysis

Little interest in exactness of constructions

Viète, Fermat, and Descartes were the main protagonists in the creation and adoption of algebraic analysis as a tool for geometry. As explained in the general introduction (cf. Section 1.5), I consider this development as the principal dynamics within the early modern tradition of geometrical problem solving. Yet Fermat figures less prominently in my story than Viète, and much less so than Descartes. The reason is that, although Fermat was strongly interested in the new analysis, and applied it in various mathematical areas, he showed little concern about geometrical constructions and their exactness.[1]

When Fermat had learned about Descartes' methods of analysis, he often took a polemical position toward them; therefore, a discussion of the relevant later writings of Fermat is better left until after the explanation, in Part II, of Descartes' approach to construction and geometrical exactness (Section 29.2). For my present purpose it suffices here to discuss Fermat's contributions to the analytical techniques of geometrical problem solving dating from the period before he became acquainted with Descartes' results.

Documents

The Fermatian documents relevant to his early ideas about geometrical problem solving are:[2] **1** a geometrical fragment concerning the "three line-locus," from the period 1629–1636;[3] **2** the restitution of Apollonius' treatise on plane loci[4] from c. 1636 of which copies are known to have circulated in Paris by the end of 1637; **3** a proposition about the parabola[5] from or just before 1636; **4**

[1] Cf. [Bos 1996b].

[2] For the dates of these documents I rely on Mahoney's chronology, [Mahoney 1994] pp. 403–411.

[3] "Loci ad tres lineas demonstratio," first published in [Fermat 1891–1922] vol. 1, pp. 87–91.

[4] "Apollonii Pergaei libri duo de locis planis restituti" first published in [Fermat 1679] pp. 12–43, also in [Fermat 1891–1922] vol. 1, pp. 3–51.

[5] "Propositio D. de Fermat circa parabolen," first published in [Fermat 1679] pp. 144–145, also in [Fermat 1891–1922] vol. 1, pp. 84–87.

the *Introduction to plane and solid loci*,[6] of early 1636, which circulated among Parisian mathematicians by the end of 1637; **5** an appendix to the previous item, concerning solid problems, from 1636 or later;[7] and **6** the solution, from 1636, of a problem proposed by Étienne Pascal.[8]

Classical analysis and loci

From these writings it appears that Fermat knew the tradition of geometrical problem solving well and could apply both the classical and the modern, Vietean, methods of analysis. The problem proposed by Étienne Pascal (item **6** above) was a triangle problem: Consider a triangle ABC with sides a, b, c and angles α, β, γ and let h be the perpendicular from A to BC; let the angle γ and the ratio $h : (b - c)$ be given; it is required to find the triangle "in kind" (the data determine the triangle up to similarity). Fermat gave an analysis in classical style (cf. Section 5.1), showing that the problem was plane, gave the construction, proposed variant triangle problems for others to solve, warned that some of them were non-plane, and showed himself to be aware of Pappus' precept by adding that that those which were plane should not be solved by solid means.[9]

Fermat's reconstruction of an Apollonian treatise (item **2** above) shows that he shared another concern common within the early modern tradition of geometrical problem solving, namely, the restitution of the classical works on analysis mentioned by Pappus (cf. Section 4.7). However, the topic of the treatise, namely, loci, had not attracted much interest earlier. In this respect also Fermat's note on the three-line locus (item **1** above) marks a novel interest within geometrical problem solving.

Fermat's interest in loci is related to one issue in which he adopted an approach different from Viète's, namely, the construction of solid problems. Before Fermat and Descartes, few early modern mathematicians practiced the construction of solid problems by conic sections (cf. Section 5.5). Viète had supplemented geometry with the neusis construction as a new postulate to deal with solid problems. Fermat returned to the classical approach in this matter, adopting construction by means of conic sections as the proper way of dealing with solid problems. In classical analysis the conics by which a solid problem should be contructed were found as loci (cf. Section 5.1 and Table 5.1). Fermat's interest in loci and conics as means of construction led him to explore the potential of Vietean algebra for dealing with loci and conic sections, and this in turn led him to what later was recognized as the principle of analytic geometry: the relation between curves and equations in two unknowns.

---

[6] "Ad locos planos et solidos isagoge," [Fermat Isagoge] first published in [Fermat 1679] pp. 1–8.

[7] "Appendix ad Isagogen topicam, continens solutionem problematum solidorum per locos," [Fermat Appendix] first published in [Fermat 1679] pp. 9–11.

[8] "Solutio problematis a Domino Pascal propositi," [Fermat 1891–1922] vol. 1, pp. 70–74 (first published in 1779 in an edition of Pascal's *Oeuvres*).

[9] [Fermat 1891–1922] vol. 1, p. 74: "Sed observandum in quaestionibus de triangulis, quoties problema poterit solvi per plana, non recurrendum ad solida. Quod quum norint viri doctissimi, supervacuum fortasse subit addiddisse."

"Isagoge"

The main document in which Fermat established the relation between curves, loci, and equations is the famous *Introduction* (item **4** above). Apart from stating the relation generally,[10] he showed explicitly in the *Introduction* that equations of first or second degree in two unknowns correspond to plane loci (i.e., straight lines and circles) or solid loci (i.e., parabolas, hyperbolas, and ellipses). For detailed discussions of the *Introduction* I refer to the standard secondary sources.[11] For my present purpose I need only describe the way in which Fermat derived the constructions of the loci from the pertaining equations in two unknowns.[12] In the case of the straight line he argued in classical fashion that if the axis and the origin are given in position and the coefficients of the equation are given in magnitude, the locus is given in position. He did not explicitly refer to Euclid's *Data*, trusting his reader to recognize from the terminology of "givens" that he based his arguments on the propositions of that book. He did not specify the further synthesis, that is, the construction. A similar tacit reference, now to Apollonius' *Conics*, underlay his treatment of the non-linear cases (he distinguished several forms of quadratic equations). Again he showed that if the coordinate axes were given in position and the coefficients of the equation were given in magnitude, the defining parameters of the conics were given in position (the vertex and the axis) and in magnitude (the *latus rectum* and the *latus transversum*). Evidently he assumed the reader to know how to find in the *Conics* the construction of the conic section on the basis of these data;[13] he did not specify these constructions.

"Appendix"

The *Introduction* explained how algebraic analysis could be used to solve locus problems. In the *Appendix* (item **5** above) to the *Introduction* Fermat applied the new technique to the solution of solid problems. He assumed that the problem was already reduced to a third- or fourth-degree equation in one unknown. As the proper solution of solid problems was their construction by the intersection of conic sections, the program of the *Appendix* required that Fermat show how, given a third- or fourth-degree equation in one unknown x, one could find two second-degree equations in two unknowns x and y, such that the conic sections represented by these equations intersected in one or more points that yielded the roots of the original equation. Fermat saw that this situation would be achieved if the unknowns x and y were taken to be the coordinates of the points of intersection of the conics. Thus the question was reduced to finding a system of two quadratic equations in x and y equivalent to the original equation in x in the sense that eliminating y from the system would result in the original equation. These quadratic equations then exhibited the plane or solid loci by

[10]The statement, which is usually considered as the first formulation of the principle of analytical geometry, occurs at the beginning of the treatise ([Fermat Isagoge] p. 91): "Quoties in ultima aequalitate duae quantitates ignotae reperiuntur, fit locus loco et terminus alterius ex illis describit lineam rectam aut curvam." ("Whenever two unknowns are found to remain in the final equation we have to do with a locus and the extremity of the second of these describes a straight line or a curve.").

[11]In particular [Mahoney 1994] pp. 76–92.

[12][Fermat Isagoge] pp. 91–103.

[13]Namely *Conics* I-52–58, cf. Note 32 of Chapter 3.

which the solid problem was to be constructed.

In the *Appendix* Fermat indeed achieved the program outlined above. In Section 5.6, Analysis 5.11, I have used Fermat's analysis of the two mean proportionals problem as an example of the resulting algebraic method of analysis involving loci. The mean proportionals problem served as a specific example in the *Appendix*. Fermat also explained a general method, applicable to all third- and fourth-order equations. He did not present this method explicitly in its full generality, because it involves case distinctions depending on whether the coefficients are positive or negative. Instead Fermat gave two examples that illustrated the two techniques he used to adjust to the various cases. I present the method here explicitly using indeterminates, which may be positive or negative:

Analysis 13.1 (Any solid problem — Fermat)[14]
Given: any solid problem reduced to a third- or fourth-degree equation in x; it is required to find the equations of two plane or solid loci whose intersections determine x.

Analysis:
1. Assume the equation reduced to the form

$$x^4 = a + bx + cx^2 ,$$

by elimination of the third-degree term.
2. Add $-2dx^2 + d^2$ to each side (d will be adjusted later):

$$(x^2 - d)^2 = a + bx + (c - 2d)x^2 + d^2 .$$

3. Equate each side to e^2y^2 (e positive, to be adjusted later) and reduce the first equation; this yields:

$$\begin{aligned} x^2 - d &= ey , \\ a + bx + d^2 &= (2d - c)x^2 + e^2y^2 ; \end{aligned}$$

The first equation represents a parabola. Take d such that $2d - c > 0$ and take $e^2 = 2d - c$; then the second equation represents a circle.
4. The problem is solved by drawing the parabola and the circle.

Some 45 years earlier Viète had achieved the reduction of all solid problems to either the trisection of an angle or the determination of two mean proportionals. Because both these classical problems could be constructed by neusis, he had proposed to supplement geometry by a new postulate legitimating the neusis as a geometrical construction procedure. However, as I noted in Section 10.3, Viète proved the reducibility in principle only, he did not provide explicit constructions for fourth-degree equations. Thus Fermat's analysis of the general solid problem constituted a marked improvement on Viète's; it provided an explicit construction, and it did not require the complicated algebraic manipulations involved in reducing fourth- to third-degree equations. Fermat was well aware of the importance of his improvement. He commented:

[14] [Fermat Appendix] pp.104–105.

> Forget, therefore, Viète's *climactic parapleroses*[15] by which he reduces quartic equations to quadratics by means of cubic equations with a squared root. For, as has been shown, quartic and cubic equations can henceforth be solved with the same elegance, ease and brevity; nor, do I think, can they be solved any more elegantly.[16]

The assessment was entirely justified. Yet Fermat was not the first to achieve this improvement; at least some 10 years earlier Descartes had found a similar procedure to construct any solid problem by the intersection of a parabola and a circle. I discuss his construction in Chapter 17.

13.2 Concluding remarks

Problem solving

The few aspects of Fermat's mathematics discussed above may suffice to characterize his position and role in the early modern tradition of geometrical problem solving. He was fully conversant with the themes and techniques current in the tradition. He solved Pascal's triangle problem by classical analytical techniques, but his main interest concerned the new Vietean analysis. Here his interest in classical locus problems and his adoption (contrary to Viète) of construction by the intersection of conics, led him to an essential improvement of the solution of solid problems. More still: this combination of interests induced him to elaborate the relation between equations in two unknowns and curves, and to state, in great generality, the central principle of analytic geometry. These fundamental contributions to algebraic analysis gave a new impetus to the principal dynamics within the tradition of geometrical problem solving. However, Descartes' new techniques, created simultaneously and publicized more effectively, came to overshadow the Fermatian innovations.

Interpretation of exactness

There are few passages in Fermat's writings dealing explicitly with the interpretation of constructional exactness. In connection with Pascal's triangle problem he reminded his readers of Pappus' precept, which he stated as a matter of course. He also adopted the construction of solid problems by the intersection of curves as self-evident without further justification. Indeed, Fermat's interest lay primarily in analysis, much less in the constructions required in the synthesis of geometrical problems; he usually left it to the reader to work out the constructions from the analysis. His attitude toward the interpretation of exactness, then, is best characterized as an appeal to authority related to indifference about the issue of construction.

Fermat's attitude was more than a mere preference for analytical questions; he accepted algebraic analysis as an autonomous mathematical topic indepen-

[15] This is Viète's term for the reduction of fourth-degree equations to third-degree ones discussed in Note 16 of Chapter 10.

[16] [Fermat Appendix] p. 107: "Abeant igitur *climacticae* illae *parapleroses* Vietaeae, quibus aequationes quadratoquadraticas reducit ad quadraticas per medium cubicarum abs radice plana. Pari enim elegantia, facilitate et brevitate solvuntur, ut jam patuit, perinde quadratoquadraticae ac cubicae quaestiones, nec possunt, opinor, elegantius."

dent from geometry or arithmetic. In the early decades of the seventeenth century this was an uncommon attitude; it rather characterizes a concern belonging to the period 1650–1750, in which (cf. Section 1.3) analysis, now including infinitesimal analysis, emancipated itself as a separate mathematical discipline, separate from geometry and with its own subject matter. In this respect Fermat's attitude was remarkably advanced.

Chapter 14

Geometrical problem solving — the state of the art c. 1635

14.1 Introduction

The field

The present chapter concludes my discussion of the early modern tradition of geometrical problem solving before Descartes. Problem solving constituted the primary context of Descartes' geometrical studies to which Part II is devoted. His contributions, however, changed the theory and practice of geometrical problem solving in so fundamental a manner that they eclipsed many of the techniques, concepts, and concerns of the earlier tradition. It is therefore appropriate to conclude Part I by a sketch of the state of the art of geometrical problem solving around 1635, that is, just before Descartes published his innovations (and also before Fermat's new techniques began to circulate among cognoscenti).

The domain of problems constituting the subject matter of the tradition of geometrical problem solving has been surveyed in Chapter 4. The list of problem types given there (cf. Table 4.1) applies for the period up to and including the mid 1630s.

Geometrical problems generated considerable mathematical activity in the period between the publication of Pappus' *Collection* in 1588 and that of Descartes' *Geometry* in 1637. To illustrate the extent of this activity I have collected in Table 14.1 the titles of books published in this period that contained substantial sections devoted to geometrical problem solving. I do not claim completeness for the list nor correctness in detail — for that reason the items from the table that are not discussed elsewhere in my study are omitted from the bibliography.

1588 ⋆Pappus *Mathematicae collectiones* Pesaro **1589** ⋆Pappus *Mathematicae collectiones* Venice **1591** ⋆Viète *In artem analyticen isagoge* Tours **1592** ⋆Viète *Canonica recensio* Tours **1593** ⋆Van Roomen *Ideae mathematicae* Antwerp ⋆Viète *Supplementum geometriae* Tours ⋆Viète *Variorum responsorum l, VIII* Tours ⋆Viète *Zeteticorum libri quinque* Tours/Paris, (1593–1600) **1594** ⋆Scaliger *Mesolabium* Leiden ⋆Scaliger *Cyclometrica elementa duo* Leiden ⋆Scaliger *Appendix ad cyclometrica* Leiden Viète *Munimen adversus nova cyclometrica* Paris **1595** Christmann *De quadratura circuli* Frankfurt Viète *Pseudo-Mesolabum* Paris ⋆Viète *Ad problema . . . responsum* Paris **1596** Van Ceulen *Vanden Circkel* Delft ⋆Van Roomen *Problema Apolloniacum* Würzburg ⋆Prado & Villalpando *In Ezechielem explanationes* Rome (1596–1605) **1597** Van Roomen *In Archimedis circuli dimensionem expositio et analysis* Würzburg **1599** ⋆Benedetti *Speculationum liber* Venice Ramus *Arithmetica . . . geometria* Frankfurt ⋆Ramus *Scholae mathematicae* Frankfurt	**1600** Sems & Dou *Van het gebruyck der geometrische instrumenten* Leiden ⋆Viète *Apollonius Gallus* Paris ⋆Viète *De numerosa potestatum ad exegesi resolutione* Paris **1602** ⋆Pappus *Mathematicae Collectiones* Pesaro **1603** Dybvad *In geometriam Euclidis . . . demonstratio linealis* Arnhem Dybvad *In geometriam Euclidis . . . demonstratio numeralis* Leiden Ghetaldi *Propositiones de parabola* Rome **1604** ⋆Clavius *Geometria Practica* Rome **1605** Dürer *Institutiones geometricae* Arnhem Van Roomen *Mathesis Polemica* Frankfurt **1607** ⋆Ghetaldi *Variorum problematum collectio* Venice ⋆Ghetaldi *Apollonius redivivus* Venice Ghetaldi *Supplementum Apollonii Galli* Ve ⋆Snellius ⟨*Peri logou apotomes . . .*⟩ *resus geometria* Leiden **1608** ⋆Roth *Arithmetica philosophica* Nürnberg ⋆Snellius *Apollonius Batavus* Leiden **1609** Van Roomen *Mathematicae analyseos Triumphus* Leuven **1611** ⋆Clavius *Opera Mathematica* Mainz (1611–1612) Clavius *Refutatio cyclometricae Iosephi Scaligeri* Mainz (1611–1612) **1612** ⋆Anderson *Supplementum Apollonii rediviv* Paris Cardinael *100 Geometrische Questien* Amsterdam

Table 14.1: Books published 1588–1635 and containing substantial sections devoted to geometrical problem solving (items marked ⋆ are mentioned in the bibliography)

1613 ⋆Ghetaldi *Apollonius redivivus* Venice **1615** ⋆Anderson ⟨*Aitologia*⟩ Paris Van Ceulen *Van den Cirkel* Leiden ⋆Van Ceulen *De arithmetische en geometrische fondamenten* Leiden ⋆Van Ceulen *Fundamenta arithmetica et geometrica* Leiden Viète *De aequationum recognitione et emendatione* Paris Viète *Ad angularium sectionum analyticen theoremata* Paris **1616** ⋆Cyriacus *Problemata duo* Paris ⋆Van Lansbergen *Cyclometria nova* Middelburg **1617** ⋆Anderson *Brevis diakrisis* Paris Snellius *Eratosthenes Batavus* Leiden **1619** ⋆Anderson *Exercitationes mathematicae* Paris ⋆Van Ceulen *De circulo* Leiden ⋆Kepler *Harmonices mundi libri V* Linz ⋆Molther *Problema Deliacum* Frankfurt **1620** Ramus *Meetkonst* Amsterdam **1621** Landroni *Instrument pour construire . . . les sections coniques* Turin ⋆Snellius *Cyclometricus* Leiden **1623** Bruni *Frutti Singolari della Geometria* Vicenza **1624** Viète *Opus restitutae mathematicae analyseos seu algebra nova* Paris	**1625** ⋆Euclid *Data* Paris Schwenter *Geometria Practica* Nürnberg **1629** ⋆Girard *Invention nouvelle en algèbre* Amsterdam **1630** ⋆Ghetaldi *De resolutione et compositione* Rome Vasset *L'algèbre nouvelle de Mr Viète* Paris Vaulézard *La nouvelle algèbre de Mr Viète* Paris Viète *Introduction en l'art analytic* Paris Viète *L'algebre nouvelle* Paris Viète *Les cinq livres des zetetiques* Paris Vaulézard *Examen de la traduction . . . des cinq livres des zetetiques de M. Viete* Paris **1631** Bruni *Dell' Armonia Astronomica et Geometrica* Vicenza Harriot *Artis analyticae praxis* London Léotaud *Elementa geometriae practica* "Dolae" Oughtred *Arithmeticae in numeribus et speciebus institutio* London Vaulézard *Examen de la traduction faicte par Antoine Vasset* Paris ⋆Viète *Isagoge . . . logistice speciosa* Paris ⋆Viète *Logistice speciosa* Paris **1634** Metius *Manuale arithmetice et geometrice practice* Amsterdam Stevin *Les oeuvres mathématiques* Leiden Herigone *Cursus seu mundus mathematicus* Paris (1634–1644) **1635** Viète *Isagoge* Leiden **1636** ⋆Mersenne *Harmonie universelle* Paris Viète *Algebre* Paris

Table 14.2: Table continued

14.2 Principal dynamics: algebraic analysis

Class and degree

The main changes within the tradition of geometrical problem solving over the period c. 1588 – c. 1635 were occasioned by what I have called the principal dynamics within the field: the creation and adoption of algebraic methods of analysis. The classification of problems into "plane," "solid," and "line-like" ones was originally based on the kinds of curves necessary for their construction. By 1635 the technique of translating geometrical problems into algebraic equations was widely known, and the classification was more often related to the degrees of these equations than to the nature of the curves needed in the construction of the problems. The relation between the geometrical constructibility of a problem and the degree of the corresponding equation is not a straightforward one (it depends on the reducibility of the equation) and by 1635 it was not fully understood.

Problems leading to equations of first or second degree were known to be plane. Thus for instance Ghetaldi noted in 1607:

> Problems which can be algebraically explained can be geometrically constructed as long as their equations remain within the confines of squares.[1]

However, sometimes the analysis of a plane problem could lead to an equation of higher degree than 2. In Section 5.4 I have argued that Ghetaldi probably was confronted with this phenomenon in dealing with a neusis problem that was known to be plane and that nevertheless led to an equation of fourth degree. Descartes was the first to realize that this discrepancy related to the reducibility of the equation.

Another sign that the role of reducibility of equations was not completely understood is the conviction expressed by some mathematicians, that problems leading to cubic equations were *ipso facto* solid. Thus when Kepler asked Van Roomen by letter about the constructibility of regular polygons with 7, 9, 11, or 13 sides, Van Roomen gave the pertaining equations and remarked:

> Of these, the second is simply cubic, the others can be resolved into cubic ones. Hence, if you accept the axiom proposed by the most expert analysts, namely that cubic equations cannot be solved by geometrical means, you have the conclusion you asked for.[2]

Apart from the question of reducibility, the construction of problems leading to cubic and quartic equations had been fully explored by Viète (cf. Section 10.3). He had shown in particular that any such problem could be reduced

---

[1] [Ghetaldi 1607] p. 5: "At problemata, quae algebraice explicari possunt, dummodo quadratorum metam aequationes non excedunt, possunt quoque et geometrica ratione construi."

[2] [Bockstaele 1976] p. 290: "Ex hisce, secunda aequatio est mere cubica, reliquae in cubicas resolvuntur. Quod si itaque axioma hoc a peritissimis analistis propositum admittas, aequationes videlicet cubicas a geometris non resolvi, jam conclusum est quod quaeris."

to a neusis and that therefore they were indeed solid in the original sense, because a neusis could be realized by the intersection of conics. As I have noted in Section 10.3 (Note 20), a proof that conversely all problems which were solid in the classical sense led to equations of degree four, was beyond Viète's interest.

Higher-order problems

Pappus had classified all problems beyond the solid ones as line-like, to be solved by curves such as the cissoid, the conchoid, the spiral, and the quadratrix. In algebraic terms this class splits into two: problems whose equation is of degree > 4 and those that lead to no algebraic equation at all. Using Cartesian terminology, I call a problem "supersolid" if it is line-like in Pappus' sense and leads to an algebraic equation. Descartes was to demonstrate (cf. Chapter 26) that such problems could be constructed by the intersection of pairs of algebraic curves (at least one of which had to be of degree > 2). I know of no higher-order algebraic curves that were introduced before Descartes to solve supersolid problems. The conchoid and the cissoid, which are higher-order curves, were originally introduced to solve solid problems, namely, the neusis and the two mean proportionals problem.

It appears that the only supersolid problems extant in the tradition around 1635 concerned higher-order angular sections and mean proportionals. The higher-order mean proportionals, leading to equations $x^n = ab^{n-1}$, were algebraically uninteresting. There was more excitement in the algebraic approach to angular sections. Several mathematicians derived algebraic equations in this respect. To give one (the most influential) example: Viète derived the following equations (in modern notation):[3]

$$\begin{aligned}
c &= x \\
d &= 2 - x^2 \\
c &= 3x - x^3 \\
d &= 2 - 4x^2 + x^4 \\
c &= 5x - 5x^3 + x^5 \\
d &= 2 - 9x^2 + 6x^4 - x^6 \\
c &= 7x - 14x^3 + 7x^5 - x^7 \\
d &= 2 - 16x^2 + 20x^4 - 8x^6 + x^8 \\
c &= 9x - 30x^3 + 27x^5 - 9x^7 + x^9
\end{aligned} \qquad (14.1)$$

in which c is the arc of a given angle within a circle of radius 1, d is the arc of the complement of the given angle, and the successive x's are the arcs of the given angle, its half, its third, its fourth, etc., parts. Viète explained the recursive relations by which the coefficients of the further equations could be found.

The relevance of these equations to geometrical construction, and hence to problem solving, was slight. Mathematicians knew that some of the higher-order angular sections were plane or solid, despite the high degree of the corresponding

[3] [Viète 1615b] pp. 294–297, cf. [Viète 1983] pp. 432–434; many variants of these equations, valid for other choices of the unknown, are mentioned in Viète's texts.

equations; but they did not study the reducibility of these equations. Nor did they try to find geometrical constructions in the supersolid cases. The algebraic study of the angular sections seems to have been inspired rather by the use of the equations for calculating rational approximate values and by the intriguing recursive relations among the coefficients of the equations. Thus the higher-order angular sections did not contribute to any innovation in the matter of exact geometrical construction beyond solid means.

The main example of a line-like but not supersolid problem was, of course, the quadrature of the circle. Others were the general angular section and the general section of a ratio (cf. Sections 4.3 and 4.4). It seems that Descartes was the first to realize that these problems were characterized by the fact that they could not be reduced to algebraic equations.

14.3 Construction

Construction of problems

Although by 1635 Pappus' precept was well known, no mathematician felt bound to the procedures of construction by the intersection of curves, which underlay this precept. The classical Greek texts known by 1635 provided many other construction procedures for non-plane problems, and new ones had been developed by early modern authors. Indeed, by the time Descartes developed his ideas about geometry, there was a confusing variety of methods of construction for non-plane problems.

In Section 4.1 (cf. in particular Table 4.2) I have classified these methods in four groups: approximate procedures, intersection of conics, intersection of other curves, and reduction to standard problems. The interest in approximate geometrical procedures (as opposed to approximate numerical ones) appears to have diminished during the first decades of the seventeenth century. Eutocius' list of constructions for two mean proportionals (cf. Section 2.4) described some special instruments devised for performing such approximate constructions, one for the construction of two mean proportionals attributed to Plato and one, attributed to Eratosthenes, for constructing any number of mean proportionals. By 1635 these instruments were well known but there was little active interest in them. Nor was there much interest in devising new such instruments; Molther (Chapter 12) was an exception, and he actually considered his procedure not as approximate but as legitimately geometrical.

The construction of solid problems by the intersection of conics was practiced very little; Van Roomen's solution of the Apollonian tangency problem (cf. Construction 5.8) was one of the few occurrences (and as it turned out an unnecessary one, because the problem was plane). When Descartes divulged his newly discovered construction of two mean proportionals by the intersection of a parabola and a circle c. 1625 (cf. Chapter 17), he was one of the first to take up this classical approach to solid problems and to reach an essentially new result.

There was more interest in the creation of special curves (rather than the conics) for solving solid or line-like problems. Villalpando's curves for mean proportionals (Definition 4.10) and Molther's spiral for constructing regular heptagons (Section 12.3) provide examples of this interest. The quadratrix and the spiral were well known by 1635 and mathematicians were aware that if one accepted these curves as means of construction, all angular sections could be performed. No new curves were devised for that purpose, nor were logarithmic curves introduced to solve in a similar line-like fashion the general problem of mean proportionals and the division of ratios.[4]

Most early modern constructions of non-plane problems consisted of a reduction to certain standard problems, often without further reference to any particular method for constructing these. As to solid problems the most common procedure was to reduce them either to two mean proportionals or to trisection. Of these it appears that the former occurred more often than the latter, but that was not, of course, because of preference, but because, apparently, problems reducible to two mean proportionals were in the majority among the solid problems that were treated. The reduction of solid problems to a neusis seems to have been little practiced, although the possibility of this reduction was known and considered significant, witness Viète's and Molther's choice of neusis as an extra postulate. In Section 4.2 I mentioned the Vietean concept of "constitutive problems" corresponding to certain standard equations. Solid problems were sometimes reduced to these constitutive problems (cf. Construction 4.22).

Construction of curves

The procedures of construction that employed the intersection of curves begged the question of how these curves themselves were to be constructed. As we have seen, several early modern geometers considered this question. By 1635 the main procedures for constructing curves were: generation by the intersection of surfaces (as the conic sections), tracing by combinations of motions (as the quadratrix), tracing by special instruments, and pointwise construction. All these methods were of classical origin.

It appears, however, that if a problem was constructed by the intersection of curves, the construction of these curves was not seen as part of the construction of the problem. Rather, the constructing curves were assumed to be given beforehand (the quadratrix, the spiral, or such new curves as Villalpando's proportionatrix), or they were assumed constructible without further explanation (as was the case with the conics).

Some mathematicians did refer to instruments for tracing the conic sections necessary in solving solid probems. The general idea about these instruments seems to have been that they were not very practicable and that in any practice, such as making sundials or plates of astronomical instruments, better precision could be reached by pointwise constructions of the curves. Mydorge, for instance, claimed that pointwise constructions were the best in practice.[5] A letter

[4]Cf., however, Section 16.5.

[5]Mydorge devoted the second book of [Mydorge 1631] to "The geometrical description of conics in the plane by points" ("De geometrica conicarum linearum in plano per puncta descriptione") pp. 83–134; in a "monitum" at the beginning of the book (p. 81) Mydorge stated that tracing conics by instruments produced unsatisfactory results.

from 1592 of Van Roomen to Clavius contains an informative reference to this practice of pointwise curve construction:

> Among other things I started to draw Gemma's astrolabe by points, but I did not finish the job because it is a general astrolabe; I started to calculate the distances of all the intersection points of meridians and parallels from the centre and the pole of the astrolabe, in order that in this way the construction of this general astrolabe would be easier. I also found a geometrical way, and indeed a very easy one, to find these intersection points.[6]

Snellius on constructing curves

Mydorge's remarks and Van Roomen's report concerned the practical side of curve tracing. In fact there was little explicit debate on the question of how in pure geometry the curves that served as means of construction should be constructed themselves. Clavius' attempt at legitimizing the pointwise construction of the quadratrix had produced some debate and had proved unconvincing. Kepler, as we have seen, altogether rejected the construction of curves other than circles or straight lines.

A somewhat more extensive discussion can be found in Snellius' work on cyclotomy of 1621. It is a good example of the level of argument on the issue at that time. In his address to the reader[7] he discussed the quadrature of the circle and used the occasion to discuss geometrical construction by the intersection of curves and Pappus' precept. In this connection he explained the generation of line-like curves: they were traced by the point of intersection of two curves that moved along a (plane or curved) surface.[8] He mentioned the classical interest in spirals, the quadratrix, and other curves for the purpose of construction. He noted that these curves were difficult to trace.

Snellius stated that motion in pure geometry was imaginary in the sense that it was conceived in the mind of the geometer. In devising such imaginary motions the pure geometer might show great cleverness, but the motions did not lead to operations by ruler and compass and hence they were not really usable in geometry. Neither could pointwise construction really be accepted as geometrical in Snellius' opinion. Mathematicians who did so compared the procedure with pointwise construction of conic sections, but that showed they didn't know the classical authors who always considered pointwise construction as a last resort, to be tried only when all legitimate plane and solid means had failed.[9]

---

[6] [Bockstaele 1976] p. 98: "Inter caetera incaepi licet non perfecerim Astrolabium Gemmae, quia catholicum est, per puncta delineare; sicque incepi calculari omnium punctorum in quibus concurrunt meridiani et paralleli, distantias, a centro et polo astrolabii, ita ut facilius sit futura hoc pacto Astrolabii illius catholici constructio. Inveni quoque rationem geometricam eamque facillimam eadem puncta concursuum inveniendi."

[7] [Snellius 1621] pp. $\star 1^v$ sqq.

[8] [Snellius 1621] p. $\star\star 2^r$: " quae . . . ex duarum linearum in una superficie sese intersecantium motu et communis sectionis vestigio delineantur."

[9] [Snellius 1621] p. $\star\star 5^v$: "At cum eam delineationem cum conicis sectionibus conferunt, audaciae etiam imprudentiam addunt, et nimium secure in veterum scriptis versati umbras

14.4 Interpretation of exactness

Various arguments

In the previous chapters I have analyzed in some detail the opinions of Clavius, Viète, Kepler, Molther, and Fermat on geometrical, and in particular constructional, exactness. In this section I mention various other contemporary opinions on the matter.

Pappus' distinction between plane, solid, and line-like problems became quickly known after 1588; from the early 1600s mathematicians referred to it as a matter of course. A characteristic example is Anderson writing on trisection:

> Pappus explains in book 4 of the *Collection* how to trisect a rectilineal angle by means of hyperbolas or the conchoid of Nicomedes, by which it is established that the proposed problem may be classified as what the ancient called solid or even line-like.[10]

Mathematicians often invoked Pappus' term "sin" ("peccatum") or equivalents in reference to solving problems by inappropriate means. Thus Anderson noted, writing about the problem of the perpendicular to the parabola:

> It was considered no light offence for someone to solve a plane problem by means of conics or line-like curves.[11]

And Mersenne referred to Pappus' precept, extending it somewhat, when he wrote in his *Universal Harmony* about the construction of mean proportionals:

> Now the ancients, as Pappus reports, were of the opinion that it was a great fault to solve by solid or line-like loci a problem which by its nature could be solved by plane loci only. Therefore, similarly, I consider it no less a fault to solve by line-like loci, or by intricate movements, or by approximative tracing, a problem which by its nature can be solved by solid loci.[12]

It was generally agreed that there was an essential difference between plane and non-plane problems. The difference was usually expressed by saying that for non-plane problems the true geometrical solution had not yet been found, and that the extant constructions were not sufficiently geometrical because they were

---

rerum non res ipsa aestimarunt. Nam illud quidem genus semper postremum est habitum, ut re desperata ad ⟨grammikas epistasis⟩ tanquam sacram anchoram confugerent. Cum enim neque per plana neque per solida quaesiti solutionem legitimam assequi possent, tum istis demum locus erat, tanquam re omnibus modis desperata."

[10] [Anderson 1619] p. 29: "Angulum vero rectilineam trisecare sive per hyperbolas, sive conchoïdem Nicomedis, docuit Pappus lib. 4. Collectionum Mathematicarum, ex quibus problema propositum ad ⟨sterea⟩ sive etiam ⟨grammika⟩ veteribus sic dicta referri posse constat."

[11] [Anderson 1619] p. 24: "Nec leviter peccatum existimasse, si quis problema planum, per conicas vel linearia absolvisset."

[12] [Mersenne 1636] p. 408 of the separately paginated *Traité des instruments a chordes*: "Or comme les Anciens, au rapport de Pappus, ont estimé que c'estoit une grande faute de resoudre par les lieux solides, ou lineaires, un Probleme, qui de sa nature pouvoit estre resolu par les seuls lieux plans: i'estime semblablement que la faute n'est pas moindre, de resoudre par des lieux lineaires, ou par des mouvements impliquez, ou par des descriptions à tastons, un Probleme, qui de sa nature peut estre resolu par les lieux solides."

mechanical. For instance, Rivault wrote, in a separate treatise on mean proportionals inserted in his 1615 edition of Archimedes' works, that all constructions of two mean proportionals found until now were manual procedures...

> But procedures performed by hand are a horror in true geometry.[13]

All these procedures should be considered mechanical

> because a large part of them is based on mechanical instruments and the others don't perform the construction in a geometrical manner, although thereafter they do provide a geometrical proof.[14]

It was often explained why certain constructions were ungeometrical. Yet positive criteria to distinguish genuinely geometrical procedures were hardly ever formulated explicitly. Snellius was an exception; at several places he mentioned ruler and compass as essential for truly geometrical construction. The quadrature of the circle, he wrote, was not yet found

> authoratively and by ruler and compass according to the rules of the art[15]

and neither, he noted elsewhere, were the regular polygons of seven or nine sides constructed

> in a geometrical way by ruler and compass.[16]

It was known from classical sources that angular sections and the quadrature of the circle could be found if the quadratrix and the spiral were accepted as given. The general opinion on these curves in the early modern period, especially after Clavius' ultimately unconvincing attempt at legitimizing the quadratrix, appears to have been that their use in constructions was ungeometrical. Moreover, they were considered as unsuitable for any practical use.[17]

An interesting dissident opinion on geometrical exactness was held by Van Lansbergen, who wrote (among other things) on cyclometry and tried to remove the image of crankiness which this field had acquired because of the many false quadratures of the circle that had been proposed.[18] Van Lansbergen also dealt with that quadrature and gave a limit process by which ever more precise bounds for π could be given. He then argued that his procedure was geometrically

[13] [Archimedes 1615] p. 92: "Abhorrent vero ⟨ta cheirourgemata⟩ à vera geometria."

[14] [Archimedes 1615] p. 92: "cum pars magna eorum mechanicis instrumentis nitatur, reliqui ⟨kataskeuen⟩ quidem non geometricè absolvunt, tamen deinceps geometricè demonstrant."

[15] [Ceulen 1615] p. 118: "apodictice et secundum artis praecepta per circinum et regulam."

[16] [Ceulen 1619] Praef. iii^r: "geometrica ratione per circinum et regula."

[17] Thus Van Roomen wrote in 1692 to Clavius about the quadratrix and its use for squaring the circle: "placet quidem, sed calculum promovere non potest, verum necessum est per inscriptionem et circumscriptionem fieri" ([Bockstaele 1976] p.93) and, on its use for constructing regular polygons: "sed cum praxin illam diligenter examinassem, eam scopo meo videlicet calculo inutilem esse inveni" (*ibid.*) p. 94.

[18] [Lansbergen 1616], the argument is in the four page dedication at the beginning of the book.

acceptable as the true solution of the quadrature.[19] His reason was that all the other approximations, though correct as approximations, left a finite margin, whereas in his procedure any degree of precision could be reached. Hence, his procedure (not the values it provided) constituted the exact geometrical solution of the problem.

14.5 Conclusion

The state of the art

In 1588 the publication of Pappus' *Collection* had provided a new impetus to the interest in geometrical problem solving. Soon afterward the writings of Viète set in motion the principal dynamics in the field: the creation and elaboration of algebraic methods of analysis. By 1635, however, the acceleration in the developments seems lost; Ghetaldi's *magnum opus* ([Ghetaldi 1630]) is characteristic for the post-Vietean, pre-Cartesian state of the art: a proliferation of special studies on special problems, several, but unstructured methods of analysis, and little interest in elaborating the new algebraic techniques beyond their application to solid problems.

Thus geometrical problem solving at that time presents the image of a field waiting for essential breakthroughs. As a very brief summary of the discussions of the previous and the present chapter (and with the hindsight of later developments) we may identify three issues as principally in need of such breakthroughs.

First, the number of available procedures of construction beyond the plane means of straight lines and circles had grown considerably, while no generally accepted order of preference or precedence among these procedures had emerged. Thus the objectives of problem solving had become opaque and the practice lacked a clear direction. Second it was realized that the use of algebraic methods of analysis involved the relations among problems, equations and constructions, but the general nature of these relations was not well understood. Third, essential progress in problem solving required a more definite and refined interpretation of the exactness of constructional procedures than had been developed within the tradition.

In 1637 Descartes' *Geometry* appeared; it provided the ideas and concepts for these breakthroughs. Part II of my study is devoted to the development of these Cartesian ideas and techniques.

[19] [Lansbergen 1616] p. 22.

Part II

Redefining Exactness: Descartes' *Geometry*

Chapter 15

Introduction to Part II

15.1 Descartes, construction, and exactness

Subject and themes

The present, second, part of my study is devoted to Descartes' ideas on geometrical construction and exactness. In the General Introduction I sketched the structure of the story of construction (Section 1.3) as consisting of two periods and one central figure, Descartes. As in Part I, my primary subject is geometrical construction, and my main objective is to understand the processes involved in the interpretation of mathematical exactness. In the early modern period Descartes was the key figure with respect to these issues. His *Geometry* of 1637 was to be the dominating influence in mathematics for more than 50 years. It was, as I will show, largely motivated by the need, as perceived by its author, for a more precise and reasoned definition of exactness in geometry.

A study of Descartes' geometrical achievements involves many themes: his mathematical ideas and results as such, their origins and their chronology, the philosophical context and motivations of his investigations, and the relation of his ideas and results to their mathematical context. These themes will be discussed in detail in the following chapters. By way of introduction, I briefly survey them here.

Note

It should be noted that the present Part II is not an overall study of Descartes' mathematics, but an inquiry into his ideas on geometrical construction and exactness. For instance, I do not deal with his study on polyhedra[1] nor with his method of tangents. Although most of the contents of the *Geometry* is covered in the next chapters, the emphasis I have placed on the different parts is determined by the special aim of my research and would probably have been different had I wanted to represent Descartes' mathematical work in its entirety.

[1] [Descartes 1987].

15.2 Mathematical ideas and results

Geometry Descartes expounded his doctrine of geometry in his *Geometry* of 1637. The work constituted a turning point in the development of the conceptions of construction and exactness in geometry. It was a strongly programmatic book, based on a distinct vision of geometry. Descartes saw geometry primarily as the art of solving geometrical problems; thus he placed his work explicitly within the early modern tradition of geometrical problem solving. The program he had set himself, and which to a large extent he fulfilled in the *Geometry*, was to provide a complete method for solving geometrical problems. The method was twofold, comprising an analytical and a synthetical part. It covered, Descartes maintained, all geometrical problems, for which he accordingly provided a full classification, extending and modifying Pappus' classification. He also gave a classification of curves. Finally, he defined and defended a strict demarcation of geometry, separating legitimately geometrical objects and procedures from non-geometrical or, as he termed them, "mechanical" ones.

Analysis Descartes adopted algebra as the principal technique for geometrical analysis. A problem was to be reduced to an equation of appropriate form; on the basis of this equation the geometrical solution, that is, the construction, was to be found. By 1637, this was a well-tested approach, especially through Viète's "specious logistics." Descartes did not take over Viète's particular choice of the letter symbols and notations in algebra, but introduced his own. He also adopted a different interpretation of the algebraic operations in geometry, introducing a unit length and thereby, in principle, avoiding the Vietean requirement that equations be homogeneous. Moreover, he realized that for achieving the proper geometrical solution of a problem, that is, its simplest possible construction, the final equation should be irreducible. Accordingly, he classified problems according to the degree of the irreducible equation (in one unknown) to which they could be reduced by his analytical method.

Construction Starting from the result of the analysis of a problem, the synthetical part of Descartes' general method provided the solution, that is, the construction of the problem. The claim to solve all problems forced Descartes to explore possibilities of construction beyond the classical "plane" and "solid" means, and to develop a canon of construction of essentially wider range than earlier ones, in particular, Viète's. Establishing such a canon required a definite choice of the means of construction, an interpretation of their hierarchy with respect to simplicity, and a clear demarcation between these means and others deemed unacceptable in geometry.

Following classical Greek usage, Descartes adopted curves as means of construction. This interpretation of the constructional procedure required a demarcation between geometrically acceptable and non-acceptable curves and a classification of the former as to simplicity. Descartes asserted that the geometrically acceptable curves were precisely those with algebraic equations; all others he called "mechanical" and he excluded them from geometry proper. He further asserted that geometrically acceptable curves were simpler in as much as the algebraic degrees of their equations were lower; accordingly, he introduced a classification of algebraic curves on the basis of their degrees. Thus Descartes' doctrine of geometrical construction prescribed that problems should be constructed by the intersection of algebraic curves of lowest possible degree. Together, these assertions constituted Descartes' interpretation of exactness in geometry. Descartes based his choices and assertions about construction on deliberate and deep considerations concerning the nature of exact knowledge in geometry, the tracing of curves by motion, and their construction by other procedures. His new approach to construction eclipsed almost all earlier ideas on the subject such as those discussed in Part I, and it became the starting point for virtually all later construction-related arguments in geometry and infinitesimal analysis.

The analytical part of Descartes' method served to reduce problems to algebraic equations, hence the synthetic part had to provide procedures for constructing the roots of these equations. Descartes implemented this technical part of the method for equations of degrees 1–2, 3–4, and 5–6. For each of these classes he decided on a certain standard form of the equation, provided methods to reduce any equation of degree ≤ 6 to the corresponding standard form, and gave standard constructions of the roots of the standard equations. He was convinced that his approach could be extended to cover equations of arbitrary degrees, but he left the extension beyond the sixth degree to his followers.

Algebra

Descartes chose algebra as a tool for analysis in geometry, but the existing algebraic techniques were insufficient for his purposes. Two requirements in his program for problem solving in particular necessitated the development of new algebraic techniques: irreducibility and the transformation of equations to standard forms. The final equation to which a problem was reduced had to be irreducible, and, for reasons related to the particular choice of constructing curves in the standard constructions of the roots, Descartes required rather special standard forms of the equations. He therefore had to elaborate a considerable number of new algebraic techniques concerning the reduction of equations and their transformation into special forms. Almost all Descartes' contributions to the algebraic theory of equations are directly related to these two issues.

Analytic geometry

In the framework of his general doctrine of solving geometrical problems Descartes developed algebraic methods for dealing with curves, in particular the technique of associating an equation in two unknowns to a curve. This technique, which made it possible to study properties of curves in terms of properties of their equations, is now generally recognized as the characteristic constituent of analytic geometry or coordinate geometry; as a result Descartes' *Geometry* is often considered and studied as the origin of analytic geometry. In retrospect such a view has some justification; it was certainly by the influence of the *Geometry* that the study of curves via their equations became a wide-spread and highly successful mathematical practice. However, the *Geometry* itself, and the ideas that shaped it, can only be understood if it is recognized that the book's primary aim was to provide a general method for geometrical problem solving and not to establish a technique for studying curves.

Episodes and periods

The Cartesian themes mentioned so far have to be considered in their chronological development. As to Descartes' own mathematical ideas the available source material suggests a distinction of a small number of episodes and periods. These are: a) The years 1618–1620 with the first very stimulating intellectual exchanges with Beeckman; b) the subsequent six or seven years with as central result the construction of solid problems by a parabola and a circle; c) the period of the final redaction of the *Rules for the direction of the mind*[2] c. 1628; d) the subsequent eight or nine years with as central event the first investigation of Pappus' problem in 1631–1632; e) the period of writing the *Geometry* in 1637; and f) the period after the *Geometry*.

The *Rules* of 1628 and the first studies on Pappus' problem of 1631–1632 have a special significance in the development of Descartes' geometrical thinking. The *Rules* document a frustration that Descartes experienced at the time of writing, namely, his inability to extend the means of geometrical problem solving beyond plane and solid problems. The studies on Pappus' problem provided the key ideas to overcome this obstacle and arrive at the doctrine of geometry formulated in the *Geometry*.

15.3 Philosophical and mathematical context

A philosopher's mathematics

Descartes' mathematics was a philosopher's mathematics. From the earliest documented phase in his intellectual career, mathematics was a source of inspiration and an example for his philosophy, and, conversely, his philosophical concerns strongly influenced his style and program in mathematics.

The question of the influence of Descartes' mathematics upon his philosophy has a long tradition in Cartesian studies. I don't address that question here; it extends by far the limitations of the present study. However, I do address the converse question of the influence of his philosophical concerns on his mathematics. Moreover, I occasionally use information that his philosophical writings provide about the development of his mathematical thought.

Method

The two main philosophical concerns that informed Descartes' mathematics were method and exactness. Descartes' only mathematical publication, the *Geometry*, was an appendix to a philosophical treatise on the method "of rightly conducting one's reason and seeking the truth in the sciences," the *Discourse on method*.[3] Thereby Descartes presented his treatise explicitly as an elaboration of such a method for a special field, namely, geometry. We may therefore expect

[2] [Descartes Rules]

[3] [Descartes 1637b]; cf. [Descartes 1985–1991] vol. 1, p. 111.

that his methodical choices in mathematics were guided to a certain extent by philosophical considerations.

Exactness

For Descartes the aim of methodical reasoning was to find truth and certainty. In geometrical context this quest concerned what I refer to by the term "exactness." Because the precise nature of exactness, especially exactness of constructions, was an unresolved issue at the time, Descartes applied all his philosophical depth and tenacity to the interpretation of geometrical exactness. As a result his ideas on the matter were of essentially higher quality and wider scope than those of his predecessors. Descartes indeed provided the most systematically reasoned interpretation of constructional exactness in geometry in the early modern period. His interpretation involved a few crucial choices such as using curves as means of construction and demarcating between geometrical and non–geometrical curves. Here, as in the case of method, it will be of interest to assess how his philosophical ideas guided these choices.

Geometrical problem solving

In Chapter 1 (Sections 1.3 and 1.5) I have identified the contexts of the developments regarding geometrical construction, and the principal dynamics within these contexts. In the period c. 1590 – c. 1650 the context was the early modern tradition of geometrical problem solving and the principal dynamics was the creation and adoption of (finite) algebraic analysis as a tool for geometry. In the period c. 1635 – c. 1750 the context was the investigation of curves by finite and infinitesimal analysis, and the principal dynamics was the emancipation of analysis from its geometrical context. The publication of Descartes' *Geometry* occurred in the overlap of the two periods, and the book was one of the main motors of the dynamics of the second period. Yet, as mentioned above, I find that in its aim, structure, and substance Descartes' book, as well as his earlier studies, belonged essentially to the context of the first period, and only by its later influence to the second. Consequently, I study Descartes' achievements concerning geometrical construction primarily in the context of the early modern tradition of geometrical problem solving, dealing with his investigations of curves mainly in as far as they were related to problem solving.

Chapter 16

Construction and the interpretation of exactness in Descartes' studies of c. 1619

16.1 The general art to solve all problems

The letter to Beeckman

In 1618 Descartes met Isaac Beeckman in Breda. From their association precious documents have survived, in particular five letters[1] that Descartes wrote to Beeckman in early 1619. In the second of these, dated 26 March 1619, Descartes formulated a program for his future investigations. The text is deservedly famous;[2] it shows how Descartes at an early age charted his scientific quest with remarkable clarity and determination. For my present purpose the text has additional value because Descartes formulated his program entirely in terms of a classification of mathematical problems with respect to their solution by calculation or construction. Thus the letter to Beeckman provides a natural starting point of my inquiry into the development of Descartes' ideas about construction.

The text

The letter opened with a reference to four inventions Descartes had made by means of what he called his new instruments. The inventions concerned angular sections and the solution of certain types of cubic equations; I return to these

[1] Descartes to Beeckman 24-I-1619 ([Descartes 1964–1974] vol. 10, pp. 151–153), 26-III-1619 (*ibid.* pp. 154–160), 20-IV-1619 (*ibid.* pp. 161), 23-IV-1619 (*ibid.* pp. 162–164), 29-IV-1619 (*ibid.* pp. 164–166); cf. also Beeckman to Descartes, 6-V-1619 (*ibid.* pp. 167–169). The text of the letters has survived because Beeckman copied them in the *Journal* he kept. This journal was discovered in 1905 by C. de Waard (cf. [Descartes 1964–1974] pp. 17–18), who later edited the the complete journal, cf. [Beeckman 1939–1953].

[2] Cf. [Shea 1991] p. 44, see also [Costabel 1969] and [Costabel 1983] on the early mathematical studies of Descartes.

results below in Section 16.4. Then followed the formulation of a scheme of what Beeckman later, when he copied the letter into his *Journal*,[3] called

> The much desired general art to solve all problems.[4]

I quote the section here in full:

> And to tell you quite openly what I intend to undertake, I do not want to propound a *Short art* as that of Lullius,[5] but a completely new science by which all questions in general may be solved that can be proposed about any kind of quantity, continuous as well as discrete. But each according to its own nature. In arithmetic, for instance, some questions can be solved by rational numbers, some by surd numbers only, and others can be imagined but not solved. For continuous quantity I hope to prove that, similarly, certain problems can be solved by using only straight or circular lines, that some problems require other curves for their solution, but still curves which arise from one single motion and which therefore can be traced by the new compasses, which I consider to be no less certain and geometrical than the usual compasses by which circles are traced; and, finally, that other problems can only be solved by curved lines generated by separate motions not subordinate to one another; certainly such curves are imaginary only; the well known quadratrix line is of that kind. And in my opinion it is impossible to imagine anything that cannot at least be solved by such lines; but in due time I hope to prove which questions can or cannot be solved in these several ways: so that hardly anything would remain to be found in geometry. This is truly an infinite task, not for a single person. Incredibly ambitious; but through the dark confusion of this science I have seen some kind of light, and I believe that by its help I can dispel darkness however dense.[6]

---

[3]Cf. Note 1.

[4][Descartes 1964–1974] vol. 10, p. 156, note: "Ars generalis ad omnes quaestiones solvendas quaesita."

[5]A reference to the *Ars brevis* (composed 1308), the strongly combinatorial philosophical method of Raymond Lull.

[6][Descartes 1964–1974] vol. 10, pp. 156–158: "Et certe, ut tibi nude aperiam quid moliar, non Lullij *Artem brevem*, sed scientiam penitus novam tradere cupio, quâ generaliter solvi possint quaestiones omnes, quae in quolibet genere quantitatis, tam continuae quàm discretae, possunt proponi. Sed unaquaeque iuxta suam naturam: ut enim in Arithmeticâ quaedam quaestiones numeris rationalibus absolvuntur, aliae tantùm numeris surdis, aliae denique imaginari quidem possunt, sed non solvi: ita me demonstraturum spero, in quantitate continuâ, quaedam problemata absolvi posse cum solis lineis rectis vel circularibus; alia solvi non posse, nisi cum alijs lineis curvis, sed quae ex unico motu oriuntur, ideoque per novos circinos duci possunt, quos non minus certos existimo & Geometricos, quàm communis quo ducuntur circuli; alia denique solvi non posse, nisi per lineas curvas ex diversis motibus sibi invicem non subordinatis generatas, quae certe imaginariae tantùm sunt: talis est linea quadratrix, satis vulgata. Et nihil imaginari posse existimo, quod saltem per tales lineas solvi non possit; sed spero fore ut demonstrem quales quaestiones solvi queant hoc vel illo modo & non altero: adeò ut pene nihil in Geometriâ supersit inveniendum. Infinitum quidem opus

Problems	About discrete magnitude	About continuous magnitude
First class	Problems (numerical equations) whose solutions are rational numbers	Plane problems solvable by straight lines and circles
Second class	Problems (numerical equations) whose solutions are "surd" (i.e., irrational) numbers	Non-plane problems solvable by curves that can be traced by one single motion
Third class	Problems (numerical equations) that can be imagined but that have no (real numbers as) solution	Problems solvable only by certain special curves that cannot be traced by one single motion

Table 16.1: The classification of problems in Descartes' letter to Beeckman

16.2 The classification of problems

Discrete magnitude

It will be useful to analyze this text in considerable detail. The classification of scientific questions that Descartes advanced in the letter may be represented schematically as in Table 16.1. In accordance with terminology introduced in Section 6.2 I use the term "magnitude" for Descartes' "quantity." I first discuss the questions concerning discrete magnitude; in the next section I turn to those concerning continuous magnitude. In formulating his classification of questions concerning discrete magnitude, Descartes restricted himself to arithmetic — he evidently assumed that all questions about discrete magnitude could be reduced to arithmetical problems. He distinguished three classes of these:[7] problems whose solutions were rational numbers, those whose solutions were "surd" numbers, that is, irrational roots of rational numbers, and finally problems that could be imagined but that had no solutions. Apparently the first two classes concerned algebraic equations in one unknown, with positive roots that might be rational or irrational.[8] So we may assume that the remaining third class also referred to equations, but now without positive, rational, or irrational roots. As it is very unlikely that Descartes knew that equations can have positive roots that are not expressible by radicals, two possibilities remain for the

est, nec unius. Incredibile quàm ambitiosum; sed nescio quid luminis per obscurum hujus scientiae chaos aspexi, cujus auxilio densissimas quasque tenebras discuti posse existimo." The translation is mine; the translation in [Descartes 1985–1991] vol. 3 pp. 2–3 is more free and I disagree with its rendering of some technical terms.

[7]Gäbe has pointed out ([Gäbe 1972] p. 117–118) the similarity of Descartes' distinction to the classification of problems as to their solvability, which Clavius gave in Chapter 14 of his *Algebra* ([Clavius 1608]).

[8]Descartes may also have thought of systems of equations reducible to one equation in one unknown. It seems unlikely that he had Diophantine equations in mind, because their solutions are essentially restricted to rational numbers.

third class: equations with negative roots and equations with complex roots. The first possibility has to be rejected because it is clear from his treatment of equations in the letter to Beeckman and in the *Private reflections*[9] that, like his contemporaries, Descartes disregarded equations that admit only negative roots.[10] The third class, then, referred to equations with complex roots.

Complex numbers The idea of a special kind of number, involving the square roots of negative numbers, had been suggested by some sixteenth-century algebrists in connection with the solution of quadratic and cubic equations. Cardano discussed in his *Great art* of 1545 the problem of finding two numbers with sum 10 and product 40, which led to the quadratic equation[11]

$$x(10-x)=40\,. \tag{16.1}$$

Applying standard rules he found as roots

$$5+\sqrt{-15} \text{ and } 5-\sqrt{-15}\,. \tag{16.2}$$

About $\sqrt{-15}$ Cardano wrote that one had to "imagine" it, and that it was a "sophistical" quantity.[12]

In his *Algebra* of 1572 Bombelli gave a more substantial discussion of number expressions involving square roots of negative numbers. He did so in connection with the so-called "casus irreducibilis" of cubic equations. In that case Cardano's method of solution[13] led to roots of negative quantities, although the examples Bombelli gave evidently admitted real solutions. Bombelli discussed in particular the equation[14]

$$x^3=15x+4\,, \tag{16.3}$$

which obviously had a solution $x=4$, and for which Cardano's algorithm led to an expression involving square roots of negative quantities:

$$x=\sqrt[3]{2+\sqrt{-121}}+\sqrt[3]{2-\sqrt{-121}}\,. \tag{16.4}$$

Taking over Cardano's terminology Bombelli called these numbers "sophistical" and investigated how one could calculate with them.[15] He noted that

$$\sqrt[3]{2\pm\sqrt{-121}}=2\pm\sqrt{-1}\,, \tag{16.5}$$

---

[9]See Note 17.

[10]Cf. [Descartes 1964–1974] vol. 10 pp. 155–156 (letter to Beeckman 26-III-1619); Descartes took over the then standard distinction of three forms of the quadratic equation, namely (in modern notation), $x^2=ax+b$, $x^2=ax-b$, and $x^2=b-ax$, with a and b positive; he did not consider the case $x^2+ax+b=0$, obviously because it had no positive solution. Nor did Descartes accept negative solutions later when he wrote the *Geometry*.

[11]Here and in the passage on Bombelli below I have modernized the notation.

[12][Cardano 1966] vol. 4, Cap. 37, Regula 2, p. 287: "ideo imaginaberis R_x.$\overline{\text{m}}$.15. . . .," ". . . huic quantitati, quae verè est sophistica . . .:" cf. [Cardano 1968] p. 219.

[13]See Note 91 of Chapter 4.

[14][Bombelli 1966] pp. 223-225.

[15][Bombelli 1966] p. 133: "la quale parerà a molti più tosto sofistica che reale."

so that the algorithm indeed yielded $x = 4$ as solution, although by a detour involving uninterpretable imagined quantities like the square roots of negative quantities.

The third class of questions (discrete)

If we compare these earlier algebraic results with Descartes' statements in his letter to Beeckman, it appears that Cardano's problem, leading to the quadratic equation 16.1, fits Descartes' description of the third class of problems well; one can imagine the problem — to find two numbers with sum 10 and product 40 — but not the solution. The more advanced treatment of complex numbers in Bombelli's argument about cubic equations does not fit Descartes' description, because Bombelli's cubic equation had a real solution. Most probably, then, Descartes had quadratic equations without real roots in mind and his term "imagine" might be an echo of Cardano's use of the same term.[16]

16.3 Problems about continuous magnitude

Geometry paradigm of continuous magnitude

Descartes assumed in his letter that problems about discrete magnitude were essentially arithmetical. Similarly, in stating that questions about continuous magnitude could be solved either by straight lines and circles or by other curves, he implicitly assumed that such questions were essentially geometrical. Thus continuous magnitude was primarily geometrical magnitude, and problems about such quantities in general required the determination, that is, the construction, of certain line segments. The ideas that line segments were the archetypal form of continuous magnitude, that consequently geometry was the paradigm science of continuous magnitude, and that construction was the essential method of problem solving, remained central elements in Descartes' philosophy of mathematics.

The classification

The identification of scientific problems about continuous magnitude with geometrical problems enabled Descartes to classify them. The letter to Beeckman shows that he did so by modifying Pappus' classification. For Pappus the essential demarcation lay between the plane and solid problems, on the one hand. and the line-like ones, on the other hand (cf. Section 3.2); Descartes shifted that demarcation line. In Descartes' and Pappus' classifications the first class was the same, it consisted of the "plane" problems, which were to be constructed

[16]It is not known whether Descartes studied Cardano's or Bombelli's works; he probably acquired his basic algebraic knowledge from Clavius' *Algebra* ([Clavius 1608]), which, however did not discuss roots of negative numbers. In the notes that Beeckman entered in his journal on occasion of Descartes' visit in October 1628, there occurs a short passage from which it appears that by that time Descartes had come to use the term "imaginary" for numbers: "Irrationales numeros, qui aliter explicari non possunt, explicat [sc. Descartes] per parabolam; nominat autem quasdam radices veras, quasdam implicatas, id est minores quàm nihil, quasdam imaginarias, id est omnino inexplicabiles; ac videt ex tabulâ vulgari, quot aliqua aequatio radices habere possit quarum una sit quaesita." ([Descartes 1964–1974] vol. 10 pp. 335.) In the *Geometry* (cf. Section 27.1) Descartes used the term "imagined" for the non-real roots of equations, stating that an n-th degree equation had n real or imaginary roots.

by straight lines and circles. But Descartes' second class covered more than Pappus'; it consisted of all non-plane problems that were solvable by the intersection of curves that could be traced by one single motion. Descartes referred in particular to curves traced by a certain kind of instruments that he called "new compasses" and that he described in his personal notes; I discuss these compasses and the curves traced by them in the next section.

As the letter makes clear, he allowed other curves than the conic sections, so his second class consisted of more problems than Pappus' "solid" ones. Descartes considered the problems in his second class, and their solutions, no less geometrical than the plane ones. For him the essential demarcation lay between these problems and others that he could not accept as certain and geometrical. The essential difference between the two kinds of problem concerned the generation of the curves that were used in their construction. The curves generated by the usual compass and by Descartes' "new compasses" were traced by one single motion; therefore, they were acceptable. If, however, the solution of a problem could only be achieved by a curve whose tracing involved several motions that were not mutually subordinated, then the problem belonged to the third class and the procedure was not certain and geometrical. Descartes deemed such curves "only imaginary:" the quadratrix was an example. In Section 16.5 I return to these imaginary curves and to Descartes' assertion that for any problem such a curve could be imagined.

The assertion implied that the classification was complete: any problem that did not belong to te first or second class, belonged to the third. The problems in each class were solvable (be it that only in the first two classes the solutions were properly geometrical) and so Descartes could claim that, if his program was completed, hardly anything would remain to be found in geometry.

Analogy between the two classifications

Descartes' words suggested that he saw the classification for continuous magnitude as analogous to that for discrete magnitude. The classifications did indeed correspond in the sense that they were both tripartite, that in both cases the first class contained the elementary problems, the second class the more advanced problems, and the third class the problems whose solution involved conceptual difficulties indicated by the term "imaginary."

But there was little more than that to the analogy. For the third class of arithmetical problems the conceptual difficulty had to do with roots of negative numbers, whereas for the third class of geometrical problems the difficulty concerned the motions by which the constructing curves were traced. With modern hindsight these two classes do not correspond either; the one has to do with complex numbers, the other (cf. Section 16.5) with transcendental curves. Nor did the first two classes in the two classifications of problems correspond; the solution of a plane geometrical problem, for instance, might well involve quadratic irrationalities so that from the arithmetical standpoint the problem would belong to the second class (cf. Section 6.4). It is not clear in how far Descartes realized the essential difference of the two ultimate classes, but he was certainly aware that the other classes did not precisely correspond. We may therefore conclude that he did not see the analogy of the two classifications as completely strict.

16.4 The "new compasses"

The "Private reflections"

In his letter to Beeckman Descartes referred to instruments for tracing curves. Another Cartesian document from the same period sheds more light on these "new compasses" and the related ideas about curve tracing. It is the text known as "Cogitationes Privatae" (*Private reflections*), which has come down to us by a somewhat tortuous route.[17] In a number of passages of these personal notes Descartes sketched instruments which he called "new compasses" and which were to be used for angular sections, mean proportionals, and the solution of certain types of cubic equations.[18] Thus we may assume that they were the same as the ones referred to in the letter to Beeckman. These instruments served to trace curves and the tracing motion they produced is well characterized as arising "from one single motion" (cf. the quotation in Section 16.1). Some other passages of the *Private reflections* concerned the tracing of curves by separate, not mutually subordinated motions. These passages may serve to explain what Descartes had in mind when he related the classification of problems about continuous magnitude to the nature of the motions that trace the curves necessary for their constructions.

The "new compass" for trisection

Three "new compasses" are mentioned in the *Private reflections*, one for trisecting angles and two for solving certain cubic equations. I first discuss the compass for angular sections. Descartes' figure and text illustrated the case of trisection, but he made it clear that he envisaged further obvious adaptations of the instrument to serve for dividing angles in 4, 5, 6, etc., equal parts.

Instrument 16.1 (Trisector — Descartes)[19]

> Four rulers (see Figure 16.1) OA, OB, OC, and OD, are connected in the point O, around which each can turn. Four equal rods EI, FJ, GI, HJ, with length a, can turn around the points E, F, G, H, which are on the four arms at distance a from O. The rods are

[17] [Descartes CogPriv]; the text, dating from 1619–1620, was among the papers of Descartes which Clerselier kept and of which Leibniz made copies in 1676. These papers are no longer extant. Leibniz' copies were published in 1859–1860 by Foucher de Careil who had found them among the Leibniz manuscripts in the Hannover library; he probably devised the title "Cogitationes Privatae." Foucher de Careil's edition of the text was particularly unsatisfactory because he had not recognized the cossic signs as such, whereby much of the mathematical content became meaningless. When the editors of the *Oeuvres de Descartes* prepared the reedition of these texts in the 1890s and 1900s, they found that Leibniz' copy was no longer in the Hannover library. In the absence of both the original and Leibniz' copy, the editors, with help of Gustaf Eneström, reconstructed the text on the basis of Foucher de Careil's text and inner mathematical logic; the resulting text is the now accepted version of the *Cogitationes Privatae*.

[18] See also [Serfati 1993].

[19] [Descartes CogPriv] p. 240: "Circinus ad angulum in quotlibet partes dividendum."

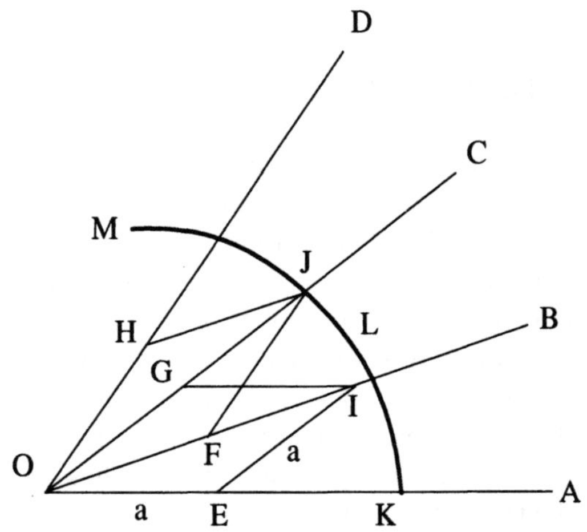

Figure 16.1: Instrument for trisection — Descartes

pairwise joined in hinges at I and J; the hinges themselves can move freely along OB and OC. It is easily seen that by this arrangement, the two arms OA and OD can form any angle within a large range and that the three inner angles AOB, BOC, and COD will always be equal; hence the instrument can serve to trisect any angle. Variants of the instrument with more intermediate arms and rods can be used for dividing angles in 4, 5, 6, etc., equal parts.

The curve $KLJM$ indicated in Figure 16.1 is the one traced by the point J when the angle DOA increases from 0 degrees (the theoretical maximum is 180°). It is a sixth-degree curve.[20] Note that, if a point J' on the curve is given, the position of the instrument (i.e., the location of the arms) for which the hinge J coincided with J' can be constructed by ruler and compass [J $(= J')$ is given, so OC is given too, OB can be found by bisecting $\angle COA$, etc.]. Thus the curve $KLJM$ in a sense incorporates all possible positions of the instrument.

Descartes did not envisage to trisect angles by directly applying his instrument. Rather he intended to use the instrument to trace the curve $KLJM$ and to use the curve for trisecting any angle. The procedure was as follows:

[20]Its equation is $4a^4x^2 = (x^2+y^2)(x^2+y^2-2a^2)$; in polar coordinates: $r = 2a\cos(\alpha/2)$.

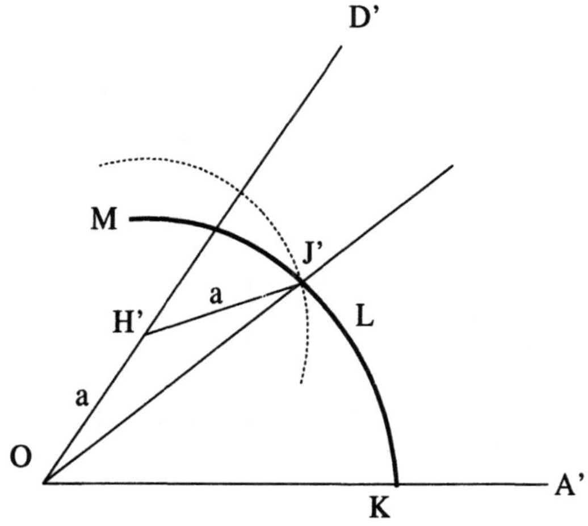

Figure 16.2: Trisection — Descartes

Construction 16.2 (Trisection — Descartes)[21]
Given: an angle $D'OA'$ (see Figure 16.2); it is required to trisect it.

> **Construction**:
> **1.** Apply the instrument with the arm OA along the axis OA'; move the arm OD from position OA outward; a tracing pin fixed at J traces the curve KLM.
> **2.** Mark off $OH' = a$ on OD'; draw a circle around H' with radius a; it intersects the curve KLM in J'.
> **3.** Draw OJ'; then $\angle D'OJ' = \frac{1}{3}\angle D'OA'$, so the angle is trisected.
> [**Proof**: Immediate by the constitution of the instrument.]

The mesolabum

The two instruments for solving certain cubic equations were adaptations of an instrument that Descartes did not directly discuss in the *Private reflections* but that, we may conclude from the text, he had already devised earlier. We know the instrument better from the *Geometry* in which he discussed it in detail. I refer to this instrument as "the mesolabum."[22] It was based on the simple

[21] [Descartes CogPriv] pp. 240–241.

[22] In the *Cogitationes Privatae* Descartes referred to a curve described by one point in the instruments for solving cubic equations as the "curve of the mesolabum compass" ([Descartes CogPriv]: p. 238–239: "linea circini-mesolabi"). It is the curve traced by the point F in Construction 16.4 below. That curve serves no function for solving the cubic equations; its use is in finding two mean proportionals by the mesolabum as explained in the *Geometry*. We may therefore conclude that in 1619 Descartes had already devised the

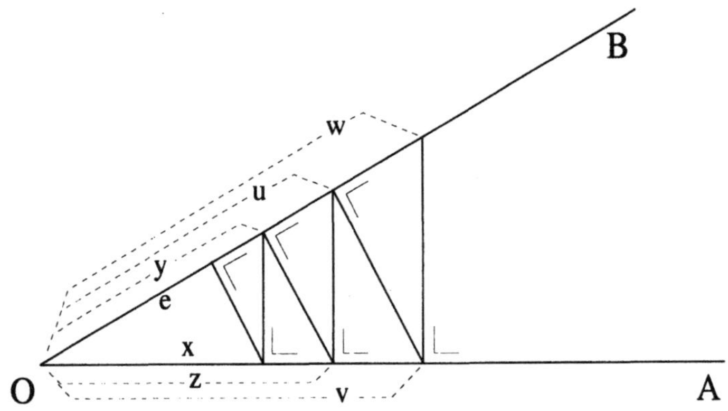

Figure 16.3: Line segments in geometrical progression

consideration that in an array as in Figure 16.3 of consecutive perpendiculars between the arms OA and OB of an angle the segments intercepted along the arms are in geometrical progression. Thus, with letters as in the figure:

$$e : x = x : y = y : z = z : u = u : v = v : w = w : \cdots . \tag{16.6}$$

Hence, if we fix e and let $\angle BOA$ increase from 0 degrees, x will take all possible values $\geq e$ and the corresponding configurations will yield all possible increasing geometric progressions starting with e. In particular, if a numerical interpretation is adopted and e is taken to be equal to 1, the geometric progression becomes 1, x, x^2, x^3, etc., and equations in x can be interpreted as relations between the line segments along the arms OA and OB. Descartes' instrument translated this idea into a mechanism. I describe the instrument on the basis of the figure from the *Geometry*.

Instrument 16.3 (Mesolabum — Descartes)[23]

> YZ and YX are rulers movable around Y (see the facsimile in Figure 16.4). At B on YX a ruler BC is fixed perpendicularly to YX. A number of rulers CD, EF, GH are adjusted along YZ; they can slide along the ruler YZ while remaining perpendicular to it; there are similar moving rulers DE and FG along YX. When YZ is fixed

mesolabum and that he devised the compasses for cubic equations as adaptations of it.

[23] [Descartes 1637] pp. 317–319, 370–371.

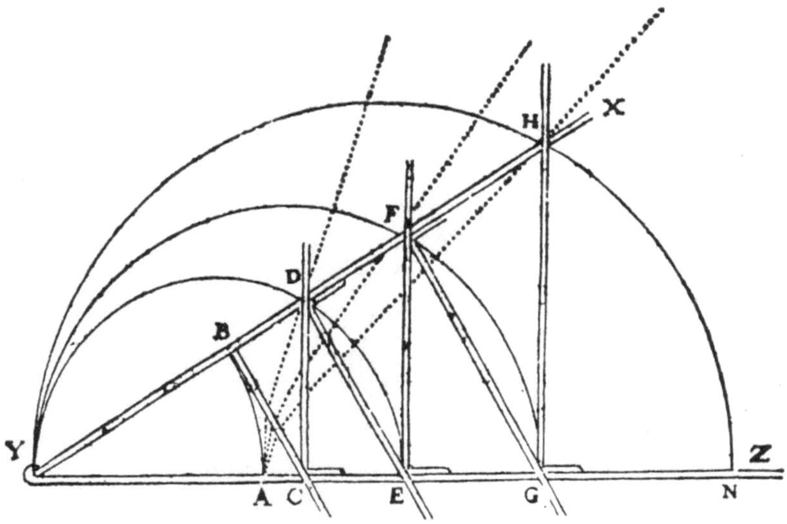

Figure 16.4: Descartes' mesolabum (*Geometry* p. 318)

and YX is turned such that the angle XYZ increases, BC is supposed to push CD along YZ; CD in turn pushes DE along YX, DE pushes EF, etc. At all instants during that motion the lines BC, CD, DE, EF, FG, GH will be connected and thus mark a series of segments YB, YC, YD, YE, YF, YG, YH, in continuous proportion along the arms YZ and YX.

The dotted curves in the figure are described by the points D, F, and H, respectively when the angle XYZ increases from 0 to c. 90 degrees. Note that, in the same way as in Instrument 16.1, each of these curves incorporates all possible positions of the instrument, in the sense that from each point on them the corresponding positions of the arms and rulers can be constructed by ruler and compass.

Similarly to the trisection procedure, Descartes used the curves rather than the instrument itself for determining mean proportionals. The procedure for finding two mean proportionals was as follows:

Construction 16.4 (Two mean proportionals — Descartes)[24]
Given: two line segments e and a, e being equal to YB in the mesolabum (see Figures 16.4 and 16.5, the letters correspond); it is required to construct the two mean proportionals x and y between e and a.

Construction:

[24] [Descartes 1637] pp. 317–319, 369–371.

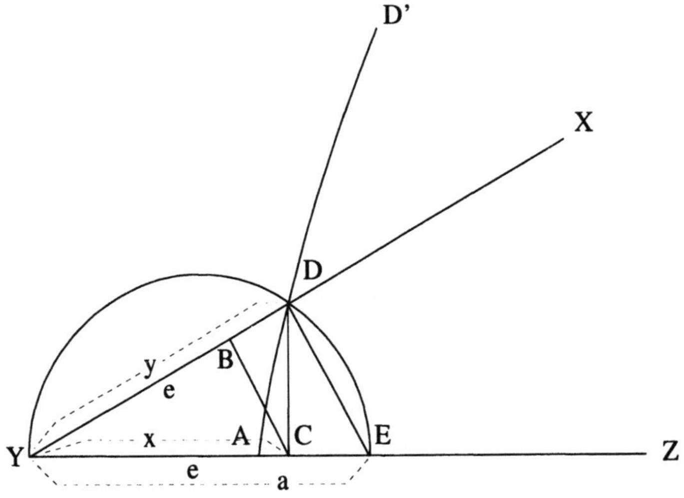

Figure 16.5: Construction of two mean proportionals — Descartes

1. Use the mesolabum to trace the curve ADD' described by the point D.
2. Mark off $a = YE$ along YZ; draw a semicircle with diameter YE; it intersects the curve ADD' in D.
3. Draw YD; mark $YB = e$ along it; draw $BC \perp YB$.
4. The segments $x = YC$ and $y = YD$ are the required two mean proportionals, i.e., $e : x = x : y = y : a$.
[**Proof**: Draw CD and DE; $DE \perp YD$ because D is on the semicircle; $DC \perp YZ$ because of the generation of the curve ADD'; the result follows from the similarity of the relevant triangles.]

Although e was fixed, this construction could also be used to find two mean proportionals between any pair of line segments f and g: Take a such that $f : g = e : a$; construct x and y as above; find u and v such that $e : f = x : u = y : v = a : g$; then u and v are the required mean proportionals between f and g.

The curves traced by points F and H in Descartes' figure of the mesolabum (Figure 16.4) could be used to construct four and six mean proportionals, respectively; the corresponding semicircles were also indicated in the figure. The further generalization of the instrument was obvious.

Universality of the two instruments

It is noteworthy that the trisector and the mesolabe have a certain universality that can hardly have escaped Descartes' notice. Viète had shown (cf. Sec-

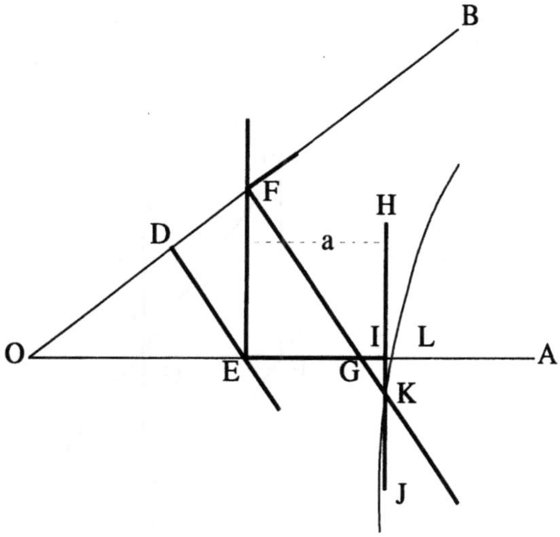

Figure 16.6: The first variant of the mesolabum

tion 10.3) that all solid problems could be reduced to either the construction of two mean proportionals or the trisection of an angle. The only problems beyond the solid ones that were discussed in the literature at the time were angular sections (beyond trisection), mean proportionals (beyond two), and the quadrature of the circle. Therefore, in their general form the two instruments, together with the ruler and the compass, took care of all known geometrical problems except the quadrature of the circle. By 1619 Descartes may well have cherished the hope or the conviction that there were no other geometrical problems, and that therefore all problems not related to the quadrature of the circle would be solvable by the curves traced by his instruments.

Variants of the mesolabum

In the *Private reflections* Descartes described two variants of the mesolabum designed for solving special cubic equations. The idea behind these instruments was the following: If e is taken to be the numerical unit, then the mesolabum marks a series of proportionals e, x, x^2, x^3, etc., along its arms (see Figure 16.3). If, for any a, the instrument can be opened so that (see Figure 16.4) $CE = a$, then we have $x^3 = YE = YC + CE = x + a$, so $x = YC$ is the solution of the equation

$$x^3 = x + a \, . \tag{16.7}$$

Descartes' first variant of the mesolabum[25] is an adaptation meant to implement this possibility. Descartes retained the fixed ruler DE (Figure 16.6) and the

[25] [Descartes CogPriv] pp. 234–237.

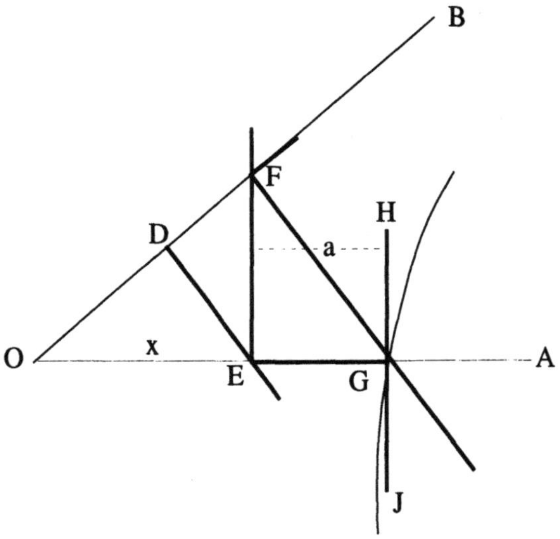

Figure 16.7: Solving $x^3 = x + a$ with the first variant of the mesolabum

second sliding ruler FG, but he replaced the first sliding ruler by a sliding system of parallel rulers FE and HIJ at distance a from each other by a beam EI, which could slide along OA (probably he meant the distance to be adjustable). While the compass was opened, the point K of intersection of FG and HJ traced a curve;[26] this curve intersected the horizontal axis in L. If now it was required to solve Equation 16.7, the angle AOB was opened whereby the points E, F, and G slid along the arms to positions E', F', and G'. The sliding process was stopped when G' coincided with L; the resulting configuration of the compass provided the required root $x = OE'$.

The second variant[27] was meant to solve the equation

$$x^3 = x^2 + a\,. \tag{16.8}$$

In this case the text is very unclear. Eneström, who provided notes to the mathematical passages in the edition of the text, suggested that Descartes made a mistake, adding a system like $FEHIJ$ in the previous case (Figure 16.6) along the arm OB and thus in fact solving $x^4 = x^2 + d$. Another interpretation, keeping to the equation Descartes mentioned but involving a more complicated instrument, seems possible to me, namely, that Descartes envisaged additional

[26]The text here has "stylo *c* mobili" ([Descartes CogPriv] p. 235 line 12), which I suggest to read as "stylo *e* mobili." The point *c*, corresponding to I in Figure 16.4, does not trace a curve as it remains on the horizontal axis; the point *e*, however, corresponding to K in my figure, does trace the curve that Descartes obviously had in mind.

[27][Descartes CogPriv] pp. 238–240.

parts of the instrument, not drawn in the figure, in order to transfer $x^3 = OG$ from the arm OA to the arm OB. This would locate the difference $x^3 - x^2$ along OB in the same way as $x^3 - x$ was located along OA in the first instrument, and thus the second instrument would serve analogously for solving the equation $x^3 = x^2 + a$.

In both cases Descartes made it clear that the instruments were used to trace curves that in their turn were used to construct the solutions. Compared with the mesolabum itself the two variants were less versatile; the curve traced by the mesolabum could serve for determining two mean proportionals between the unit e and any other magnitude, whereas in the case of the two variants each value a required its own curve to be traced. However, Descartes did not comment on this lack of generality of the variants.

The *Private reflections* also contain passages on the algebraic solution of certain classes of cubic equations. Descartes wrote the equations by means of cossic symbols in the style of Clavius' *Algebra*, adding a sign for an undetermined coefficient. In particular he used substitutions of the form $x = \alpha y$ to make one coefficient equal to a given number or two coefficients equal to each other.[28] In these explorations he made many mistakes both conceptual and calculational; in fact these passages show that by 1619 Descartes was comparatively a stranger to algebra.

16.5 The geometrical status of curve tracing

The second class of problems (continuous)

The "compasses" from the *Private reflections* illustrate what Descartes meant by curves "that originate from one single motion." Evidently the single motion was the turning of one arm of the compass while the other arm remained fixed. This motion determined the motions of the other parts of the instrument, in particular that of the pin tracing the curve. The letter to Beeckman shows that Descartes considered this tracing procedure to be as exact and geometrical as the tracing of straight lines and circles. Thus all curves traced by the mesolabum, the trisector, and their variants could be used for solving problems from Descartes' second class. These curves were meant to determine higher-order angular sections and mean proportionals. Hence, as I noted earlier, Descartes' second class of problems about continuous magnitude was larger than the classical class of solid problems.

Tracing "imaginary" curves

Descartes claimed in his letter to Beeckman that all remaining problems could be solved by special curves, which, however, could not be traced by motions such as provided by the new compasses. These curves were traced by separate motions not subordinated to each other; they were "only imaginary." Descartes mentioned the quadratrix[29] as an example. The passage in which he mentioned

[28] [Descartes CogPriv] pp. 234, 236–237, 244–245; the sign for an undetermined coefficient was O. Descartes discussed such substitutions later in the *Geometry*, cf. Section 27.2.

[29] It seems likely that Descartes knew about the quadratrix from Clavius' treatise on it (cf. Section 9.2), if not from Pappus' *Collection* itself (cf. Section 3.2), or from [Viète 1593b]

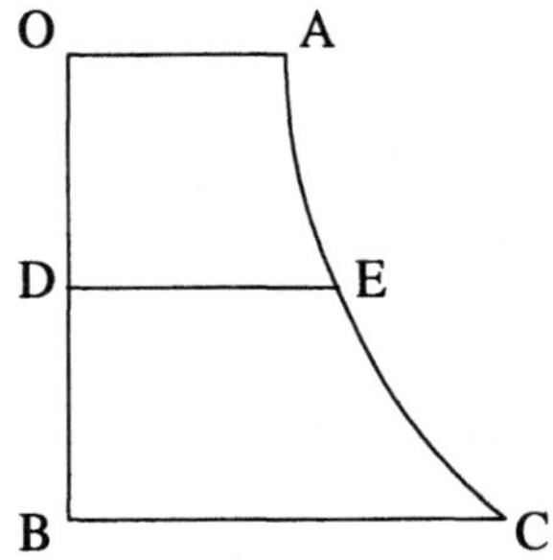

Figure 16.8: The "linea proportionum" — Descartes

the quadratrix[30] concerned another curve, which he called the *linea proportionum*. Descartes studied the latter curve in connection with free fall and with compound interest. In the figure which Descartes considered (cf. Figure 16.8), OB is a time axis, OA is a debt, which after time OD has grown to DE and after time OB to BC. The curve AEC is the linea proportionum. Evidently Descartes had the curve in mind with the property that any sequence of equally spaced ordinates (dividing the line segment OB in equal parts) is a geometric sequence increasing from OA to BC, for that is how a debt increases under compound interest.[31] Presumably Descartes realized that the curve could be used for finding mean proportionals; the equally spaced ordinates are mean proportionals between OA and BC. In modern terms the linea proportionum is

(cf. Section 10.4); in a later letter of 13-XI-1629 to Mersenne ([Descartes 1964–1974] vol. 1 pp. 69–75) he referred to Clavius with respect to the quadratrix. See also [Gäbe 1972] pp. 113–132.

[30] [Descartes CogPriv] pp. 222–223.

[31] This interpretation is based on Descartes' text: "Ad talia pertinet quaestio de reditu redituum. G.v., mutuo accepi OA; post tempus OB, debeo BC; post tempus OD, debebam tantùm DE, si AEC ducta sit linea proportionum." ([Descartes CogPriv] pp. 222–223. I have changed the letters; they now correspond to those in Figure 16.8.) The figure in the edition of the text in [Descartes 1964–1974] is not in full agreement with the interpretation as the curve is drawn with its convex side toward the axis, contrary to the graph of a debt under compound interest. However, in view of the history of the text (cf. Note 17) I think that this discrepancy may well have arisen somewhere in the editing process.

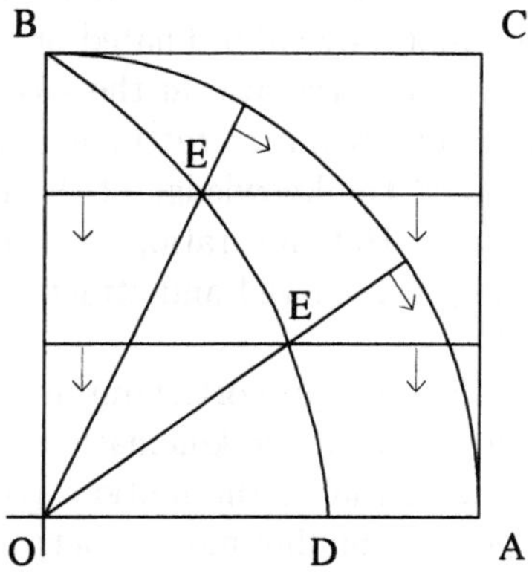

Figure 16.9: The quadratrix

the exponential curve with equation

$$y = a^x . \tag{16.9}$$

Like the quadratrix, the linea proportionum is a transcendental curve. As we will see (cf. Chapter 24), Descartes later arrived at the conclusion that algebraic curves were acceptable in geometry, and that non-algebraic curves were not. But by 1619 Descartes had not yet developed the idea of a correspondence between a curve and an equation; his arguments about curves like the quadratrix and the linea proportionum, and about their difference from other curves, concerned the manner in which they were traced.

Descartes wrote:

> The linea proportionum should be associated with the quadratrix; the latter curve indeed arises from two motions that are not subordinated to each other, one circular and one straight.[32]

Descartes referred here to the usual generation of the quadratrix by motion as given by Pappus (cf. Construction 3.3). The "circular" and the "straight" motions were those of the radius OB and the side BC, respectively (cf. Figure 16.9), whose intersection E traced the quadratrix $BEED$. These motions were uniform and such that in the time radius needed to sweep a quarter of the

[32] [Descartes CogPriv] pp. 222–223: "Linea proportionum cum quadratrice conjungenda: oritur enim [quadratrix] ex duobus motibus sibi non subordinatis, circulari et recto."

circle, the side traversed the full square. Descartes claimed that these motions could not be subordinated to each other, and that similarly the linea proportionum could not be described by two mutually subordinated motions. In the 1619 documents Descartes did not argue explicitly why the two tracing motions for the quadratrix could not be mutually subordinated or be generated by one single motion. But he discussed the curve also in the *Geometry* and there he explained that the subordination of circular to rectilinear motion necessary for tracing the quadratrix presupposed the knowledge of the ratio of the circumference and the diameter of a circle. But that ratio, he claimed, could never be known because it was a ratio between curved and straight lines.[33] I return to this argument in Section 24.2 below.

Descartes did not return to the linea proportionum in later studies and it is not clear in how much detail he analyzed the kinematical tracing of the curve. However, he may well have realized that if the motion along the axis (OB in Figure 16.8) was uniform, the corresponding motion perpendicular to the axis should have a velocity that varied proportionally to the corresponding abscissa. It is difficult to see how an instrument could effectuate this combination of motions. So Descartes may well have concluded that the linea proportionum could not be traced by instruments with the same certainty as the curves for multisecting angles and for determining mean proportionals were traced by his "new compasses."

Sections of angles and ratios

The mesolabum, its variants, and the linea proportionum concerned mean proportionals; the trisector, its variants, and the quadratrix concerned angular sections. There was a certain analogy between the two problem types of which Descartes certainly was aware. Angular section meant dividing a given angle; similarly, the determination of mean proportionals was seen as a division problem. Finding n mean proportionals between two line segments a and b meant dividing the ratio $a : b$ in $n+1$ equal parts in the sense explained in Section 4.4. The analogy between the two problems has its counterpart in the two corresponding curves that feature two analogous properties: If (cf. Figure 16.10) a segment OF along the axis corresponds, in the case of the quadratrix, to an angle θ or, in the case of the linea proportionum, to a ratio σ, then any division of OF in equal parts induces a division of θ or σ in as many equal parts (angles or ratios, respectively).

Apart from the section of an angle in n equal parts (which includes the section in parts with a rational ratio $p : q$), the quadratrix can also be used for the "general section" of an angle in two parts with a given, not necessarily rational, ratio α. This construction was explained in Pappus' *Collection* (cf. Construction 3.4). It is possible that Descartes considered the analogous general section of ratios and the possibility to construct such sections by means of the linea proportionum. However, the passage on the linea proportionum in the notes of 1619 is too brief to conclude whether or not he did so. Hence I do not include

[33] [Descartes 1637] p. 317.

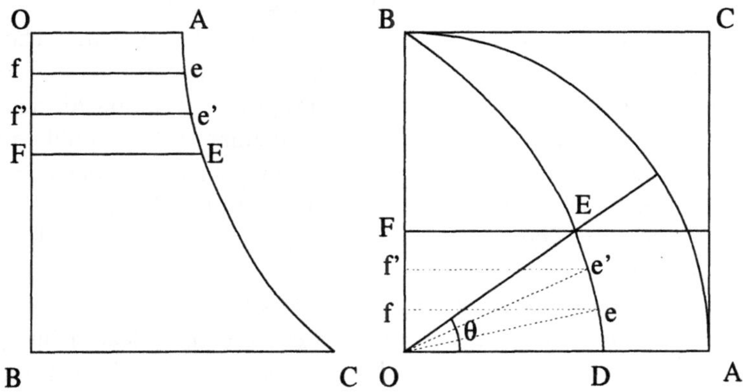

Figure 16.10: Quadratrix and linea proportionum. $Of = ff' = f'F$; $OA : fe = fe : f'e' = f'e' : FE$ (linea proportionum); $\angle AOe = \angle eOe' = \angle e'OE$ (quadratrix).

this problem in my further discussion of Descartes' third class of problems.[34]

The third class of problems (continuous)

We have seen that Descartes' mesolabum, the trisector, and their variants were intended to draw curves by means of which any number of mean proportionals between two given lines and any number of equal parts of a given angle could be constructed. As these curves were traced in an exact and geometrical way, these problems belonged to Descartes' second class. With respect to angular sections the border line between the second and the third class of problems lay between the divisions of an angle in equal parts and the general section of the angle. (In the case of sections of a ratio the borderline would also lay between the mean proportionals and the general section of a ratio but, as said above, we do not know whether Descartes considered this general section.) None of the variants of the trisector provided a curve for the general section of an angle; this problem required the quadratrix or the spiral for its geometrical construction. But Descartes considered the quadratrix insufficiently exact and precise and classified problems solved by this curve, in particular the general section of an angle, in his third class of problems. I have argued in Section 3.2 that the generation of the quadratrix (and the spiral) by combined motions was probably found as a direct translation of the angular section problem in kinematic terms: the motions install a direct relation (an isomorphism one might say) between the divisions of the side of a square and the division of the quarter arc within the square. The curve thus as it were embodies all possible instances of angular sections.

The situation resulting from Descartes' classification is a remarkable one: dividing an angle in equal parts required ever more complicated but (according to Descartes) sufficiently exact and precise curves traced by variants of the trisector. The general angular section requires the quadratrix, a curve that straightforwardly represented the relation involved in dividing an angle according to a given ratio. Yet this problem was relegated to the third class. Here it is illuminating to recall Descartes' statement, in his letter to Beeckman, that

[34] In analogy with the general section of the angle one may also consider a general section of a ratio. In the interpretation which saw compounding ratios as addition, the sum of $a : x$ and $x : b$ was $a : b$ and the determination of x was seen as a section of $a : b$ in two parts. In some cases the parts of a ratio were considered to have themselves a ratio; thus (cf. Section 4.4 Equation 4.14) if

$$a : x_1 = x_1 : x_2 = x_2 : x_3 = x_3 : x_4 = x_4 : x_5 = x_5 : x_6 = x_6 : b \, , \qquad (16.10)$$

x_3 was said to divide the ratio $a : b$ in parts of which the one, $a : x_3$ was 3/7-ths of $a : b$, and the other, $x_3 : b$, 4/7-ths, so the ratio of the ratios $a : x_3$ and $x_3 : b$ was $3 : 4$. In the 14th century Oresme had pioneered a generalization of this concept of a ratio of two ratios to cover all ratios (rather than only those occurring in sequences of continued proportionals). It is easily seen that the concept, if made explicit, would be a logarithmic one, namely $(a : b)* : *(c : d) = \frac{\log a - \log b}{\log c - \log d}$, where $* : *$ denotes the ratio of ratios. With such an extended conception of division, a general section of ratio can be envisaged, analogous to the the general section of an angle. These sections of ratios can be effectuated by means of the linea proportionum: the section of the ratio $OA : BC$ (see Figure 16.8) according to a given ratio α can be performed by constructing a point f'' on OB such that $Of'' : f''B = \alpha$ and drawing the ordinate $f''e''$ of f''; then $OA : f''e''$ and $f''e'' : BC$ are the required parts of the ratio $OA : BC$.

nothing can be imagined "that cannot at least be solved by such lines."[35] Evidently he meant that, in the same way as the general angular section, any problem could be translated in terms of a correspondence embodied by a curve. Hence any problem was solvable if the pertaining curve were given. But the case of the quadratrix also showed that, although such a curve could readily be imagined, it might be difficult to conceive a method to trace or construct it and unacceptable to assume it as given. Precisely here Descartes had to articulate the demarcation between his second and third classes of problems, because for solving a problem he required more than merely imagining a constructing curve; in order to qualify as proper means of geometrical solution, the curves should be available at least as effectively as the curves traced by the "new compasses."

There is another consideration that may illuminate Descartes' demarcation between his second and third problem classes. If he accepted the quadratrix as means of construction for angular sections, the "new compasses" would be superfluous. Moreover, the whole intricate inner structure of the bisection, trisection etc., of angles would be lost if they were reduced, via the quadratrix, to the simple bisection, trisection etc. of line segments. The compasses for angular sections, on the contrary, retained a suggestive differentiation within each problem type; the increasing complexity of the curves they traced related to the increasing difficulty of the corresponding angular sections. A similar loss of structure would result from discarding the variants of the mesolabum and using the linea proportionum for all mean proportionals. I am inclined to believe that Descartes' rejection of the quadratrix and the linea proportionum as means of construction was caused in part by a tacit, or perhaps subconscious, realization that their acceptance would, so to speak, spoil the game of geometrical problem solving by trivializing intricate questions.

16.6 Conclusion — Descartes' vision of geometry c. 1619

Geometry

The letter to Beeckman and the relevant passages from the *Private reflections* enable us to characterize the first documented phase of Descartes' thinking about geometry, construction, and the interpretation of exactness. He held that in geometry constructions should be performed by the intersection of straight lines and curves. The curves were acceptable as means of construction if they could be traced by instruments such as the ruler, the compass, or the "new compasses." These new compasses featured combinations of motions that were subordinated to each other in such a way that the resulting tracing could be characterized as one continuous motion; the subordination was the key criterion for the acceptability of the resulting curves. The quadratrix and the linea proportionum did not satisfy this criterion.

This conception of construction entailed a more precise classification of problems than the classical one formulated by Pappus. Descartes, as it were, ordered

[35] Cf. Note 6.

and subdivided Pappus' class of "line-like" problems. He realized that there was a large class of problems more complicated than the solid ones but still geometrically solvable by means of curves such as those traced by the new compasses. For Descartes the essential demarcation lay between these problems and those that were only solvable by the quadratrix and similar curves.

The 1619 documents do not yet show a particular interest in analysis, in methods for finding the constructions. Nor was algebra as prominent as it would become later in Descartes' thinking. He devised instruments for the geometrical solution of equations, but there are no signs that he considered equations as prototype for all geometrical problems. It is difficult to say in how far Descartes at the time was aware of Viète's analysis, but if so, he had not realized the force of its central idea, namely, to reduce problems to algebraic equations.

Philosophical aspects

The influence of Descartes' philosophical ideas on his mathematics primarily concerned the programmatic and methodological aspects. We will see in the following chapters how these interests were later combined with the concerns for certainty and exactness.

The letter to Beeckman shows that Descartes saw mathematical problems as a paradigm for all scientific problems. Those about discrete magnitude were analogous to the problems in arithmetic and those concerning continuous magnitude to the problems in geometry. His statements mainly concerned the latter category and thus his program for a "completely new science" was modeled to a large extent upon the early modern tradition of geometrical problem solving. In particular, he argued about the aims and the methodological questions of the new science from analogy with geometry, whose primary aim was problem solving by construction.

The analogy to mathematical, primarily geometrical, problems, with their full classification, entailed a particular feature of Descartes' vision of the new science, namely its completeness. It was a science "by which all questions in general may be solved." Descartes' programmatic vision was not open-ended; the new science, although "an infinite task, not for a single person," was in an important sense finite and could in principle be completed.

Originality

Although it is difficult to determine precisely the influences Descartes underwent in his early mathematical formation, we may assess globally the novelty of his ideas at the time. The notion of solving equations by instruments was not particularly new and Descartes' algebraic technique was decidedly below the standard of the time. The instruments for trisection and mean proportionals do not strike one as particularly novel within the tradition of geometrical problem solving. The former translated an obvious idea into a mechanism and for the latter Descartes might have been inspired by a passage in Clavius' Euclid edition in which a similar arrangement of a variable angle and adjustable perpendiculars is suggested for an approximate construction of mean proportionals.[36] More

[36] [Euclid 1589] pp. 778–779; I have discussed the procedure above, Construction 4.8.

novel was his insistence that the problems were not to be constructed directly by the instruments but indirectly by the curves that the instruments traced.

The methodological ideas Descartes formulated in his letter to Beeckman were decidedly new. His criteria and his classification of geometrical problems and procedures went beyond the classical ideas and made these more precise. His arguments about the difference between curves like the quadratrix, on the one hand, and the curves traced by the new compasses, on the other hand, were also new; he was to work them out more fully later.

Interpretation of exactness

By 1619 Descartes adopted the attitude to the interpretation of geometrical exactness, which in Section 1.6 I characterized as appeal to authority and tradition; he took over the classical construction by curves. He did, however, formulate a conviction that he would keep throughout the further development of his geometrical thinking, namely, that in geometry curves were to be accepted or rejected on the basis of the motions or combinations of motions by which they were traced (and not on the basis of other ways of generating curves such as the intersection of surfaces or the definition as loci).

Chapter 17

Descartes' general construction of solid problems c. 1625

17.1 The construction of roots of third- and fourth-degree equations

Sources

While in Paris in 1625–1626 Descartes communicated to some mathematicians a construction (without proof) of two mean proportionals by means of a parabola and a circle.[1] One of the recipients was Mydorge, who devised a proof of the construction and showed it to Descartes. Later, in 1632, Mersenne sent Descartes another proof, provided by Roberval.[2] Mersenne published Descartes' construction (without mentioning his name) and Roberval's proof in his *Universal harmony* in 1636.[3] Meanwhile Descartes had met Beeckman again in 1628. At that occasion Descartes showed him the construction, the proof by Mydorge, and a general construction of the roots of any third- or fourth-degree equation also by means of a parabola and a circle. Beeckman copied these in his Journal.[4] About the general construction he noted:

> Mr Descartes values this invention so much that he avows never to have found anything more outstanding, indeed that nothing more outstanding has been found by anybody.[5]

---

[1]Cf. especially [Mersenne 1933–1986] vol. 1 pp. 256–259 and [Descartes 1964–1974] vol. 10 pp. 651–659.

[2]Cf. Descartes to Mersenne, end June 1632, [Descartes 1964–1974] vol. 1, pp. 254–257.

[3][Mersenne 1636] Livre IV, pp. 407–412; the Latin edition of the book contained a less elaborated version ([Mersenne 1636b], Liber IV, Prop. II, pp. 146–147). The French text of the passage is published in the sources mentioned in Note 1 above.

[4][Beeckman 1939–1953] pp. 136–139, the passage is also published in [Descartes 1964–1974] vol. 10, pp. 342–346.

[5][Beeckman 1939–1953] p. 139, cf. [Descartes 1964–1974] vol. 10 p. 346: "Hanc inventionem

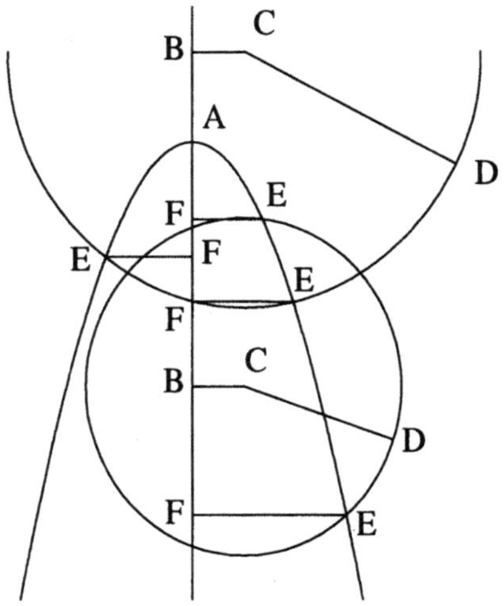

Figure 17.1: Construction of the roots of third- and fourth-degree equations — Descartes

It seems that Beeckman was the only one to whom Descartes showed the general construction, until, in 1637, he published it in the *Geometry* as the standard construction of the roots of third- and fourth-degree equations (Construction 26.1).

The construction

The construction that Descartes showed to Beeckman was as follows:

Construction 17.1 (Roots of third- and fourth-degree equations — Descartes)[6]
Given a third- or fourth-degree equation:[7] $x^4 = \pm px^2 \pm qx \pm r$; *it is required to construct its roots.*

Construction:
1. Draw a parabola with vertical axis, vertex A as highest point, and *latus rectum* 1 (see Figure 17.1).
2. If the sign of p is $+$, then take $AB = \frac{1+p}{2}$ from A down along the axis; if that sign is $-$ and $p < 1$, then take $AB = \frac{1-p}{2}$ from A

tanti facit D. des Chartes, ut fateatur se nihil unquam preastantiùs invenisse, imò a nemine unquam praestantiùs quid inventum."

[6]As transcribed by Beeckman in 1628–29, [Descartes 1964–1974] vol. 10. pp. 344–346. I have corrected a mistake in the case distinctions with respect to the sign of the coefficient p.

[7]The text in Beeckman's *Journal* has no algebraic symbols and gives the equation in words: "Primo praeparetur aequatio ita ut remaneat biquadratum aequale + vel minus certo numero quadratorum, + vel − certo numero radicum, et plus vel minus certo numero absoluto." [Descartes 1964–1974] vol. 10, p. 344.

down along the axis; if the sign is $-$ and $p > 1$, then take $AB = \frac{p-1}{2}$ upward along the axis; finally if the sign is $-$ and $p = 1$, take $B = A$.[8]
3. Take $BC = \frac{q}{2}$ perpendicular to the axis either to the right or to the left (the choice is left to the geometer executing the construction).
4. If the sign of r is $+$, then take a line segment $CD = \sqrt{CA^2 + r}$; if it is $-$, then take[9] $CD = \sqrt{CA^2 - r}$.
5. Draw a circle with center C and radius CD; it intersects the parabola in points E; draw perpendiculars EF to the axis.[10]
6. If the sign of q is $-$, then the segments EF for which EC intersects the axis represent the positive roots, the others the negative ones; if the sign of q is $+$, then the segments EF for which E is at the same side of the axis as C represent the positive roots and the others the negative ones.
[**Proof** (modern): The parabola has equation $y = x^2$. Let the center C of the circle have coordinates (a, b) and let its radius be c; its equation is $(x-a)^2 + (y-b)^2 = c^2$, and the x coordinates of the points of intersection satisfy the equation $x^4 = (2b-1)x^2 + (2a)x + (c^2 - a^2 - b^2)$. Hence if we chose a, b, and c such that $2b - 1 = \pm p$, $2a = \pm q$, and $c^2 = \pm r + a^2 + b^2$, the points of intersection of the circle and the parabola will yield the roots; Descartes' construction does precisely that.]

Descartes was aware that any third- or fourth-degree equation could be reduced to the standard form presupposed in the construction; this reduction was at that time a standard algebraic technique.[11] Thus his construction was indeed general.

Origin of the construction

If this general procedure is applied to the problem of determining two mean proportionals, the resulting construction is the same as the one[12] that Descartes showed to some mathematicians in 1625–1626. However, we do not know when Descartes found the two constructions nor whether he found the general one earlier or later than the special one. Because of the close relationship between the general and the special constructions I am inclined to believe that they were found at approximately the same time, which I indicate as "c. 1625."[13]

[8] Here I have corrected a mistake of Descartes or a copying error of Beeckman. In the text the case that the sign of p is $-$ is subdivided as to whether the "difference between the unit and the number of squares," i.e. $|1-p|$, is < 1, > 1 or $= 1$. That distinction is incorrect, as can be checked easily; reading p instead of $|1-p|$ leads to the correct distinction.

[9] Descartes noted that $CA^2 - r$ will be positive because otherwise all roots would be imaginary — apparently he assumed that at least one root was real.

[10] Here Descartes noted that there are as many intersections as there are roots, not counting an intersection in the vertex.

[11] Viète explained the technique in [Viète 1615] pp. 127–132 (tr: [Viète 1983] pp. 236–246).

[12] Cf. Note 3.

[13] Describing Descartes' meeting with Faulhaber in Ulm, Lipstorp suggested in 1653 that as early as 1620 Descartes had found the general construction (cf. [Descartes 1964–1974] vol. 10 pp. 252–253). De Waard (cf. [Mersenne 1933–1986] vol. 1 p. 258 note 4) and Milhaud ([Milhaud 1921] pp. 75–76) concluded that Descartes had actually shown the construction to

It is unlikely that Descartes explained the way he found the construction of two mean proportionals when he showed it to mathematicians in Paris. At any rate the published texts by Mydorge and Roberval[14] do not provide clues. Their proofs are synthetical, and it appears to be impossible to reconstruct a feasible analysis corresponding to the lines of these synthetic proofs.

It may be that Descartes arrived at the general construction by the method of indeterminate coefficients. The proof which I have added above shows that such a technique leads directly to the construction. Moreover, in his notes to the 1659 Latin edition of the *Geometry* Van Schooten added a derivation of the construction by indeterminate coefficients.[15] However, it may also be that Descartes found the general solution by successive generalizations of his construction of two mean proportionals.

Importance

According to Beeckman[16] Descartes saw his general construction at the time as his most significant achievement in geometry. Mersenne prefaced the publication of Descartes' construction of two mean proportionals by parabola and circle by an argument[17] that may well have reflected Descartes' own reasons for valuing the result so highly. First Mersenne quoted Pappus' statement that it was an error in geometry to construct with inappropriate means. He stated that few contemporary mathematicians constructed solid problems by conic sections. Menaechmus had provided two such constructions of two mean proportionals, one by parabola and hyperbola, the other by two parabolas. Descartes' construction used not two but only one conic, and a most simple one at that, namely, the parabola. It was therefore, Mersenne concluded, an improvement as compared with the classical ones. We may well assume that Descartes considered these arguments applicable even more to his general construction of roots of third- and fourth-degree equations, because with that construction he had found at one stroke the simplest possible construction for any solid problem.

Viète's and Fermat's constructions

When Descartes' general construction was published in 1637, Fermat had also found a general construction of third- and fourth-degree equations by means of a parabola and circle, but he had not published it. In dealing with Fermat's result (cf. Section 13.1, Analysis 13.1) I have compared the construction of solid problems by a parabola and circle to the earlier Vietean solution of solid problems and listed the advantages of the former method. Viète had shown that

---

Faulhaber. Later commentators have treated Lipstorp's statement with caution. Shea does not want to date the invention before 1620 but considers it possible that Descartes referred to the construction when he wrote in March 1620 "I have begun to understand the foundations of a marvellous invention" ([Descartes 1964–1974] vol. 10 p. 179; [Shea 1991] p. 57). Costabel has suggested that the construction of two mean proportionals could have been found as a modification of Menaechmus' construction; he gave 1628 as probable date for the general construction, cf. [Descartes 1977] pp. 309–313.

[14] Cf. Note 2.

[15] [Descartes 1659–1661] vol. 1 pp. 323–324; it is a note to the passage on p. 391 of [Descartes 1637].

[16] Cf. Note 5.

[17] [Descartes 1964–1974] vol. 10 pp. 653–655, cf. Note 3.

all geometrical problems that were reducible to equations of third or fourth degree could be solved by neusis. Thus Viète's result covered the same class of problems as Fermat's and Descartes'. However (cf. Section 10.3), Viète gave explicit constructions only for third-degree equations; for fourth-degree ones he asserted the constructibility on the basis of algebraic manipulations (in particular the methods to reduce the algebraic solution of fourth-degree equations to third-degree ones); he did not translate the transformations into constructional procedures, probably because he realized that these would be unwieldy. Thus his result was for a large part an abstract proof of constructibility, not an actual construction. Fermat realized that his own approach avoided all the algebraic complications of Viète's, but he gave only an analysis and left it to his readers to work out the construction. By 1625 Descartes had already found such an explicit method of construction. When it was published in 1637, it was clearly recognizable as new and as more powerful and simpler than the methods available in print.

17.2 Descartes' geometrical ideas c. 1625

Algebra and instruments

A comparison of the construction of the roots of third- and fourth-degree equations with the results from c. 1619 discussed in the previous chapter reveals some important changes in Descartes' ideas about geometry. The most evident of these was that by 1625 algebra had become more important and instruments less. Although we do not know exactly how Descartes found his general construction, it is clear from the result that he had become much more familiar with algebraic manipulations than when, in the *Private reflections*, he speculated about instruments to solve cubic equations.

No doubt the constructions also convinced Descartes of the importance of reducing problems to equations. Thus a central element of Descartes' doctrine of geometry, not yet explicitly present in the 1619 texts, had now entered: problems should be reduced to equations; the equations should then provide the constructions. This tenet was not new in 1625, Viète had advocated it since c. 1590. By adopting it Descartes linked up with what I have called the principal dynamics within the early modern tradition of geometrical problem solving: the creation and adoption of algebraic analysis as a tool for geometry.

Generality

In 1619 Descartes hoped to achieve generality via instruments: the trisector, which could be modified to serve for other angular sections, and the mesolabum, which could be extended so as to provide any number of mean proportionals. Although the new construction by parabola and circle did not extend beyond the solid problems, it did provide the solution of all solid problems, and its simplicity and effectiveness may well have suggested to Descartes that the means for further generalization lay in algebra rather than in instruments. Moreover, the construction suggested an essential unification in the art of geometrical problem solving compatible with Pappus' classification of problems. Linear and quadratic equations corresponded to plane problems solvable by straight lines and circles; third- and fourth-degree equations corresponded to solid problems and were now covered by one simple general construction employing the parabola as the only non-plane means. The result clearly suggested to proceed by searching for an equally elegant and convincing concord of geometry and algebra for higher-order problems and equations. In the years after 1625 Descartes indeed proceeded in this direction, but the simplicity of his general construction for solid problems proved unattainable in the case of higher-order problems.

Interpretation of exactness

The texts relating to the construction by parabola and circle contain no explicit remarks about its exactness. It seems that Descartes accepted without question the use of a parabola in constructing solid problems. Apparently the authority of Pappus' canon of construction was sufficient legitimation for this approach. Already in Descartes' earlier ideas about construction with the new instruments there was a move from the instruments themselves as means of construction to the curves traced by the instruments. In the construction with the parabola and circle, the question how the parabola was traced or otherwise produced was not discussed. Later Descartes devoted much attention to the proper methods of tracing the curves that were used in constructions.

Chapter 18

Problem solving and construction in the "Rules for the direction of the mind" (c. 1628)

18.1 The *Rules*

The "Rules" and mathematics

I now turn to the *Rules for the direction of the mind* (*Regulae ad directionem ingenii*),[1] Descartes' unfinished attempt to formulate rules of reasoning, dating, in its final form, from c. 1628. The *Rules*, written in Latin, were not published during his lifetime. The work has great relevance for the understanding of Descartes' mathematical thought because the rules he formulated were to a large extent inspired by mathematics. The question in what ways mathematics, and in particular the idea of a "universal mathematics" inspired the *Rules* has been treated extensively in the literature on Descartes[2] and I don't deal with it here. Rather I discuss a more restricted, and in a way reverse question, namely: what do the *Rules* tell us about Descartes' mathematical ideas at the time, in particular concerning geometrical construction and the interpretation of exactness.

Structure of the text

Descartes' treatise contains 21 rules. Apart from the last three, each rule is formulated in one or a few sentences and is followed by a lengthy explanation. The first 12 rules deal with the actions of the mind necessary for dealing methodically with questions and for achieving solid and true judgments; they are

[1] [Descartes Rules].

[2] See in particular the notes by J.L. Marion in [Descartes 1977] pp. 156–158, 160–164 and 302–309, and further: [Crapulli 1969], [Israel 1990], [Mittelstrass 1978], [Mittelstrass 1979], [Pasini 1992], [Schuster 1980], [Serfati 1994], and [Tamborini 1987].

not specifically mathematical. At the end of the explanation of the twelfth rule Descartes sketched the structure he had in mind for his treatise. He characterized the first 12 rules as dealing with what he called "simple propositions," whose truth could be directly intuited by a well-prepared mind. Two further sets of 12 rules were to follow, one about problems that "can be understood perfectly, even though we do not know the solutions to them," another about problems that "are not perfectly understood."[3] Descartes added that questions of the former kind occurred primarily in arithmetic and geometry. However, the *Rules* break off in the middle of the second set of 12 rules; Rules 19–21 lack the explanatory texts.

Mathesis universalis The *Rules* can be seen as an elaboration of the ideas explained in the letter to Beeckman of March 1619 (cf. Chapter 16). The "new science" which Descartes envisaged in that letter unified all scientific problems by noting that they all dealt with quantities, either discrete or continuous. Arithmetic and geometry then functioned as the prototype sciences for the two kinds of problem, respectively, and they provided a classification for each kind. In the *Rules* Descartes also reduced reasoning to problem solving, and he made a further step toward the unification of all scientific problems. He explained (in the commentary to the fourth rule) that arithmetic, geometry, and such sciences as astronomy, music, optics, and mechanics, had in common that they studied "numbers, shapes, stars, sounds, or any other object whatever," and that therefore they dealt with questions of "order and measure." Thus the foundation of these sciences lay in a "universal mathematics" (mathesis universalis) that dealt abstractly with "all the issues that can be raised concerning order and measure irrespective of the subject matter."[4]

Geometry In Descartes' vision of the "mathesis universalis" (or at least in the method of reasoning he outlined in the *Rules*) the unifying element was geometry as the theory of spatial extension. All problems could be reduced to problems about magnitude, and geometry provided the best frame for representing the so-reduced problems. If they concerned discrete quantity, they could be represented as problems about plane configurations of points; if continuous quantity was involved, lengths and areas were the appropriate means of representation. This spatial (essentially two-dimensional) representation served the mental faculty of the "imagination," which Descartes essentially saw as a screen on which the mind could model figures, operate on them, and inspect the results of such operations. Thus the fourteenth rule was:

[3] [Descartes Rules] pp. 428–429: "propositiones simplices:" "intelliguntur perfectè, etiamsi illarum solutio ignoretur:" "non perfectè intelliguntur:" translations from [Descartes 1985–1991] p. 50.

[4] [Descartes Rules] p. 378: "in numeris, vel figuris, vel astris, vel sonis, aliove quovis objecto:" "ordo et mensura:" "mathesis universalis:" "id omne . . . quod circa ordinem et mensuram nulli speciali materiae addictam," translations from [Descartes 1985–1991] vol. 1 p. 19.

> The problem should be re-expressed in terms of real extension of bodies and should be pictured in our imagination entirely by means of bare figures. Thus it will be perceived much more distinctly by our intellect.[5]

And in the explanation of the rule Descartes wrote:

> The final point to note is this: if we are to imagine something, and are to make use, not of the pure intellect, but of the intellect aided by images depicted in the imagination, then nothing can be ascribed to magnitudes in general which cannot also be ascribed to any species of magnitude.
>
> It is easy to conclude from this that it will be very useful if we transfer what we understand to hold for magnitudes in general to that species of magnitude which is most readily and distinctly depicted in our imagination. But it follows from what we said in *Rule* Twelve that this species is the real extension of a body considered in abstraction from everything else about it save its having a shape.[6]

Algebra

Thus Descartes' ideas about the constitution of the imagination and about its role in human understanding gave geometry a central place as paradigm science of order and measure. Viète and others had recognized algebra as the key to analytical methods in geometry, and Descartes himself had realized its power in his general construction of solid problems. Naturally, then, algebra also played a crucial role in the *Rules*; a large part of the treatise may be characterized as Descartes' endeavor philosophically to understand the application of algebraic methods in solving problems about magnitudes in general.

18.2 The arithmetical operations

From problem to equation

The second part of the *Rules* dealt with the preparation of problems for analysis by algebra. Rules 13–15 taught how to strip a problem of its superfluous aspects, to find and enumerate its simplest constituent parts, to reduce it to a problem about extension, that is, about geometrical magnitude, to use figures for representing it distinctly to the mind, and, if necessary, to draw the figures on paper as aids to the imagination.

---

[5] [Descartes Rules] p. 438: "Eadem est ad extensionem realem corporum transferenda, et tota per nudas figuras imaginationi proponenda: ita enim longè distinctius ab intellectu percipietur," translation quoted from [Descartes 1985–1991] vol. 1 p. 56).

[6] [Descartes Rules] pp. 440–441: "Ut verò aliquid etiam tunc imaginemur, nec intellectu puro utamur, sed speciebus in phantasiâ depictis adjuto: notandum est denique, nihil dici de magnitudinibus in genere, quod non etiam ad quamlibet in specie possit referri.

Ex quibus facilè concluditur, non parùm profuturum, si transferamus illa, quae de magnitudinibus in genere dici intelligemus, ad illam magnitudinis speciem, quae omnium facillimè et distinctissimè in imaginatione nostrâ pingetur: hanc verò esse extensionem realem corporis abstractam ab omni alio, quàm quod sit figurata, sequitur ex dictis ad regulam duodecimam," translation quoted from [Descartes 1985–1991] vol. 1 p. 58.

The next rules (16–21) described in general terms the technique of translating a problem into an equation. Descartes explained the successive steps of this process: use short symbols to denote the elements of a problem that have to be kept in mind (16); disregard whether the terms are known or unknown and find their interrelations (17); use the four operations addition, subtraction, multiplication, and division in noting down these interrelations as equations (18); search for equations, as many as there are unknown terms (19); apply (20) a further procedure (Descartes noted that he would explain this procedure later, but the extant version of the *Rules* does not contain such an explanation; most probably he envisaged a method for testing whether the equation was reducible); reduce the equations to a single one of lowest possible degree (rule 21) .... And here the *Rules* break off. The sequel one would expect, namely, rules for deriving the solution of the problem from the equation arrived at in Rule 21, is absent.

Interpretation of the arithmetical operations

The extant text suggests that by the time Descartes broke off writing the *Rules* he had become aware of two fundamental questions concerning the use of algebra he was exploring, namely, the interpretation of the arithmetical and algebraic operations for general magnitudes, and the derivation of the solution of a problem from its equation. In the *Rules* he answered the former question partially, but left the latter unanswered.

As to the introduction of arithmetical and algebraic operations in geometry, Descartes faced the question of the legitimacy of introducing numbers and algebra into geometry. This question was much discussed around 1600 (cf. Chapter 7); the most extreme answer had been given by Stevin, who claimed that number was continuous quantity and that there was no essential difference in subject-matter between arithmetic and geometry (Section 7.3). Descartes however, having given to geometry the role of paradigm science of continuous quantity, could not adopt such an extreme position, so he found himself faced with the question Viète had earlier attacked, namely, to reinterpret the arithmetical and algebraical operations so as to apply to geometrical magnitude.

Viète (cf. Sections 8.2 and 8.3) had adopted a dimensional interpretation of the operations; the product of two line segments was a rectangle, the product of three a rectangular block, and he allowed higher abstract dimensions for products of more than three factors. Later, in his *Geometry*, Descartes introduced a unit line segment and gave a non-dimensional interpretation in which the product of two or more line segments was again a line segment (cf. Section 21.1). In the *Rules* he made use of a unit but did not fully remove the dimensional aspects of magnitudes.

Descartes introduced[7] a unit length and a unit square (see Figure 18.1); I denote them by e and E, respectively, the side of E is e. He represented

[7] Rules 15 and 18, [Descartes Rules] pp. 453–454, 461–468, cf. [Descartes 1985–1991] pp. 65–66, 71–76. After Descartes' visit to him in 1628 (cf. Section 17.1), Beeckman wrote in his Journal a note "Algebrae Des Cartes specimen quoddam" ([Beeckman 1939–1953] vol. 3 pp. 95–97, also published in [Descartes 1964–1974] vol. 10, pp. 333–335). The first part of this note explains the representation of numbers and magnitudes by means of line segments or rectangles in the same way as in Rules 15 and 18.

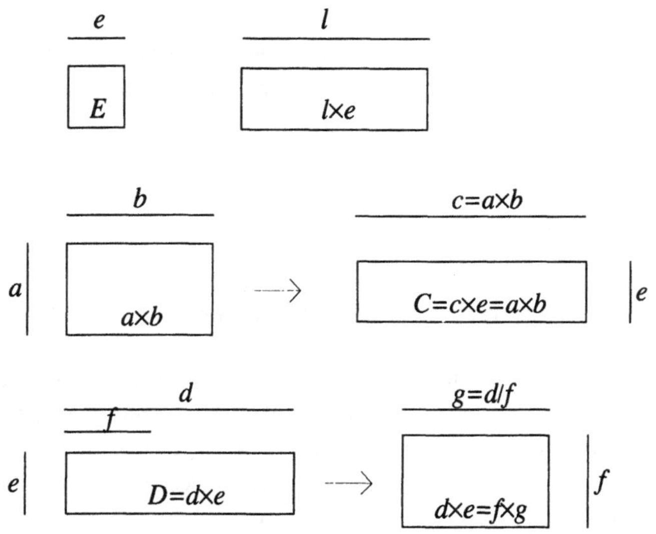

Figure 18.1: The interpretation of arithmetical operations in the *Rules*

magnitudes either as line segments or as rectangles with width e and considered these representations as equivalent. Thus the line segment l and the rectangle with length l and width e were interchangeable. Addition and subtraction were performed by joining or removing either the line segments or the rectangles. For multiplication, the factors had to be interpreted as line segments, say a and b. To find their product the rectangle with sides a and b had to be transformed into a rectangle C with width c and equal in area to rect (a, b). The length c of C, or equivalently the rectangle C itself, was the product of a and b. Similarly for dividing two magnitudes, the dividend had to be interpreted as a rectangle and the divisor as a line segment. Thus dividing d by f (with d and f line segments) one should first form the rectangle $D = d \times e$, then transform this rectangle into rectangle of equal area and width f; the other side of this rectangle, say g, was the required quotient d/f, as indeed $D = d \times e = f \times g$.

Descartes noted that these operations depended on transformations of rectangles:

> In this way, the entire business is reduced to the following problem: given a rectangle, to construct upon a given side another rectangle equal to it. The merest beginner in geometry is of course perfectly familiar with this; nevertheless I want to make the point, in case it should seem that I have omitted something.[8]

[8] [Descartes Rules] p. 468: "Ita enim totum hoc negotium ad talem propositionem reducitur: dato rectangulo, aliud aequale construere supra datum latus.

And there the commentary ends, it does not contain the promised explanation, only the enunciations of rules 19–21 follow. It is clear, however, which method Descartes had in mind: the construction of the fourth proportional by *Elements* VI-12 (cf. Construction 4.1). In the multiplication of a and b above, the product c is such that $\text{rect}\,(a,b) = \text{rect}\,(c,e)$, hence (by *Elements* VI-16) c is the fourth proportional of e, a and b. Similarly in the division of d by f, the quotient g is such that $\text{rect}\,(f,g) = \text{rect}\,(d,e)$, hence (again by *Elements* VI-16) g is the fourth proportional of d, e, and f.[9]

18.3 The algebraic operations

Root extraction At the end of the extant *Rules*, the preparations were completed for the application of the primary arithmetical operations to general magnitudes. The obvious next step in the program was the extension to the algebraic operations: root extraction and solving equations generally. Descartes considered root extraction as a kind of division, and therefore he briefly discussed it after his explanation of division in the commentary to Rule 18. He wrote:

> As for those divisions in which the divisor is not given but only indicated by some relation, as when we are required to extract the square root or the cube root etc., in these cases we must note that the term to be divided, and all the other terms, are always to be conceived as lines which form a series of continued proportionals, the first member of which is the unit, and the last the magnitude to be divided. We will explain in due course how to find any number of mean proportionals between the latter two magnitudes. For the moment we must be content to point out that we are assuming that we have not quite done with these operations, since in order to be performed they require an indirect and reverse movement of the imagination, and at present we are dealing only with problems which are to be treated in the direct manner.[10]

Quod etiamsi vel Geometrarum pueris sit tritum, placet tamen exponere, ne quid videar omisisse." Translation quoted from [Descartes 1985–1991] vol. 1 p. 76.)

[9] Anticipating the explanation in Section 21.1 I note that if we add the Euclidean construction of the fourth proportional to the explanation given in the *Rules*, and remove the double interpretation of magnitudes as both line segments and rectangles, we arrive precisely at the definitions of multiplication and division as given later in the *Geometry*.

[10] [Descartes Rules] p. 467: "In illis autem divisionibus, in quibus divisor non est datus, sed tantùm per aliquam relationem designatus, ut cùm dicitur extrahendam esse radicem quadratam vel cubicam etc., tunc notandum est, terminum dividendum et alios omnes semper concipiendos esse ut lineas in serie continuè proportionalium existentes, quarum prima est unitas, et ultima est magnitudo dividenda. Quomodo autem inter hanc et unitatem quotcumque mediae proportionales inveniendae sint, dicetur suo loco; et jam monuisse sufficiat, nos supponere tales operationes hîc nondum absolvi, cùm per motus imaginationis indirectos et reflexos faciendae sint; et nunc agimus tantùm de quaestionibus directè percurrendis." Translation quoted from [Descartes 1985–1991] vol. 1 pp. 75.)

Thus Descartes interpreted the n-th root of a line segment a as the first of $n-1$ mean proportionals between the unit e and a. If we call the root x, we have

$$e : x = x : x^2 = \ldots = x^{n-1} : a \,, \tag{18.1}$$

so $x^n = a$. His terminology shows that he considered finding x as a kind of division of a, one in which the divisor itself was not given, but a relation was indicated which the divisor had to satisfy. In the case of the square root of a, the unknown divisor was x and it had to satisfy the relation $e : x = x : a$. Descartes' text admits a still wider interpretation of "division:" he may have had in mind any given relation between the unknown divisor and its powers, that is, any equation in x. Thereby both root extraction and the solution of equations became a kind of division. A related conception can be found in Stevin's work, who considered equation solving as a generalized procedure of taking fourth proportionals.[11]

Root extraction different from arithmetical operations

Although he saw root extraction as a kind of division, Descartes also did indicate one crucial difference between the two operations: contrary to division, root extraction required "an indirect and reverse movement of the imagination," and he added that "at present we are dealing only with problems which are to be treated in the direct manner."[12] It may be that Descartes had in mind to treat root extraction and the solution of equations in the third set of 12 rules and considered them as problems that "are not perfectly understood" (cf. Note 3 above). This would be in keeping with the passage in Rule 20 where Descartes referred to certain operations whose treatment he postponed (cf. Section 18.1). These operations probably served to reduce the equations to lower degree by splitting off factors; they were therefore divisions by unknown divisors.

It should be noted that the distinction which Descartes intended to make between the second and the third dozen of rules was an arithmetical rather than a geometrical one. Addition, subtraction, multiplication, and division belonged to the second dozen rules. Root extraction and the determination of mean proportionals required "an indirect and reverse movement of the imagination" and were therefore postponed to the third. This division corresponds to rationality and irrationality in arithmetic: the four primary arithmetical operations produce only rational numbers, but root extraction generally leads to irrationals. In geometry, where the operations are constructions, the first meaningful division is between operations that can or cannot be performed by straight lines and circles. Some root extractions (square roots in particular) remain inside this boundary, others (cube roots for instance) are outside. It was around this distinction that Descartes met with an essential obstacle[13] in his program.

[11] [Stevin 1585] pp. 264–266, cf. [Stevin 1955–1966] vol. 2 pp. 581–583; the passage is entitled "La raison pourquoi nous appellons reigle de trois, ou invention de quatriesme proportionel des quantitez; ce que vulgairement se dict equation des quantitez."

[12] Cf. Note 10.

[13] Cf. Section 6.4.

18.4 Comparison with Viète's "new algebra"

Differences and similarities

At this point it is appropriate to compare Descartes' program with Viète's "new algebra," which claimed "to leave no problem unsolved" (cf. Section 8.2 and Chapter 10) and was thereby similar to Descartes' project of the *Rules*. The two scholars differed both in their motivation and in their approach to mathematics. Viète started from algebra, he did not attempt to move outside mathematics, and, despite the "nullum non problema solvere," he was not overly concerned about the completeness of his method. Descartes started from scientific problems and their complete classification; only later he acknowledged algebra as the means to classify and analyze; and later still, by the time he wrote the *Geometry*, he arrived at, in principle, a complete method for geometry.

On the other hand, with respect to the structure of their programs, there are notable similarities between Viète's and Descartes' endeavors. Both took symbolic algebra (Viète's *specious logistics*) as the general method. Both realized the necessity of reinterpreting the algebraic operations in geometrical context, and although their interpretation differed with respect to dimensionality, they both rejected a simple identification of number and continuous magnitude. Descartes' procedure for translating a problem into an equation (*Rules* 16–21) corresponded to Viète's Zetetics; his geometrical interpretation of the arithmetical operations and the abandoned interpretation of root extraction corresponded to geometrical exegetics that Viète had worked out in his treatises on plane and solid geometrical constructions. We will see (Section 20.2) that in the final form of Descartes' geometrical doctrine (as expressed in the *Geometry*) the structural analogy with Viète's tripartite analysis (Zetetics, Poristics, and Exegetics) was even more marked.

18.5 Obstacles in the program of the *Rules*

The significance of non-plane constructions

We do not know how far Descartes had elaborated the program of the *Rules* before he abandoned it, but it seems likely that he proceeded to some extent in the direction suggested by the text we have. I surmise that in doing so he was confronted with the question how the mind could perform operations like root extraction, and in general the solution of equations, with appropriate clarity and distinctness. Here arithmetic gave no guidance; numerical procedures only provided approximations. Geometry, however, did provide methods, namely, constructions, and Descartes had realized this; when he wrote that the "merest beginner" could deal with the transformation of areas,[14] he implicitly referred to a Euclidean construction. Moreover, geometrical constructions matched Descartes' conception of the imagination, because they operated upon line segments in the two-dimensional plane. But the classical means of construction, straight lines and circles, were known to be insufficient for general mean proportionals, and *a fortiori* for the general solution of equations. Thus the argument underlying the *Rules* led, naturally and cogently, to the geometrical question: how

[14]Cf. Note 8.

to construct beyond the power of straight lines and circles? For Descartes, following this line of argument, the question, already central in the early modern tradition of geometrical problem solving, now acquired an additional philosophical significance: it appeared that a general set of rules of reasoning based on geometry as paradigm science of magnitude, required a convincing canon for constructions beyond straight lines and circles.

Crucial issues in higher-order construction

We have seen in the previous chapter that Descartes had achieved an important result in solving non-plane problems: his general construction of solid problems by the intersection of a parabola and a circle. But the realization of the significance of higher-order construction would make him the more aware that his procedure for solid problems, however advanced with respect to the then available methods in geometry, left a principal methodological question unanswered. He still lacked two essential ingredients for extending his result to a general method for constructing all problems that could be reduced to an equation. In the first place there was no argument, apart from classical authority, why construction with parabola and circle could pass the stern criteria of clarity and distinctness that Descartes required. And secondly it was not at all obvious how one should proceed for higher-order equations, that is, for the problems that did not belong to Pappus' class of solid problems and that yet should legitimately belong to geometry. Which curves should be used for their construction? Not the conics because they were not potent enough to serve beyond solid problems. Nor the quadratrix, the spiral, or the like, because Descartes kept to his opinion that these curves were to be rejected from geometry.[15] What was needed was an effective demarcation between acceptable and unacceptable curves. Descartes' remarks about "an indirect and reverse movement of the imagination" involved in root extraction suggest that he would try to link the acceptability of constructions to the motions involved in their execution, that is, to the tracing of the constructing curves. But the *Rules* contain no arguments on this issue; in 1628 Descartes left the questions about the legitimacy of higher-order constructions open. Indeed, it appears that their difficulty formed the main obstacle in the program for the *Rules* and thereby the principal reason for Descartes not to pursue his project. A few years later, however, the Pappus problem provided the occasion for Descartes to take up these questions again and at that time (as we will see in the next chapter) he came further.[16]

Interpretation of exactness

Especially in their incompleteness the *Rules* show how strongly Descartes' philosophical concerns influenced the development of his mathematics. It was precisely his philosophical program, with its definite aims of clarity, method, and completeness, that made him realize the lacunae both in his mathematical techniques and in his understanding of the interrelation between algebra,

[15] Cf. Descartes' letter to Mersenne of 13-XI-1629, [Descartes 1964–1974] vol. 1 pp. 69–75, in particular pp. 70–71.

[16] A similar explanation of the obstacle which prevented Descartes from finishing the *Rules* was proposed by Schuster, cf. [Schuster 1980], pp. 78–79.

geometrical problem solving, and problem solving in general. One of these lacunae concerned the interpretation of geometrical exactness: how to construct beyond the power of straight lines, circles, and conic sections. His theory of the imagination provided the criteria that such operations had to satisfy: they were to be performed by the imagination, through direct or indirect movements in a two-dimensional space, in such a way that the mind can gain certain and indubitable cognition of them.[17]

By 1628 Descartes' view was that constructions in geometry should be performed by means of curves. The philosophical ideas of the *Rules*, in particular the "movement of the imagination" involved in higher-order algebraic operations, suggested the importance of the motions by which these curves could be generated; these motions were to be submitted to the test of mental clarity that would demarcate between acceptable and non-acceptable curves. Henceforth the acceptability of tracing motions was the key element of Descartes' interpretation of geometrical exactness.

Thus the *Rules* illustrate a strong interaction between Descartes' ideas in philosophy and those in mathematics. The 1620s, with the *Rules* as culmination, may have been the period in the development of Descartes' thought during which this interaction was strongest. In Descartes' later formulation of rules for the guidance of the mind in his *Discourse*,[18] the analogy with the mathematical method is much less strict than in the *Rules*. It may well be that in working out the answer to the questions that were left unanswered in the *Rules*, Descartes realized that such a strict analogy was untenable and that the doctrine of geometrical construction was less easily generalized to serve general philosophical problem solving than he had hoped earlier. Thus we may date in the late 1620s the beginning of a gradual separation of the ways of Descartes the mathematician and Descartes the philosopher.

---

[17] Cf. *Rule* 2, [Descartes Rules] p. 362: "Circa illa tantùm objecta oportet versari, ad quorum certam et indubitatam cognitionem nostra ingenia videntur sufficere." ("We should attend only to those objects of which our minds seem capable of having certain and indubitable cognition" [Descartes 1985–1991] vol. 1 p. 10.

[18] [Descartes 1637b].

Chapter 19

Descartes' first studies of Pappus' problem (early 1632)

19.1 Golius' challenge

Texts Sometime in late 1631 the Dutch mathematician and philologist Jacob Van Gool (Golius) suggested to Descartes that he should try his new method in solving the problem, mentioned by Pappus, of the locus to three, four, or more lines. This is the problem that, through Descartes' treatment of it in the *Geometry*, has become famous as "Pappus' problem." The texts of two letters[1] of Descartes to Golius about the problem have survived. From the one, written January 1632, it appears that he had sent Golius an "écrit" (manuscript) containing his solution of the problem, and that he hoped to receive comments. The *écrit* itself is lost, but Descartes' first letter contained an addition to it. In the second letter, dated February 2, 1632, Descartes merely expressed his gratitude for Golius' "favorable judgment" upon his solution of the problem.

As will become clear in the next section, it is probable that Descartes achieved most of his results about the problem in early 1632. In the *Geometry* he very effectively used these results to show the technical power of his new method in solving a difficult geometrical problem of great classical standing. But the confrontation with Pappus' problem gave Descartes more than just an occasion to develop his new mathematical techniques and convince himself of their power. The letters of 1632, together with the passages on Pappus' problem in the *Geometry*, strongly suggest that Golius' challenge gave him the ideas by which he could overcome the obstacles blocking his progress at the time he left the *Rules* unfinished, and that enabled him to achieve a complete doctrine of

[1] Descartes to Golius, January 1632, [Descartes 1964–1974] vol. 1 pp. 232–236; Descartes to Golius, 2 February 1632, *ibid.* pp. 236–242.

geometrical construction.

A reconstruction In Chapter 23 I give a detailed analysis of Descartes' treatment of Pappus' problem in the *Geometry*. In the present chapter I discuss some results about special cases of Pappus' problem, which, I conjecture, Descartes found in early 1632. I also suggest how these results could have given him the principal ideas of his later doctrine of construction. My conjectures[2] are based on indirect evidence mostly from technical aspects of Descartes' published solution of the problem. These technical aspects will be dealt with in Chapter 23, so for some of the statements in the present chapter the evidence will be presented later. Moreover, I discuss the content of the addition to Descartes' letter in connection with the question of the demarcation of geometry in Chapter 24 (Sections 24.5 and 24.6). I adopt this fragmented presentation because otherwise either the chronology would be broken, or much technical material would have to be explained here, which is more naturally dealt with in the chapters about the *Geometry*.

19.2 Pappus' problem

The problem Pappus' problem was a locus problem. That is, it required the determination of a curve all of whose points shared a certain given property. Pappus mentioned the problem in his *Collection* noting that mathematicians at his time had not achieved a full solution.

The problem is as follows:[3]

Problem 19.1 (Pappus' problem)[4]
Given n straight lines L_i in the plane (see Figure 19.1), n angles θ_i, and a line segment a. For any point P in the plane, the oblique distances d_i to the lines L_i are defined as the (positive) lengths of segments that are drawn from P toward L_i making the angle θ_i with L_i. It is required to find the locus of points P for which a certain ratio, involving the d_i and depending on the number of lines, is equal to a given constant ratio δ. The relevant ratios are:

For 3 lines: $$d_1^2 : d_2 d_3 \quad (19.1)$$

For 4 lines: $$d_1 d_2 : d_3 d_4 \quad (19.2)$$

For 5 lines: $$d_1 d_2 d_3 : a d_4 d_5 \quad (19.3)$$

For 6 lines: $$d_1 d_2 d_3 : d_4 d_5 d_6 \quad (19.4)$$

In general for an even

[2]In [Bos 1992] I have elaborated these conjectures in more detail than I will present them here or in the subsequent chapters.

[3]I use modern notation here to represent the problem; Descartes (and of course Pappus) did not use indices and expressed the coefficients explicitly with respect to a figure. In his formulation Descartes certainly meant the generality which modern notation can express.

[4][Pappus Collection] pp. 507–510; cf. [Pappus 1876–1878] vol. 2, pp. 676–681 and [Pappus 1986] vol. 1 pp. 118–123.

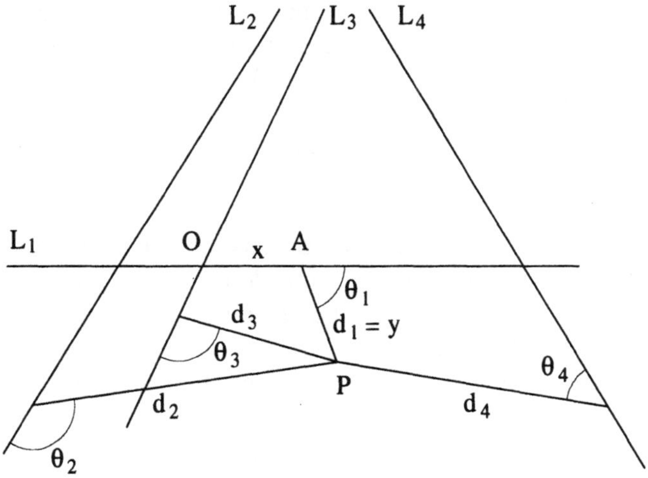

Figure 19.1: Pappus' problem

number $2k$ of lines: $\quad d_1 \ldots d_k \quad : \quad d_{k+1} \ldots d_{2k}$, (19.5)

and for an uneven

number $2k+1$: $\quad d_1 \ldots d_{k+1} \quad : \quad ad_{k+2} \ldots d_{2k+1}$. (19.6)

I call the loci of Pappus' problem "Pappus curves" or "Pappus loci," and special instances of Pappus' problem "Pappus problems," if necessary indicating the number of given lines ("five-line Pappus problem," "Pappus problem in k lines").[5]

Deriving the equations

In the *Rules* Descartes had explained that the first step in solving a problem was to derive its equation (Rules 19–21, cf. Section 18.2), which in the case of a locus problem would involve two unknowns. Doubtlessly, then, his first action in solving the problem presented by Golius was to try to derive such equations. In the *Geometry* Descartes described the general procedure of deriving the equations; it was as follows (see Section 23.2 for a more detailed discussion):[6]

Analysis 19.2 (Pappus' problem)[7]
Given and required: see Problem 19.1.

[5]Note that in Pappus' presentation the three-line locus is actually a four-line locus in the special case that two lines coincide. In analogy with the five-, seven-, etc., line loci one would expect the ratio $d_1 d_2 : ad_3$.

[6]The remarks in Note 3 apply also here.

[7][Descartes 1637] pp. 310–314, 323 sqq..

Analysis:

1. Assume a coordinate system (see Figure 19.1) with its origin at the intersection of L_1 and one of the other lines (L_3 in the figure), its X-axis along L_1 and its ordinate angle equal to θ_1; with respect to this system $d_1 = y$.

2. By employing the similarity of the relevant triangles, for any point P with coordinates x and y, the corresponding d_i can be written as

$$d_i = \alpha_i x + \beta_i y + \gamma_i \,, \qquad (19.7)$$

in which the coefficients α_i, β_i and γ_i are constant ratios expressed in terms of appropriate constant segments along the L_i determined by the given position of these lines and the given angles θ_i; hence, the α_i, β_i, and γ_i are known.

3. The constancy of the given ratio δ can now be expressed as an equation:

$$\begin{aligned} y(\alpha_2 x + \beta_2 y + \gamma_2) \cdots = \\ = \delta(a)(\alpha_l x + \beta_l y + \gamma_l)(\alpha_{l+1} x + \beta_{l+1} y + \gamma_{l+1}) \cdots \end{aligned} \qquad (19.8)$$

($l = k+1$ if there are $2k$ lines, $l = k+2$ if there are $2k+1$). The factor a on the right-hand side only occurs if the number of lines is uneven; δ is the given constant value of the ratio.

The techniques of this derivation are straightforward enough to assume that Descartes achieved it as early as in 1632. This supposition is corroborated by the fact that Descartes explicitly referred to equations in his letter to Golius.[8] Moreover, the classification he undertook in that letter (cf. Section 24.5) suggests that he had seen the general relation between the degree of the equation and the number of given lines. This relation follows immediately from the analysis sketched above: if the number of lines is increased by two, then the degree is increased by one.

Special cases

We may *a fortiori* assume that in 1632 Descartes was able to derive the equations in special cases of the problem arising for simple regular configurations of the given lines. It seems likely that he studied such special cases, both because that is a sensible strategy and because in the *Geometry* one such special case was extensively discussed (cf. Section 23.4). Descartes called it the simplest case in five lines; it featured four parallel equidistant lines and one perpendicular. Analogous configurations of two or three parallel lines and one perpendicular will also appear in the sequel of my argument.

19.3 Descartes' earliest solution of the five-line problem reconstructed

The five-line problem

In the letter to Golius Descartes stated explicitly that all Pappus curves could

[8][Descartes 1964–1974] vol. 1 p. 234.

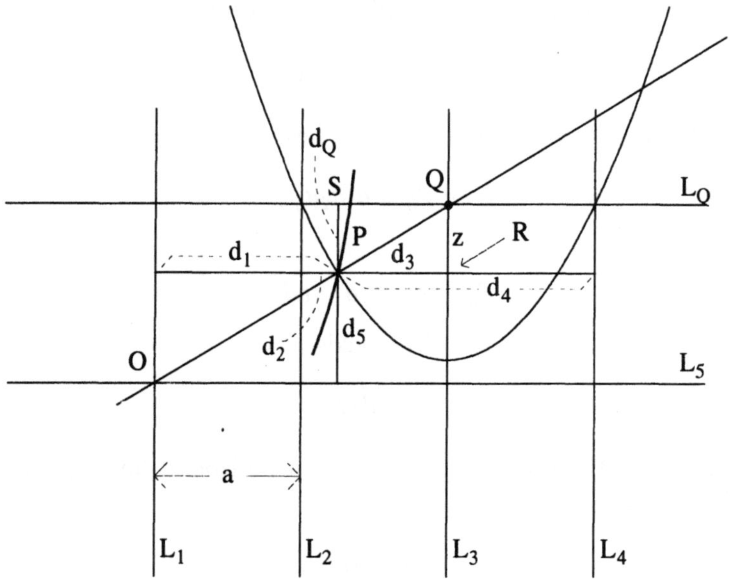

Figure 19.2: Pappus' problem in five lines

be traced by certain well-regulated motions.[9] What results could have inspired this statement? Neither in the letter nor in the *Geometry* did Descartes provide an explicit method for tracing all Pappus curves by motions. As we will see, the statement that Pappus curves (or, in general, algebraic curves) can be so traced recurs in the *Geometry*, but the argument given there was not directly tied to Pappus' problem. I conjecture that Descartes came to the statement on the basis of a solution, found in 1632, of the Pappus problem with respect to four parallel and one perpendicular lines. The problem was as follows:

Problem 19.3 (Pappus' problem in five lines)[10]
Given (see Figure 19.2) four parallel, equidistant lines $L_1, \cdots, L_4$ (for easy reference assumed to be vertical and with distance a), and one line L_5 perpendicular to them. It is required to find the locus of points whose perpendicular distances d_i to L_i satisfy

$$ad_3d_5 = d_1d_2d_4\ . \tag{19.9}$$

A motion tracing the curve

For reasons to be explained more fully in Chapter 23 below,[11] I conjecture that in 1632 Descartes argued as follows: Let P be a point on the locus and consider the line OPQ with Q on L_3. Let R be the intersection of L_3 and the

[9] Cf. Note 16
[10] [Descartes 1637] pp. 335–338.
[11] Cf. also my article [Bos 1992].

line through P parallel to L_5. If we call $QR = z$, we have

$$d_5 : d_1 = z : d_3 \,. \tag{19.10}$$

Moreover, the condition of the problem implies

$$ad_3d_5 = d_1d_2d_4 \,, \tag{19.11}$$

or

$$d_5 : d_1 = d_2d_4 : ad_3 \,. \tag{19.12}$$

Combining Equations 19.10 and 19.12 yields

$$az = d_2d_4 \,. \tag{19.13}$$

Now z is equal to the distance PS of P to a horizontal line L_Q through Q. Calling this distance d_Q, we have

$$ad_Q = d_2d_4 \,, \tag{19.14}$$

which means that the point P lies on a three-line Pappus locus[12] with respect to the three lines L_2, L_4, and L_Q. Such a three-line locus is (as we will see) a parabola with axis along L_3. If now OQ is conceived as a ruler turning around O and forcing Q to move along L_3, the line L_Q moves up or down and, because L_2, and L_4 are verticals, one may conceive the system of three lines L_Q, L_2 and L_4 as moving up and down with Q. Hence so does the three-line locus: the parabola moves up and down together with L_Q, and throughout the motion the point P on the Pappus curve is at the intersection of the ruler and the parabola.

The moving three-line locus

As noted, the argument supposes that the three-line locus in question is a parabola. We may well assume that Descartes studied the relevant three-line loci, either in their own right (because he may naturally have started with studying simple special cases) or once he had seen that he could reduce the five-line locus to a three-line one. The relevant three-line loci were those with two parallel lines L_1 and L_2 and one perpendicular L_3. Descartes would realize that there were essentially two types of such three-line loci, namely:

$$\text{type 1:} \qquad d_1d_2 = cd_3 \,, \tag{19.15}$$

$$\text{type 2:} \qquad d_1d_3 = cd_2 \,. \tag{19.16}$$

Moreover, he could easily find (with or without analytic methods) that the first type (see Figure 19.3) yields a parabola through the intersections of L_3 with L_1 and L_2; its axis is the vertical equally distant from L_1 and L_2. Similarly, the second type yields a rectangular hyperbola through the intersection of L_3 with L_2; L_1 is its vertical asymptote and its horizontal asymptote lies at distance c from L_3. In the case of Equation 19.14 the three-line problem is of the first type with distance between L_1 and L_2 equal to $2a$ and $c = a$.

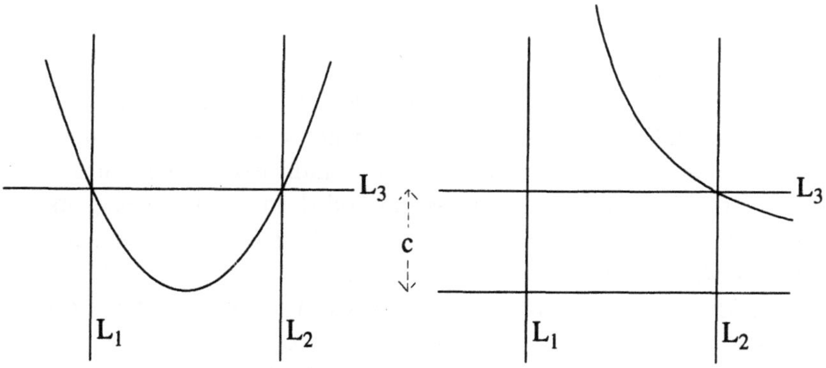

Figure 19.3: Simplest three-line loci

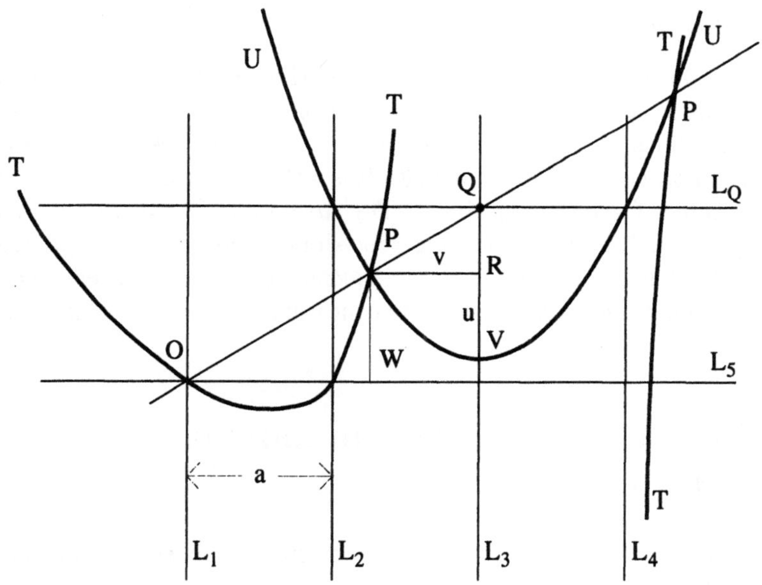

Figure 19.4: Tracing the five-line locus

The construction by tracing

The arguments about the five-line locus sketched above imply the procedure for tracing the locus, which Descartes explained in the *Geometry*:

Construction 19.4 (Five-line locus — Descartes)[13]
Given and required: see Problem 19.3.

Construction:
1. Consider (see Figure 19.4) a parabola UVU with vertical axis along L_3 and *latus rectum* equal to a (which means that its equation in rectangular coordinates u and v as indicated in the figure is $au = v^2$). The parabola can move up and down while keeping its axis along L_3. Moving with it is a point Q on the axis inside the parabola with distance a to the vertex V.
2. Consider also a straight line OQ that can turn around the intersection O of L_1 and L_5 while Q moves along L_3.
3. During the combined motion the points P of intersection of the parabola and the straight line move over the plane; they trace a new curve $TOPT$ TPT (consisting of two branches); this curve is the required five-line locus.
[**Proof**: During the process the coordinates u and v satisfy $au = v^2$. Now $a + v = d_4$ and $a - v = d_2$ from which it follows that $v^2 = a^2 - d_2d_4$ (1). Moreover, because the triangles PRQ and OWP along OQ are similar and $QR = QV - VR = a - u$, it follows that $(a - u) : d_3 = d_5 : d_1$, whence $au = a^2 - ad_3d_5/d_1$ (2). Equating the expressions (1) and (2) for au and v^2, respectively, yields $ad_3d_5 = d_1d_2d_4$.]

As mentioned, this tracing procedure is precisely the one Descartes presented in the *Geometry* as the procedure to trace the solution of the special case of Pappus' problem in five lines. The curve played a crucial role in Descartes' theory of geometrical construction (cf. Section 26.3); it is a third-degree curve which later acquired the name "Cartesian parabola." Descartes himself gave no explanation at all of how he had found the tracing procedure. I consider the appearance of precisely this tracing procedure in the *Geometry* as the main evidence for my reconstruction of Descartes' endeavors concerning Pappus' problem in early 1632.

19.4 The "turning ruler and moving curve" procedure

The procedure

The method of curve tracing used in the construction above is an instance of what henceforth I call the "turning ruler and moving curve" procedure. Special cases of this procedure occur at several places in the *Geometry*. All these cases

[12] In the sense of the analogon to the fifth-, seventh-, etc., line case; cf. Note 5.
[13] [Descartes 1637] p. 337.

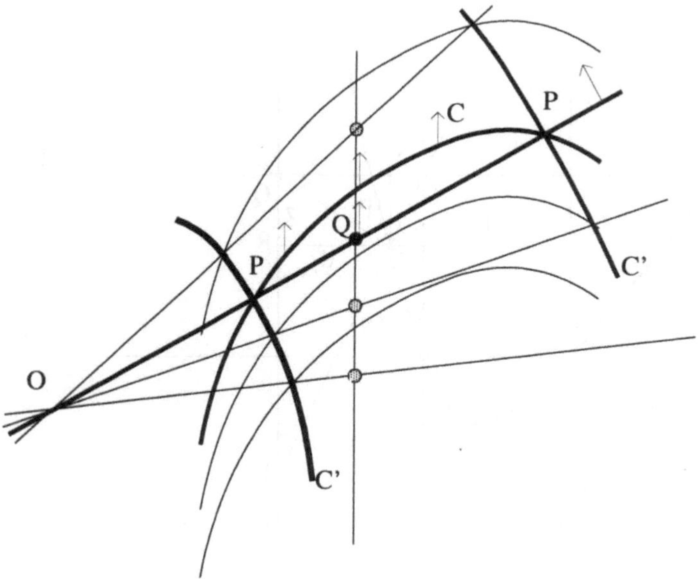

Figure 19.5: The "turning ruler and moving curve" procedure

involve a curve C moving in one fixed direction and a ruler turning around a fixed point O (see Figure 19.5). The two motions are interrelated via a point Q whose position with respect to C is fixed, which means that Q partakes in the rectilinear motion of C; the ruler connects O with Q. During the combined motion the point or points P of intersection of the ruler and the curve trace a new curve C' (possibly consisting of several branches).

A four-line problem

There is another instance of the turning ruler and moving curve procedure in the *Geometry* that, although Descartes did not mention this, very probably had a direct relation with a Pappus problem. This is the case in which the moving curve is a straight line. Descartes showed[14] that the resulting curve is a hyperbola. In the light of my reconstruction of the solution of the five-line locus, it is of interest to note that also in this case the procedure may have arisen in the solution of a Pappus problem, namely, the problem for four lines, three of which are equidistant and parallel, and the fourth perpendicular:

Problem 19.5 (A Pappus problem in four lines)
Given three equidistant lines L_1, L_2, L_3, one line L_0 perpendicular to these, and a ratio δ (cf. Figure 19.6). It is required to determine the curve whose points P satisfy

$$d_0 d_1 = \delta d_2 d_3 \,. \tag{19.17}$$

[14] [Descartes 1637] pp. 319–322.

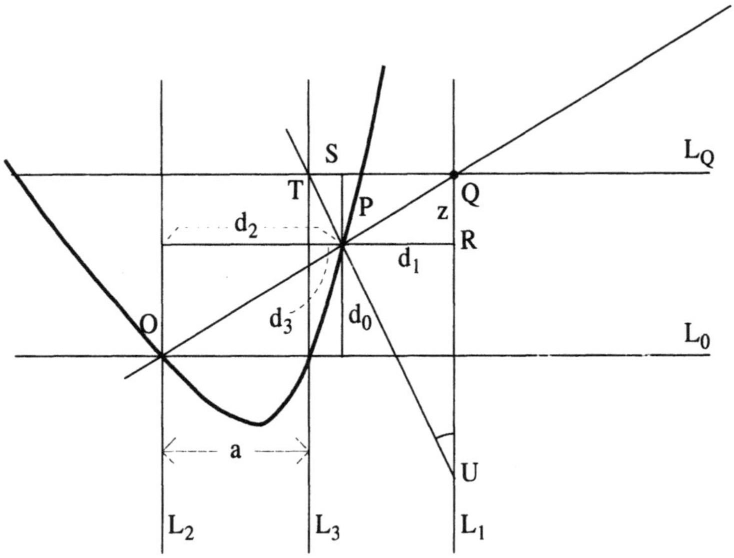

Figure 19.6: A Pappus problem in four lines

Here an argument analogous to the one in the case of the special five-line locus proceeds as follows: The point O can be chosen on either L_2 or L_3, assume it is on L_2, Q must then be on L_1. Setting $z = QR = PS$ we have $z : d_1 = d_0 : d_2$; eliminating d_1 and d_2 leads to

$$z = \delta d_3 \; . \tag{19.18}$$

This is, one could say, a two-line locus, with z equal to the distance d_Q of P to a line L_Q through Q parallel to L_0. In other words, Equation 19.18 implies that P is on the locus of points whose distances from L_Q and L_1 have a constant ratio δ. This locus is a straight line intersecting L_3 in T, T being at the same height as Q. This straight line moves up and down with Q. Thus if U is its intersection with L_1, the distance QU is constant and so is the angle at U. The four-line locus is therefore traced by a turning ruler and moving curve procedure with OQ as the ruler and the straight line UT as moving curve. This is precisely the generation of an hyperbola by the turning ruler and moving curve procedure that Descartes gave in the *Geometry*.[15]

Nature and origin of the procedure The two solutions reconstructed above, whose end results occurred in the *Geometry* as curve tracing procedures, essentially consist in reducing the given Pappus problem to a related Pappus problem in a smaller number of lines. The latter problem is recognized by considering the proportionalities that arise when a line is drawn from a point on the locus to an intersection of the transversal

[15]See Note 14.

with one of the other given lines. The turning ruler and moving curve procedure is the direct kinematic interpretation of that reduction.

It should be added that, although Descartes probably used algebraic analysis from the beginning in attempting to solve Pappus' problem, the reconstructed arguments in the solutions above do not depend on the knowledge of the equations of the curves. They can be achieved entirely without the help of algebra. Moreover, I have not been able to devise a purely algebraic line of arguments leading in a natural way from the problems to the turning ruler and moving curve procedures for tracing the solution curves.

19.5 The significance of Pappus' problem

Classification and demarcation

As I have argued, there is good reason to assume that Descartes solved some special four- and five-line loci by the reduction explained above and its kinematic interpretation. I suggest that this combination of a reduction of the problem to one of a simpler type, and the kinematic method to generate intricate curves from simpler ones, has been decisive in the formation of Descartes' ideas on geometry, on its proper demarcation, and on its legitimate methods of construction.

Here the letter to Golius, with the addition to the "écrit," provides important evidence. I discuss that text in more detail in Section 24.1, suffice it here to note that it concerned Pappus loci and the classification of curves. Descartes wrote that all Pappus loci could be traced by "one single continuous motion completely determined by a number of simple relations:"[16] he claimed that curves which could be traced in this way were acceptable in geometry, whereas curves as the spiral and the quadratrix could not be traced in this way and were therefore excluded; and he stated that the Pappus curves could be classified according to their complexity, which depended on the number of given lines. These statements are readily explained on the basis of my conjecture on the nature of Descartes' early solution of Pappus' problem. The turning ruler and moving curve procedure would suggest the idea of a hierarchy of curves, higher-order loci being generated iteratively by the motion of lower-order ones. It is likely that in early 1632 Descartes knew the equations of the Pappus loci, and noted the parallelism between the degree of the equation, the number of given lines, and the complexity of the tracing motions. The freedom of choice in the position of the given lines and his understanding of the formation of the equations of the resulting loci would convince Descartes of the great extent of the class of Pappus curves; quite possibly he could have formed the idea (actually formulated in the *Geometry*[17]) that any curve equation could occur among the equations for Pappus curves.[18] Thus I conjecture that the confrontation with Pappus' problem in early 1632 suggested to Descartes a number of crucial ideas

[16][Descartes 1964–1974] vol. 1 p. 233: ". . . describi possunt unico motu continuo, et omni ex parte determinato ab aliquot simplicibus relationibus."

[17][Descartes 1637] p. 308.

[18]The idea is incorrect; cf. Section 24.5, Note 36.

about geometry, namely: (1) that curves should be accepted in geometry in so far as they were traced by geometrically legitimate motions; (2) that these legitimately traced curves were precisely the Pappus curves; (3) that Pappus curves were precisely the ones that admitted polynomial equations; and (4) that therefore the totality of geometrically acceptable curves could be classified equivalently by the complexity of the tracing motion, the degree of the equation, and the number of given lines in the pertaining Pappus problem. In Section 24.6 I discuss how these ideas relate to the ones expressed later in the *Geometry*.

Later developments

I conjecture that Descartes arrived at these ideas by optimistic generalization of the tracing methods he found for the simple four- and five-line loci. I consider it very unlikely that Descartes actually found a general method for tracing Pappus loci. The reason is that it seems very difficult to find such a method[19] and, perhaps more important, that no such method occurs in the *Geometry*, where it would, as we will see, have been very functional.

In the years between 1632 and 1637 Descartes probably realized that he could not find a general method for tracing Pappus curves, and as a result he downplayed the connection between curve tracing and Pappus' problem when he came to write the *Geometry*. However, he retained the conviction that acceptable curves should be traceable by definite acceptable motions. We will see below (Section 24.1) how he argued for this conviction in the *Geometry*.

Hope and disillusion

My reconstruction above of Descartes' findings in early 1632 may also uncover something of the thrill that he must have experienced while dealing with Pappus' problem. No doubt the first results seemed very exciting to him: Pappus loci had algebraic equations and, because of the variability of the data, any algebraic equation would correspond to some Pappus locus. It seemed also that all Pappus curves were traceable by coordinated motions. Thus the class of traceable, acceptable curves turned out to be large, but at the same time classifiable both in terms of the special nature of the tracing motions and in terms of their equations. Curve tracing provided a sound reason for abandoning a restriction to circles and straight lines and extending the arsenal of constructing curves far beyond the conic sections and the special curves the ancients used in solving solid problems. At the same time the curves that Descartes had rejected from the very beginning, the spiral and the quadratrix, were indeed excluded; they solved no Pappus problem; moreover, they were traced by motions whose mutual relation was indeterminate, and that was exactly what made their generation different from that of the Pappus curves.

Descartes' later realization that the tracing methods for the simple four- and five-line loci could not effectively be generalized to arbitrary Pappus problems must have been a considerable disappointment. A failure to establish tracing methods for all Pappus curves meant that an essential element of Descartes' theory of geometry lacked direct proof, namely, the assertion that all algebraic

[19]The first proof that any algebraic curve can be traced by a linkage was given by Kempe more than two centuries after Descartes ([Kempe 1876]).

curves were traceable by acceptable continuous motion and that conversely all curves so traceable were algebraic. In Chapter 24 I discuss how Descartes later dealt with this defect.

Catalyst

Pappus' problem was extensively treated in the *Geometry*, but there it served primarily as a didactic device to explain Descartes' new methods, with the additional advantage of classical appeal and prestige. I claim that in the development of Descartes' thinking on geometry the problem was much more than an appealing example. It was the crucial catalyst; it provided him, in 1632, with a new ordered vision of the realm of geometry and it shaped his convictions about the structure and the proper methods of geometry.

At the time Descartes wrote the *Rules*, the analytical part of the program for geometry was clear; problems should be translated into algebra and reduced to equations. But the synthetical part, how to arrive from the equation to the construction, was far from transparent, while the philosophical setting, with its emphasis on completeness of the method, only made the matter more urgent. Construction should be by curves. But the acceptable curves should be convincingly demarcated from the unacceptable ones, they should be ordered as to simplicity and there should be enough of them to construct all problems. The study of Pappus' problem in 1632 provided the ingredients for the answers to these questions: Curve tracing as exemplified by the turning ruler and moving curve procedure was to be the criterion for acceptability; it excluded curves like the spiral and the quadratrix. Acceptable curves were precisely those that had algebraic equations; the degree of the equations corresponded to the simplicity of the tracing motion. The iteration of the turning ruler and moving curve procedure ensured the availability of sufficient curves to solve all problems. We will see that in the *Geometry* Descartes indeed suggested using the solution curve of the five-line locus as means to construct the next class of problems beyond the solid ones (cf. Section 26.3); no doubt the choice was inspired by the fact that this curve was generated, through the turning ruler and moving curve procedure, from the parabola, which itself served for the construction of solid problems.

A breakthrough

Thus the solution of Pappus' problem, with its suggestive correspondence between the algebraic and the geometrical properties of the acceptable curves, held great promise for the completion of the synthetic part of the program. Until c. 1630 the programmatic side of Descartes' ideas was still largely in keeping with the classical ideas on construction and the more recent ones on analysis by means of algebra. The episode of Pappus' problem provided the necessary ingredients for a breakthrough. In elaborating these Descartes developed truly independent and original ideas. The result was Descartes' mature doctrine of geometry, which we find in the *Geometry*.

Chapter 20

The *Geometry*, introduction and survey

20.1 Descartes' geometrical ideas c. 1619–1637 — a recapitulation

The documents from c. 1619 discussed in Chapter 16 make clear that from early on Descartes viewed the scientific enterprise with a strong programmatic interest. His program encompassed the whole of science, which he saw primarily as a problem solving endeavor, with arithmetical and geometrical problems as paradigms. Thus he formulated his programmatic ideas (in the letter to Beeckman) by giving a classification of problems concerning continuous and discrete quantity and by defining the nature of the solutions to be achieved in each class. We may consider the program as Descartes' earliest interpretation of what it meant to solve scientific problems exactly. The starting point of his interpretation of exactness was classical Greek geometry; geometrical problem solving meant construction by the intersection of curves and his classification of problems can be seen as a modification of Pappus'. He singled out the manner of tracing by motion as the primary criterion for the acceptability of curves; regular motions such as those provided by the "new compasses" (cf. Section 16.4) were acceptable; other motions, such as the ones generating the quadratrix or the linea proportionum, were, if not rejectable, at least of lower status. *c. 1619*

Although by 1620 Descartes had some algebraic interests and even endeavored to find new means of solving cubic equations, algebra had no central place in his conception of science or even of mathematics. In 1625 this situation had changed. By means of algebraic techniques he had achieved a phenomenal result: the general solution of third- and fourth-degree equations by the intersection of a parabola and a circle. Probably this result, rather than the writings of contemporary practitioners of *specious algebra*, convinced him of the importance *c. 1625* of algebra as analytical method: to reduce problems to equations. As to the synthesis, i.e., the solution of geometrical problems, he kept to the classical conception of construction by the intersection of curves.

c. 1628 The *Rules* of c. 1628 mark the stage in the development of Descartes' thought in which his mathematical and philosophical interests were linked most closely. He took up his programmatic ideas of c. 1619, included algebra as principal means of analysis, and attempted to work out the program with special concern for method, completeness, and exactness. He realized that a serious enquiry into the applicability of algebra to general fields of knowledge required an interpretation of the algebraic operations independently of the number concept and valid for magnitudes in general. Taking extension as the prototype of continuous magnitude, he worked out such an interpretation of the primary arithmetical operations for geometrical magnitudes (in particular line segments and areas). His interpretation was different from the one accepted in the Viètean school. For the primary arithmetical operations the Euclidean constructions with straight lines and circles sufficed, but for general root extraction and the construction of roots of equations other means were necessary. In the *Rules* Descartes did not proceed to the interpretation of these higher algebraic operations; he broke off the project precisely at the point where he was forced to discuss these.

The project of the *Rules*, then, confronted Descartes with the problem of construction beyond the usual plane means. For the first step, third- and fourth-degree equations, he had a construction by the intersection of a parabola and a circle; but because equations could have any degree, he evidently needed a more general procedure, not impeded by restrictions on the degree. If the circle and parabola construction was to be incorporated in it, that general procedure should involve construction by the intersection of curves, and this raised the question of which curves beyond the parabola should be chosen for constructions. Apparently by 1628 Descartes had yet no answer to that question.

c. 1632 Writing the *Rules* had led Descartes in a natural way to questions about curves, their acceptability for use in constructions, and their classification. I have argued above that the study of Pappus' problem in 1632 supplied new and effective ideas on these matters. The tracing methods that Descartes found for some Pappus curves linked up with his earlier ideas about acceptable and non-acceptable curve tracing. Moreover, the Pappus problem, the equations it gave rise to and the tracing procedures he found, suggested a hierarchy of curves in which the degree of the equation was a measure of the complexity of the tracing motion and thereby of the complexity (or simplicity) of the curve. Descartes found one curve in particular, the Cartesian parabola, which, because the parabola was employed in its tracing and because he considered its tracing particularly simple, was the obvious candidate for serving as the constructing curve for the class of problems after the solid ones. In sum, the study of Pappus' problem provided Descartes with the last missing ingredients for the interpretation of exactness of geometrical constructions, which he later presented in the

Geometry.

20.2 The questions still open before 1637

The *Geometry*[1] served as an illustrative essay accompanying the *Discourse on the method*. Descartes did not explicitly discuss the links between the method of the *Geometry* and the general rules of methodical thinking expounded in the *Discourse*. Yet, for instance, the second and third of the four rules expounded in Part 2 of the *Discourse*[2] might easily be seen as exemplified by the procedures of analysis and synthesis, respectively, as detailed in the *Geometry*. *Method*

Indeed the method of the *Geometry* consisted of:

A. An *analytic* part, using algebra to reduce any problem to an appropriate equation;

and

B. A *synthetic* part, finding the appropriate construction of the problem on the basis of the equation.

For both parts Descartes developed new ideas and techniques. Some of these he had acquired in the period before c. 1632, but a number of questions were still open. In the present section I use the analysis–synthesis division to survey and discuss the methodological questions whose final answers Descartes had to forge before or during the writing of the *Geometry*.

The final elaboration of the analytic part of Descartes' geometrical method posed mainly technical questions. Translating problems into algebraic equations involved recognizing geometrical relations as algebraic ones. That is, it implied the question whether algebraic operations as addition and multiplication could be interpreted so as to apply to geometrical objects such as line segments. Here Descartes could start from the interpretation of the primary arithmetical operations that he had elaborated in the *Rules* (cf. Section 18.2). The interpretation presented in the *Geometry* differed from the earlier one in that the dimensional aspect was removed and that it included the algebraic operations as well. I discuss it in Chapter 21. *Analysis: from problem to equation*

Descartes' doctrine of construction stipulated (see below) that the construct- *Analysis: preparing the equations*

[1]For valuable further information and alternative opinions on Descartes' *Geometry* I refer (without attempt at completeness) to: [Boyer 1959], [Costabel 1969], [Forbes 1977], [Freguglia 1981], [Galuzzi 1985], [Giusti 1990], [Grosholz 1991], [Israel 1997], [Itard 1956], [Jullien 1996], [Mancosu 1992] pp. 65–91, [Lachterman 1989] pp. 124–205, [Molland 1976], and [Molland 1991].

[2][Descartes 1637b] pp. 18–19; cf. the translation in [Descartes 1985–1991] vol. 1, p. 120: "The second, to divide each of the difficulties I examined into as many parts as possible and as may be required in order to resolve them better. The third, to direct my thoughts in an orderly manner, by beginning with the simplest and most easily known objects in order to ascend little by little, step by step, to knowledge of the most complex, and by supposing some order even among the objects that have no natural order of precedence." Cf. also Note 7 of Chapter 20.

ing curves should be simplest possible in the sense of having lowest possible degree. To achieve this, the problem itself should be reduced to an algebraic equation (in one unknown) of lowest possible degree. Moreover, Descartes' standard constructions presupposed the problems to be reduced to equations of certain standard forms. He therefore needed algebraic techniques to ascertain whether equations in one unknown could be reduced to equations of lower degree, and to transform such equations into certain standard forms. Several of these techniques had to be elaborated anew for the completion of the analytic part of his method. Descartes was probably aware of the need for such techniques when he composed the *Rules*, but his terminology there[3] suggests that he had not yet fully appreciated the complexity of the matter. In the *Geometry* he did provide the necessary algebraic techniques; they are discussed in Chapter 27 below.

Equations are not solutions

The analytic part of the method translated a problem into an equation, but an equation was not a solution. Mathematicians of the sixteenth century had developed explicit formulas, or algorithms, such as those of Cardano[4] and Ferrari,[5] for calculating the roots of third- and fourth-degree equations. But in the so-called "casus irreducibilis" the interpretation of these formulas was still dubious because they involved square roots of negative quantities. No general formulas were available for the roots of equations of degrees higher than four; the search for these seems to have started late in the seventeenth century (and stopped in the early nineteenth with Abel's proof that they do not exist). Moreover, even if formulas for the roots were available and interpretable, these, in general, did not provide geometrical constructions. For example (as mentioned earlier in connection with Viète, Section 8.2), the problem of two mean proportionals could readily be reduced to an equation, namely, $x^3 = a^2b$; this equation had an explicit algebraic solution, $x = \sqrt[3]{a^2b}$, but the cubic root sign did not give any guidance about how such a root could be geometrically constructed. Algebraically the problem might be considered solved by the explicit formula, geometrically it was not.

Algebra does not provide constructions

The fact that algebra does not provide geometrical constructions merits emphasis because too often Descartes' contribution to geometry is presented as the brilliant removal of cumbersome geometrical procedures by simply applying algebra.[6] In fact, algebra could only do half of the business, it could provide the analysis and reduce problems to equations. The other half of the job, the synthesis, the geometrical construction of the roots of the equations, remained to be done.

Synthesis: constructional exactness

The synthetic part of Descartes' program presented the most profound ques-

[3] In particular the conception of reduction as a kind of division, cf. Section 18.3.

[4] See Note 91 of Chapter 4.

[5] See Note 18 of Chapter 10.

[6] Thus Kline, for instance, writes: ". . . Descartes solved geometric construction problems by first formulating them algebraically, solving the algebraic equations, and then constructing what the solutions called for . . ." [Kline 1972] p. 317.

tions. They concerned the conception of geometrical construction itself, in other words the interpretation of constructional exactness. That interpretation required a demarcation of the class of curves acceptable for use in constructions and a criterion to judge the simplicity of these curves. As we have seen, Descartes' final ideas about this matter were probably formed during or shortly after his investigation of Pappus' problem: acceptable curves were traced by acceptable motions; they were precisely those that had algebraic equations; they were simpler in as much as their degree was lower. But the precise arguments for this position were still lacking, while possibly an early optimistic expectation that the techniques for solving Pappus' problem would straightforwardly provide these arguments later proved untenable. Thus Descartes had to forge new arguments on this matter; they are dealt with in Chapter 24 below.

Synthesis: standard constructions

After finding and justifying the demarcation of the class of constructing curves and the criterion of simplicity, Descartes had to provide the constructions themselves. He divided the problems into classes defined by the degrees of their corresponding equations. He gave (at least in principle) for each class a standard form of the equation and a standard construction by which any problem of that class could be constructed. But Descartes also had to decide which constructing curves were to be used in the standard constructions. The plane problems, leading to linear or quadratic equations, were easily incorporated; the standard constructions, by straight lines and circles, which Descartes gave for these problems were appropriate but not particularly new. Obviously he wished to incorporate his result of 1625, the construction of all third- and fourth-degree equations by a parabola and a circle as one of the standard constructions. That construction covered all solid problems. The challenging question was how to proceed beyond the solid problems. Here higher-order curves than the conics were needed. These had to be acceptable as constructing curve and simplest possible with respect to the class of problems they were to solve. From the *Rules* we may conclude that by 1628 Descartes did not realize the difficulty of this question, if indeed he was aware of it. Considerable intellectual effort was still needed to arrive at the doctrine of construction presented in the *Geometry*. In Chapters 25 and 26 I discuss this doctrine and the standard constructions.

20.3 The structure of the *Geometry*

Survey of contents

As the presentation of Descartes' achievements in the next chapters does not follow the order he chose himself in composing the *Geometry*, it is useful to present here a schematic survey of the contents of the book. The survey makes clear how strongly the structure of the *Geometry* is determined by Descartes' program of redefining exactness in geometry and providing a complete doctrine of construction in accordance with his new interpretation of exactness. For further clarification I have presented the content of the *Geometry* in tabular form — see Table 20.1.

The *Geometry* consists of three books. Book I is about geometrical analysis and construction in the methodologically unproblematical case of plane problems. It introduces the geometrical interpretation of the algebraic operations $+$, $-$, $\times$, $\div$, and square root extraction, explains the full canon of problem solving in the case of plane problems, and presents the general analysis of Pappus' problem.

Books II and III are about higher-order problem solving; they contain Descartes' new interpretation of exactness of geometrical construction. This interpretation involves a demarcation between acceptable and non-acceptable curves and a criterion of simplicity. Accordingly, Book II is about curves and their acceptability in geometry, and Book III is about the criterion of simplicity and its technical implications.

Book II opens with the explanation of the demarcation between curves that are acceptable in geometry and curves that are not. Then follows a full solution of Pappus' problem in three and four lines, and a discussion of two special cases of the problem in five lines. After this, Descartes returns to the acceptability of curves and discusses in that context the various methods of tracing curves. The remaining part of Book II is devoted to the use of curve equations in finding normals and tangents, the study of ovals, and three-dimensional geometry. These are important passages for their later influence on the development of analytic geometry and the calculus. They are, however, not closely related to construction and exactness, and therefore I do not discuss them in the present study.

Book III deals with simplicity of problems, solutions, and curves, and gives Descartes' standard non-plane constructions. In order that the constructions be simplest possible, that is, that the degrees of the constructing curves be lowest possible, the equations in one unknown have to be reduced to their irreducible components. Moreover, Descartes' standard constructions require these equations to be of certain standard forms. Consequently, Descartes provides in the first part of Book III an extensive algebraic study of properties and transformations of equations in one unknown (among them the famous "rule of signs"). In the second part of Book III he presents the standard constructions for equations of third and fourth degree (by the intersection of a parabola and a circle) and for those of fifth and sixth degree (by the intersection of a Cartesian parabola and a circle). He ends his essay with the (overconfident) statement that it should now be clear how to extend the canon of construction to equations of ever higher degree.

The table Table 20.1 illustrates the structure of the *Geometry*.[7] Descartes himself gave margin titles of sections, thus dividing his three books in 9, 19, and 32 sections, respectively. In the table I adopt a thematic subdivision of the books in fewer parts. This subdivision and the characterizations in Column 2 of the table (and the codes in Column 1) are mine. Columns 3 and 4 give the page references for the parts in the *Geometry* itself ([Descartes 1637]) and in its edition in the *Oeuvres* ([Descartes 1964–1974] vol. 6), respectively.

[7] I have published the table earlier in [Bos 1990] p. 357 (p. 44 of ed. in [Bos 1993c]).

	Book I: Plane problems	*Géométrie* pp:	*Oeuvres* vol. 6 pp:
I-A	Geometrical interpretation of the operations of arithmetic	297–300	369–372
I-B	Problems, equations, construction of plane problems	300–304	372–376
I-C	Pappus' problem; deriving the equation, cases in which the problem is plane	304–315	377–387
	Book II: Acceptability of curves		
II-A	Acceptable curves, their classification	315–323	388–396
II-B	Pappus' problem continued, solution of the three- and four-line problem, plane and solid loci, simplest case of the five-line locus	323–339	396–411
II-C	Acceptability of pointwise construction of curves and construction by strings	339–341	411–412
II-D	Equations of curves, their use in finding normals	341–352	412–424
II-E	Ovals for optics	352–368	424–440
II-F	Curves on non-plane surfaces	368–369	440–441
	Book III: Simplicity of curves and of constructions		
III-A	Acceptability of curves in constructions, simplicity	369–371	442–444
III-B	Equations and their roots	371–380	444–454
III-C	Reduction of equations	380–389	454–464
III-D	Construction of the roots of third- and fourth-degree equations, solid problems	389–402	464–476
III-E	Construction of the roots of fifth- and sixth-degree equations, "supersolid" problems	402–413	476–485

Table 20.1: The structure of the *Geometry*

Chapter 21

Algebraic operations in geometry

21.1 Descartes' interpretation

Quadratic operations

In the first two pages of the *Geometry* Descartes presented his method for applying algebraic operations to line segments. The method was new, although it combined elements that had been extant for some time. Descartes introduced a unit line segment, and defined the quadratic operations[1] for line segments by constructions in such a way that the results of the operations were again line segments. For addition and subtraction the obvious corresponding procedures were the joining and removing of line segments. Multiplication, division, and square root extraction were to be performed by combining the Euclidean constructions of fourth and mean proportionals (*Elements* VI-12, 13, cf. Constructions 4.1 and 4.2) with the adoption of a unit. This led to the following definitional constructions:

Construction–Definition 21.1 (Multiplication of line segments)[2]
Given a unit length e and two line segments a and b (see Figure 21.1), a line segment c is constructed, equal (by definition) to the product of a and b.

Construction:
1. Draw two lines intersecting in O under any angle; mark off $OE = e$ on one of the lines.
2. Mark $OA = a$ along the line on which OE is marked; mark $OB = b$ along the other line.
3. Draw EB and draw a line through A parallel to EB, it intersects the other line in C.

[1]In the sense of Section 6.2: addition, subtraction, multiplication, division, square root extraction, and the solution of quadratic equations.

[2][Descartes 1637] p. 298.

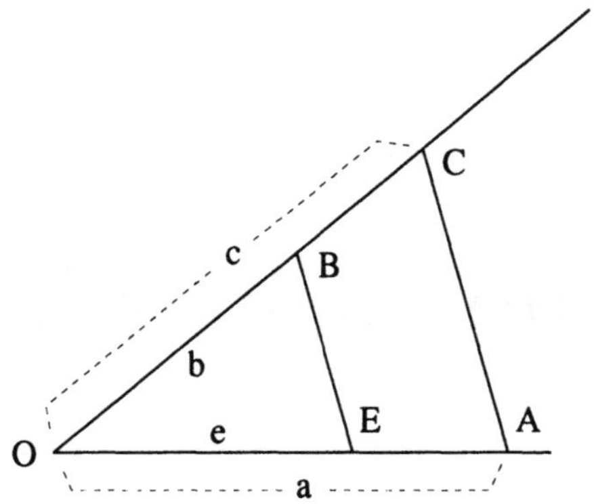

Figure 21.1: Multiplication of line segments

4. The product ab of a and b is now defined to be the segment $c = OC$.

Construction–Definition 21.2 (Division of two line segments)[3]
Given a unit length e and two line segments f and g (see Figure 21.2), a line segment h is constructed, equal (by definition) to the quotient of f and g.

Construction:
1. Draw two lines intersecting in O under any angle; mark off $OE = e$ on one of the lines.
2. Mark $OG = g$ along the line on which OE is marked; mark $OF = f$ along the other line.
3. Draw FG and draw a line through E parallel to FG, it intersects the other line in H.
4. The quotient f/g of f and g is now defined to be the line segment $h = OH$.

Construction–Definition 21.3 (Square root of a line segment)[4]
Given a unit length e and a line segment a (see Figure 21.3); a line segment b is constructed, equal (by definition) to the square root of a.

[3] [Descartes 1637] p. 298.
[4] [Descartes 1637] p. 298.

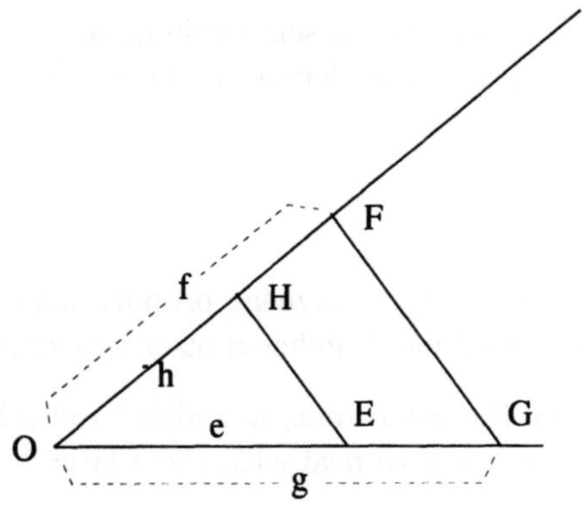

Figure 21.2: Division of two line segments

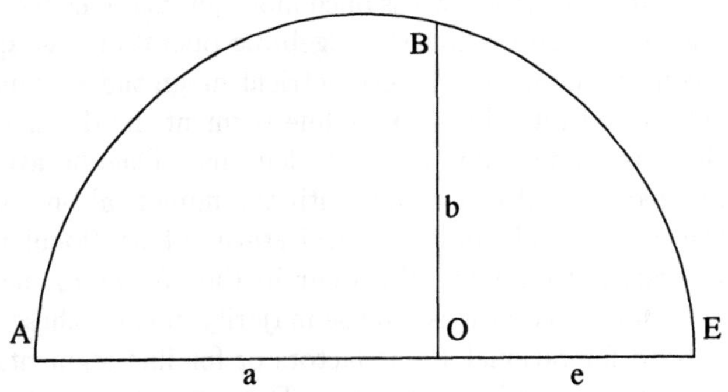

Figure 21.3: Square root of a line segment

Construction:
1. Mark points A, O, E along a line (with O between A and E), such that $OE = e$ and $OA = a$.
2. Draw a semicircle with diameter AE; draw a line through O perpendicular to AE, it intersects the semicircle in B.
3. The square root $\sqrt{a}$ of a is now defined to be the line segment $b = OB$.

Higher-order root extraction postponed

Descartes interpreted higher-order roots as mean proportionals.[5] However, he postponed the geometrical effectuating of higher root-order extraction, writing:

> I say nothing here about the cubic root, nor about the others, because it will be more convenient to deal with them later.[6]

This crucial sentence is usually overlooked when Descartes' interpretation of the algebraic operations is expounded in the secondary literature; thus the impression is given that at the outset of the *Geometry* Descartes effected, by a few simple constructions, a complete correspondence between algebra and geometry.[7] He did not and he was aware of it. Obviously, the reason for the postponement was that cubic or higher-order roots could not in general be constructed by straight lines and circles. Hence, Descartes could deal with the root extractions only after having explained how to construct beyond the Euclidean means of construction. Indeed it was only in the last book of the *Geometry* that Descartes found it "convenient" to deal with cubic and higher-order roots (cf. Chapter 26).

Line segments, not numbers

With the risk of being repetitive I stress once more (cf. Sections 6.4, 8.2, and 18.2) that in Descartes' interpretation the algebraic operations as applied in geometry did not concern numbers but geometrical magnitudes, namely, line segments. Although he did introduce a unit line segment, he did not identify line segments with their numerically expressed lengths. Thus he avoided the identification of the geometrical continuum with the numerical one, undoubtedly because of the conceptually problematical status of irrational numbers. Both rational and irrational numbers did occur in the *Geometry*, namely, as factors, terms, or solutions of equations. In the majority of cases these numbers were rational and stood for rational scalar factors or for line segments defined as rational multiples of the unit.[8] Occasionally Descartes discussed equations

[5] [Descartes 1637] p. 298: "... trouver une, ou deux, ou plusieurs moyennes proportionelles entre l'unité, et quelque autre ligne; ce qui est le mesme que tirer la racine quarrée, ou cubique, etc."

[6] [Descartes 1637] p. 298: "Ie ne dis rien icy de la racine cubique, ny des autres, à cause que i'en parleray plus commodement cy aprés."

[7] Cf. Chapter 20 Note 6.

[8] It is easily seen that Descartes' use of the unit in the definition of the multiplication of line segments ensures that $3a$, the result of joining three copies of the line segment a, is the same as $(3e)a$, the result of multiplying the line segment $3e$ with the line segment a.

whose coefficients or roots were irrational numbers, but these usually illustrated the solution of geometrical problems.[9] Needless to say that there are no transcendental numbers in the book.

21.2 Comparison with earlier interpretations

Classical geometry and Viète's interpretation

At this point it is useful to compare Descartes' approach to the use of algebra in geometry with some earlier interpretations of the quadratic algebraic operations when used outside numerical calculations.

In classical Greek geometry (cf. Section 6.2) line segments and areas could be combined in manners analogous to the ways in which the primary arithmetical operations combine numbers. Thus two line segments a and b could be joined (analogous to addition) or the one cut off from the other (analogous to subtraction), they could form a rectangle with sides a and b (analogous to multiplication), or a ratio $a : b$ (analogous to division in the case of factors of equal dimension). An area A could be "applied" to a line segment a, which meant finding a line segment b such that $\text{rect}(a, b) = A$ (analogous to division of factors of unequal dimension); similarly an area could be made into a square, which meant finding a line segment c such that $\text{sq}(c) = A$ (analogous to root extraction). Early modern mathematicians generally accepted this dimensional interpretation of the arithmetical and quadratic algebraic operations in geometry.

Viète (cf. Chapter 8) generalized this classical interpretation to apply for the letter symbols of his *specious logistics.* These letter symbols represented magnitudes of various dimensions (in analogy with line segments, areas, solids, but extended to abstract higher dimensions as well). Consequently, the expressions and equations in Viète's algebra had to be homogeneous and of integer dimension.[10] As in the classical interpretation, the meaning of an expression in Viète's *specious logistics* did not depend on the choice of a unit, indeed the device of a unit line segment did not occur in Viète's algebra.

Homogeneity eliminated

The main and crucial difference between Descartes' interpretation (as given in the *Geometry*) and the classical and Vietean ones was that Descartes, by introducing a unit, removed the necessity for formulas to be homogeneous. In his new interpretation the formula $ab + c$, for line segments a, b, c, did not denote a non-interpretable sum of an area and a line but simply the sum of two

[9] A characteristic example was the equation $x^4 - 17x^2 - 20x - 6 = 0$, with solutions $2 \pm \sqrt{7}$ and $-2 \pm \sqrt{2}$, [Descartes 1637] pp. 385–386. Here the context was clearly geometrical; the example illustrated the phenomenon that a problem may lead to a fourth-degree equation and still be plane, that is, solvable with straight lines and circles. In another example, *Ibid.* p. 379, the coefficients were irrational numbers: $x^3 - \sqrt{3}x^2 + \frac{26}{27}x - \frac{8}{27}\sqrt{3} = 0$. The example showed a method to remove the irrationals from the coefficients by the substitution $y = \sqrt{3}x$. This substitution ($y^2 = 3x^2$) is more likely to arise in a geometrical than in a numerical context.

[10] Expressions with non-integer dimension, such as $\sqrt{a}$ and $\sqrt[3]{a}$ (with reference to a line segment a) did not occur in Viète's algebra because k-th roots could only be extracted from k-dimensional quantities. Descartes' interpretation did allow these expressions; see below.

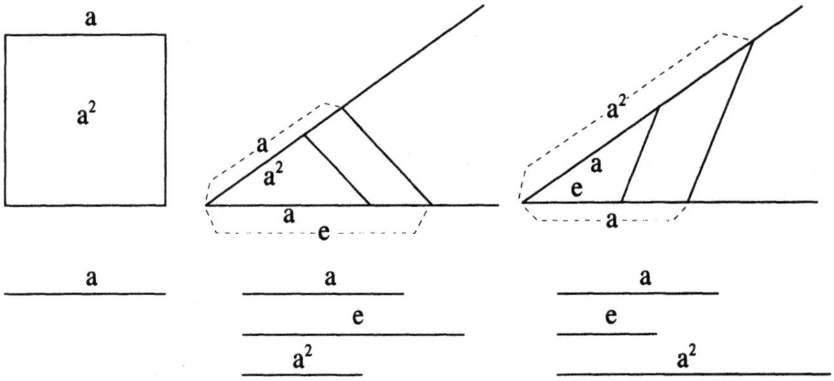

Figure 21.4: Descartes' interpretation of a^2

line segments, because by the definition of multiplication ab was a line segment as well as c. By removing the homogeneity requirement Descartes avoided the conceptual difficulties of a dimensional interpretation that had caused Viète to introduce infinitely many abstract higher dimensions for each kind of quantity.

Role of the unit Thus the introduction of a unit simplified the use of algebra in geometry conceptually in so far as dimensions higher than three were avoided. It should be remarked, however, that it introduced a complicating factor of its own, namely, that the operations were no longer uniquely determined. In the classical interpretation the result of a multiplication, a division, or a root extraction did not depend on the choice of a unit. If a, b, and c were line segments, a^2 was the unique square with side a. $\sqrt{ab}$ was the uniquely determined side of a square equal in area to a rectangle with sides a and b; that side was known to be equal to the mean proportional of a and b. For Viète $\frac{ab}{c}$ was the uniquely determined side of a rectangle that was equal in area to the rectangle with sides a and b and whose other side was c; it was known that, consequently, $\frac{ab}{c}$ was the fourth proportional of c, a, and b.

This uniqueness was lost in Descartes' new interpretation. Consider, for instance, the meaning of a^2 for a line segment a. If the unit e happened to be chosen smaller than a, then Descartes' construction yielded an a^2 that was larger than a (cf. Figure 21.4), but if the unit e was chosen larger than a, then

$a^2 < a$.[11] Descartes explained that the choice of the unit was arbitrary. Most often a geometrical situation (in particular the kind of configurations occurring in the early modern geometrical problems) did not suggest an obvious unique choice of a unit. Thus Descartes' new interpretation introduced an essential arbitrariness. Descartes did not explicitly discuss this feature. In fact, as will become clear in Section 21.3, he mostly avoided the choice of a unit and worked with homogeneous formulas.

Interpretation in the "Rules"

In the *Rules* (cf. Section 18.2) Descartes had interpreted the product of two line segments a and b as the rectangle $\mathrm{rect}(a,b)$. But by considering a rectangle with unit width e as equivalent to its length, $\mathrm{rect}(e,l) \simeq l$, he could in principle reduce any result of the algebraic operations to either an area or a line segment. Thus at that time his interpretation was still dimensional and required homogeneity in some sense, but it avoided the necessity of introducing (as Viète had done) magnitudes of dimensions higher than three. Moreover, in the *Rules* Descartes used the relation (cf. Figure 18.1)

$$\mathrm{rect}(a,b) = \mathrm{rect}(e,c) \simeq c\,, \tag{21.1}$$

in which c is the fourth proportional of e, a, and b. In the *Geometry* Descartes defined the product of a and b immediately as equal to this fourth proportional c, constructed according to the standard Euclidean construction of the fourth proportional. Thus the interpretation of the operations given in the *Geometry* can be seen as the natural successor to the one in the *Rules*, derived from it by incorporating the reduction of rectangles to line segments in the definitional constructions.

Van Ceulen

The introduction of a unit in situations where calculations were needed in geometry was not at all new. We have seen (cf. Section 7.2) how, for instance, Regiomontanus employed such a unit in his theory of triangles. However, in most cases the introduction of the unit implied the introduction of numbers as well and that was not what Descartes did. As to the use of a unit in non-numerical context there was, as we have seen in Section 8.6, at least one precursor of Descartes, namely, Van Ceulen, whose posthumous *Foundations*,[12] published in 1615, contained an interpretation of the quadratic algebraic operations that used a unit length and thereby considered products and quotients of line segments as line segments. Van Ceulen based his constructions on *Elements* III-35 (Cf. Construction 8.1). Moreover, Van Ceulen dealt only with line segments constructible by Euclidean means, thus restricting himself to line segments corresponding to quadratic irrational numbers. The main difference between his and Descartes' interpretation was that Descartes based his operations on different Euclidean constructions, and that he realized the necessity of extending the interpretation beyond quadratic algebra.

[11] The phenomenon was noted by Debeaune; see Note 14.
[12] [Ceulen 1615] and [Ceulen 1615b].

It is quite possible that Descartes knew Van Ceulen's book. It is less likely that his interpretation of the operations owed much to Van Ceulen. The idea of using a unit length was readily suggested by practical geometry (cf. Section 7.2) and the earlier form of Descartes' interpretation as extant in the *Rules* much more suggests a direct elaboration of this idea than an inspiration from Van Ceulen's somewhat idiosyncratic geometrical interpretation of the operations with quadratic irrationals.

21.3 The actual interpretation of the algebraic operations in the *Geometry*

Boldness Descartes' reinterpretation of the algebraic operations has deservedly become famous in the history of mathematics. However, that fame is based primarily on the boldness of his removal of homogeneity and dimensionality, not on the way he applied his new approach. In fact, despite the emphatic presentation of the new interpretation of the algebraic operations at the beginning of his book, he hardly ever applied it.[13] When dealing with actual geometrical problems he seldom introduced a unit and as a result the formulas he arrived at were dimensionally homogeneous. This applied, for instance, throughout his long treatment of Pappus' problem (cf. Section 23.3).

Compatibility The fact that Descartes used two different interpretations of the algebraic operations in geometry raises the question whether these interpretations were compatible. The compatibility is not obvious because the one interpretation depended on the choice of a unit and the other did not. Descartes himself did not mention this question. However, one of the first commentators of the *Geometry*, Debeaune, devoted a considerable section of his "Notes" to the dependence of products of line segments on the unit. He showed by examples that, despite this dependence, the results in Descartes' interpretation of the operations were compatible with the results of the dimensional interpretation. He concluded with the advice in general to leave the unit undetermined and calculate with homogeneous formulas unless a unit was explicitly provided at the outset.[14] The advice precisely summarized Descartes' practice in the *Geometry*.

[13]The interpretation of multiplication and division by means of a unit was not used in the *Geometry* at all. The interpretation of root extraction occurred occasionally in the third book, but alternative interpretations were used much more often. Indeed the first square root occurring after the explanation of root extraction concerned the solution of the equation $x^2 = ax + b^2$ (which Descartes denoted homogeneously, writing b^2 rather than simply b, cf. Construction 22.1). Its solution involved the root $\sqrt{\frac{1}{4}a^2 + b^2}$. Descartes did not interpret this root by means of the unit, but classically (and indeed much more naturally) as the hypothenuse of a right-angled triangle with sides $\frac{1}{2}a$ and b.

[14][Debeaune 1649], pp. 107–112 in the edition [Descartes 1659–1661] vol. 1. Debeaune discussed the products b^2, bd, and d^2 (for line segments b and d) and showed that if these were determined in Descartes' manner by means of a unit, their ratios were the same as in the classical dimensional interpretation. He also presented an example involving two different units, showing geometrically that the line segments representing the squares a^2, b^2, and c^2 of the sides a, b, and c of a rectangular triangle did depend on the choice of the unit, but that

The two interpretations are indeed compatible. To be precise: whenever a formula denotes a line segment in the classical dimensional interpretation, then Descartes' interpretation will yield the same line segment, independently of the choice of the unit. The reason is that in these cases the constructions always consist of pairs of operations each involving the unit, and in these pairs the dependence of the unit in the one operation is canceled in the other. Consider, for instance, $\sqrt{ab}$. Dimensionally interpreted this is a well-defined line segment, the mean proportional between a and b; it is independent of the choice of a unit. Interpreted according to Descartes $\sqrt{ab}$ is the square root, to be constructed by Construction 21.3, of a line segment, which itself is the product of two line segments a and b, constructed by 21.1. Both constructions involve the unit e and in each of them the result of the construction is dependent on the choice of e. But in their combination the unit e cancels.[15]

Significance

It is difficult to assess the significance of Descartes' silence about the compatibility of his new interpretation of the algebraic operations and the classical one. It seems that he was simply not aware of the problem, perhaps because in practical geometry different units were used apparently without giving rise to problems of compatibility. I find Descartes' omission striking and suggestive of a certain carelessness about the matter; it suggests that he considered his interpretation of the quadratic algebraic operations as a necessary preliminary formality for his method rather than as an important contribution to geometry and algebra. The fact that he himself hardly ever used his new interpretation points in the same direction. Descartes probably saw the other achievements of the *Geometry*, notably the geometrical interpretation of higher-order root extraction and the solution of equations generally, as much more momentous than his interpretation of the quadratic algebraic operations.

---

the relation $a^2 + b^2 = c^2$ remained valid whichever unit was chosen.

[15] According to Construction 21.1, the product c of a and b is the fourth proportional of e, a and b, so $e : a = b : c$, hence $\mathrm{rect}(e, c) = \mathrm{rect}(a, b)$. According to Construction 21.3, the square root d of c is the mean proportional of e and c, that is, $e : d = d : c$, hence $\mathrm{sq}(d) = \mathrm{rect}(e, c)$. So the two definitions involved in the expression $d = \sqrt{ab}$ imply $sq(d) = \mathrm{rect.}(a, b)$, which means that d does not depend on e. A similar argument applies in the case of $\frac{ab}{c}$ and in general to all formulas that, dimensionally interpreted, denote line segments.

Chapter 22

The use of algebra in solving plane and indeterminate problems

22.1 Problem, equation, construction

From problem to equation

Having settled the geometrical interpretation of the elementary and the quadratic operations at the beginning of the *Geometry*, Descartes dealt with the application of these operations in solving plane problems and in dealing with indeterminate problems that could be reduced to plane ones. It is instructive to analyze in some detail the interplay of algebra and geometry in his approach to these problems.

He first explained how any problem could be translated into an equation. He did so in the sixth section of the first book of the *Geometry*, which was entitled

> How one should arrive at the equations that serve for solving the problems.[1]

The section[2] contained an elaboration of the procedure described in Rules 16–21 of the *Rules* (cf. Section 18.2): One should start by assuming that the problem was solved[3] and consider a figure incorporating the solution. One should denote line segments in that figure by letters, chosing $a, b, c, \cdots$ for the given or otherwise known ones, and $z, y, x, \cdots$ for the as yet unknown ones. Then one should collect the given and required relations between these line segments and express these as equations. One should try to find as many equations as there were unknowns, which in the usual cases would be possible. If that appeared impossible, it was a sign that the problem was not fully determined. I return

[1] [Descartes 1637] p. 300: "Comment il faut venir aux Equations qui servent a resoudre les problesmes."

[2] [Descartes 1637] pp. 300–302.

[3] The classical opening of an analysis procedure; cf. Section 5.2.

to that case in Section 22.3. Then one had to eliminate all but one of the unknowns, arriving at an equation in the one remaining unknown, such as, for instance (they are Descartes' own examples,[4] note that he chose homogeneous equations and considered the letters to denote positive quantities):

$$\begin{aligned} z &= b\,, \\ z^2 &= -az + bb\,, \\ z^3 &= +az^2 + bbz - c^3\,, \\ z^4 &= az^3 - c^3z + d^4\,. \end{aligned} \tag{22.1}$$

Descartes stressed that one should use "all the divisions that are possible" to arrive at the simplest possible equation.[5] These divisions referred to methods to reduce equations by splitting off factors, which Descartes explained later in book III; I return to them in Section 27.3. With the final equation thus reached, the analytical part of the method was completed, what remained was the synthesis, that is, the construction.

From equation to construction

Having explained the general procedure to translate a geometrical problem into an equation, Descartes proceeded to the obvious next step: to translate an equation into a solution, that is, into a geometrical construction of its roots. In book I he restricted himself to plane problems for which, he asserted, the final equation was linear or quadratic. Descartes did not deal explicitly with the construction of roots of linear equations; he probably considered the procedure to be obvious from the constructions for multiplication and division. He distinguished three types of quadratic equations, depending on the signs of the coefficients, namely (note again that he wrote the equations homogeneously and that he assumed the given line segments a and b to be positive):[6]

$$\begin{aligned} x^2 &= ax + b^2 \\ x^2 &= -ax + b^2 \\ x^2 &= ax - b^2\,; \end{aligned} \tag{22.2}$$

in each case he gave a construction of the root or roots (he disregarded negative roots). For illustration, here is his construction for the first case:

Construction 22.1 (Root of $x^2 = ax + b^2$)[7]
Given two line segments a and b (see Figure 22.1), it is required to construct a line segment x satisfying $x^2 = ax + b^2$.

[4] [Descartes 1637] p. 301, it is not clear why he added the plus sign to the term az^2.

[5] [Descartes 1637] p. 302: ". . . pourvû qu'en demeslant ces Equations on ne manque point a se servir de toutes les divisions, qui seront possibles, on aura infalliblement les plus simples termes, ausquels la question puisse estre reduite."

[6] [Descartes 1637] pp. 302–303. Note that Descartes implicitly assumed that any positive two-dimensional constant could be considered as a square. In geometrical problems such constants would generally arise as rectangles, which indeed, by Construction 21.3, could be equated to a square.

[7] [Descartes 1637] pp. 302–303.

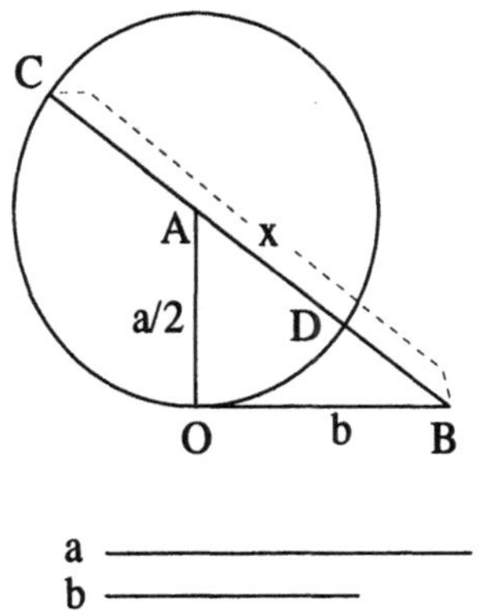

Figure 22.1: Construction of the root of $x^2 = ax + b^2$

Construction:
1. Draw a right angled triangle AOB with $OA = \frac{1}{2}a$, $OB = b$ and $\angle AOB = 90^o$.
2. Draw a circle with center A and radius $\frac{1}{2}a$.
3. Prolong AB; the prolongation intersects the circle in C.
4. $x = BC$ is the required line segment.
[**Proof**: BA intersects the circle in D; by *Elements* III-36 (cf. Construction 4.3) $BC \cdot BD = OB^2$, i.e., $x(x-a) = b^2$, so $x^2 = ax + b^2$.]

The constructed x is the positive root of the equation; the other root is negative and therefore plays no role. Descartes provided no separate proof of this construction. In the tradition of problem solving it was known; it occurred in Clavius' Euclid edition and I have discussed it above (Construction 4.3 of a segment x satisfying $x(x-a) = b^2$). Descartes added that x could be expressed as

$$x = \frac{1}{2}a + \sqrt{\frac{1}{4}a^2 + b^2}\,. \qquad (22.3)$$

It is noteworthy that Descartes presented this relation as a corollary of the construction (it follows from $CB = CA + AB$) and not as the formula resulting from an algebraic procedure for solving the equation.

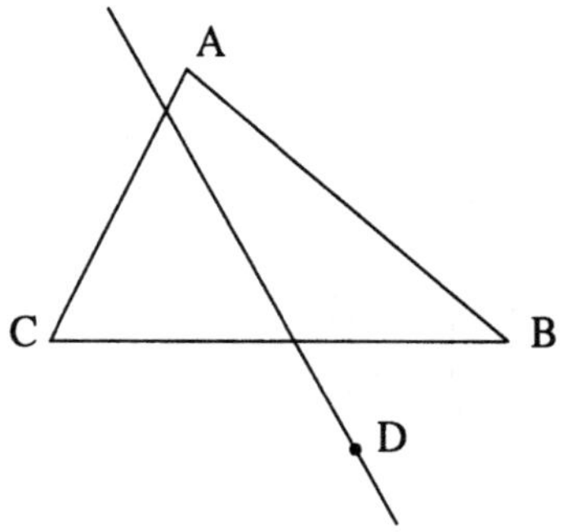

Figure 22.2: Division of a triangle

22.2 An example

Clavius' division problem

With the constructions for the quadratic equations Descartes had given a complete description of his method of problem solving as far as plane problems were concerned. He gave no special example but went on directly to discuss the Pappus problem, which is indeterminate; its solutions form a locus. It will be instructive, however, to pause and apply Descartes' method to a characteristic plane problem. For that purpose I take the triangle division problem whose solution by Clavius we saw in Section 4.8 (Construction 4.18). We have no evidence that Descartes ever discussed or solved this problem. What follows is my own application of Descartes' method to the triangle division problem. I follow the procedures that Descartes prescribed:

Problem 22.2 (Triangle division)[8]
Given a triangle ABC (see Figure 22.2) and a point D outside the triangle, it is required to draw a line through D dividing the triangle in two equal parts.

The application of Descartes' procedure to this problem consists of an analysis and a construction:

Analysis 22.3 (Triangle division)
Given etc.: cf. Problem 22.2.

[8] I have published the following Cartesian solution of the triangle division problem in [Bos 1990] pp. 353–356 (pp. 40–43 in ed. [Bos 1993c]) and [Bos & Reich 1990] pp. 206–212.

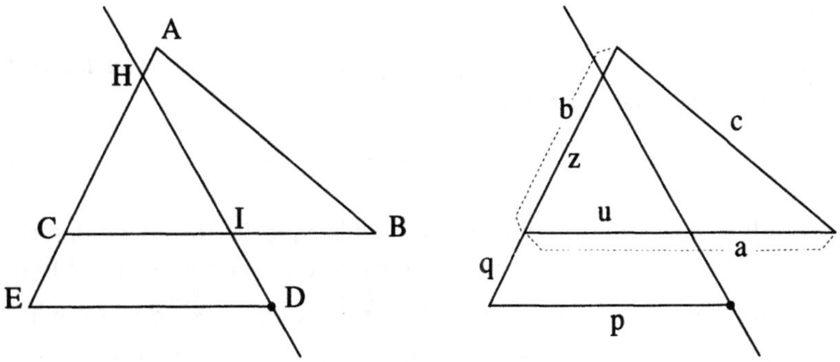

Figure 22.3: Triangle division — analysis

Analysis:
1. Assume the problem solved (see Figure 22.3 — for easier comparison I use the same figure and lettering as in my rendering of Clavius' solution, cf. Figure 4.14) and let the required line through D intersect AC in H and BC in I (we study the case that the dividing line through D intersects the sides AC and BC; the other cases can be treated in the same way); draw a line through D parallel to BC, it intersects AC prolonged in E.
2. Give names (letters) to the known elements in the figure: $CA = b$, $CB = a$, $AB = c$, $ED = p$, $CE = q$; and to the unknown elements: $CI = u$ and $CH = z$.
3. Identify the given and required relations: Because DE is drawn parallel to BC and D is given we have, by similar triangles, $u : p = z : (q + z)$, which can be written as an equation:

$$pz = u(q + z) ; \tag{22.4}$$

furthermore, it is required that the line through D divides the triangle in equal parts, hence $\Delta CHI = \frac{1}{2}\Delta ABC$, which implies, by *Elements* VI-23, that $uz : ab = 1 : 2$ or, written as an equation,

$$uz = \frac{1}{2}ba ; \tag{22.5}$$

we have now as many equations as unknowns.

4. Eliminate one unknown, namely u; this leads to

$$z^2 = \frac{\frac{1}{2}ba}{p} z + \frac{\frac{1}{2}ba}{p} q \,, \tag{22.6}$$

an equation in one unknown of second degree.

We have now arrived at the end of the analysis; the problem is translated into an equation in one unknown of the type

$$z^2 = fz + g^2 \,. \tag{22.7}$$

Descartes' standard construction of the root of this equation has been described above (Construction 22.1). In order to apply that construction we need first to construct the line segments f and g. Now $f = \frac{\frac{1}{2}ba}{p}$ and $g^2 = \frac{\frac{1}{2}ba}{p} q = fq$, whence $g = \sqrt{fq}$. If we should follow Descartes' prescribed method to the letter, we should choose a unit and apply the Constructions 21.1–21.3 to find f and g. However, the right-hand sides of the equations

$$f = \frac{\frac{1}{2}ba}{p} \tag{22.8}$$

$$g = \sqrt{fq} \,, \tag{22.9}$$

dimensionally interpreted, denote line segments and therefore the result of their construction is independent of the choice of the unit (cf. Section 21.3 and Note 15 of Chapter 21). For f the easiest construction is obtained by choosing the unit e equal to p, for g by choosing $e = q$. In fact the constructions then reduce to the ones to which the classical interpretation of these formulas would lead: $\frac{\frac{1}{2}ba}{p}$ is the fourth proportional of p, $\frac{1}{2}b$, and a, and $\sqrt{fq}$ is the mean proportional of f and q. A strict adherence to Descartes' procedures would force us to choose one unit (after all he did not explain that in these cases any unit leads to the same result) so that either f or g or both would be obtained by an unnecessarily complicated construction. It seems likely that neither Descartes nor his readers would stick to the letter here but that they would adjust the units (or follow the classical interpretation), which leads to the following construction:

Construction 22.4 (Triangle division)
Given etc: cf. Problem 22.2.

Construction:
1. Bisect b to find $\frac{1}{2}b$.
2. Determine the fourth proportional of p, $\frac{1}{2}b$, and a, or, equivalently, the product of $\frac{1}{2}b$ and a according to Construction 21.1 taking p as unit; that is (see Figure 22.4): draw two lines L and M intersecting in a point O; take p and $\frac{1}{2}b$ along L and a along M; connect the endpoints of p and a by a line; draw a parallel line through the endpoint of $\frac{1}{2}b$, its intersection with M marks the endpoint of the

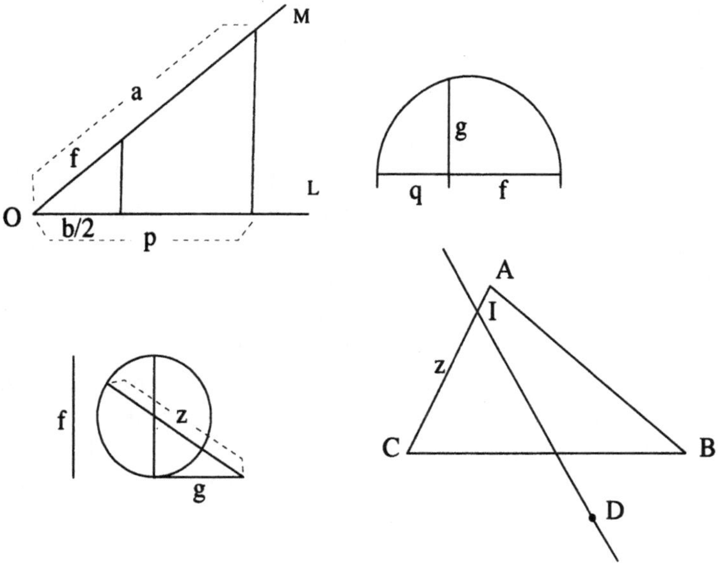

Figure 22.4: Triangle division — construction

required line f.
3. Determine the mean proportional g of f and q, or, equivalently, construct $g = \sqrt{f}$ by taking q as unit; that is: join q and f along a straight line; draw a semicircle with diameter $q + f$; draw a perpendicular to the diameter from the point where q and f meet; the intercept between that point and the circle is the required segment g.
4. Having constructed f and g we can now apply Descartes' Construction 22.1 for finding the roots of Equation 22.7 $z^2 = fz + g^2$: Mark f and g along two perpendicular lines from the point of intersection (see Figure 22.4); draw a circle with f as diameter; connect the endpoint of g and the center of the circle by a line and prolong that line until it meets the circle; the resulting line segment is the required root z.
5. Draw $CI = z$ in the original triangle and connect I and D; DI is the required line dividing the triangle in two equal parts.
[**Proof**: I omit the proof, which follows easily from the analysis.]

Comments on the example

We may now gather the conclusions to be drawn from the exercise of applying Descartes' method of analysis to a characteristic plane problem from the tradition of geometrical problem solving. The method indeed provides a construction, in fact, it leads to precisely the same construction as Clavius had found (cf. 4.18).[9] I have noted that if Descartes' approach is followed to the letter, the resulting construction is more complicated than Clavius' because we are then obliged to introduce a non-functional unit length in the construction of the multiplication or the square root.

The correspondence to Clavius' construction would be less marked had we chosen other unknowns in the analysis or eliminated differently. However, one soon convinces oneself that the choice of u and z as unknowns is the most natural one; and eliminating z rather than u leads to a quadratic equation of the second type in Equation 22.2 with coefficients differing from those in Equation 22.6 merely in that the roles of p and q are interchanged. Hence, the construction in that case is not much different from the one given here.

Our example also shows a particular difference between the analytical and the synthetical part of Descartes' procedure. The analysis, leading from problem to equation, was not determinate; there was some room for choice in selecting the unknowns and the order of their elimination. The synthesis, however, was completely determined and automatic, it consisted of applying standard constructions for the algebraic operations and standard constructions of the roots of equations.

Descartes could extend the analytical part of the procedure to higher-order problems without essential change. But the synthetic part of the solution of higher-order problems (equations) was much more difficult; it required new standard construction procedures for all higher-order equations, and these procedures in turn required a new interpretation of constructional exactness. Much of the contents of Books II and III of the *Geometry* is devoted to that new interpretation; it will be discussed in Chapters 24–26.

22.3 Indeterminate problems

Reduced to pointwise construction

In the majority of problems from the early modern tradition of geometrical problem solving Descartes' analysis would indeed lead to an equation in one unknown whose roots would correspond to the finitely many solutions of the problem. However, for some problems the final equation involved two or more unknowns. In that case the problem was indeterminate; it admitted an infinity of solutions. Descartes' approach to these problems was to reduce them to determinate ones by choosing arbitrary values for some of the unknowns:

> And one should find as many such equations as there are supposed to be unknown lines. But if so many cannot be found, and nothing of what is required in the problem has been left out of consideration, this shows that the question is not entirely determined. In such a case one may choose at will known lines for each unknown line to which there corresponds no equation.[10]

[9] Step **1** of Clavius' construction corresponds to the introduction of the line segments p, q, and $\frac{1}{2}b$, step **2** to the determination of f, step **3** to the determination of g, and step **4** to the determination of z; the techniques applied for these determinations are exactly the same.

[10] [Descartes 1637] p. 300: "Et on doit trouver autant de telles Equations, qu'on a supposé

Of such indeterminate problems Descartes considered primarily those with two unknowns.[11] In that case the solutions form a one-dimensional locus in the plane, a straight line or a curve, and Descartes' procedure supplied two coordinates (one chosen and one constructed) of a point on the curve. In this way arbitrarily many points on the curve could be found, and thus the curve could be constructed point by point.[12] Only in the case of the three- and four-line Pappus problem and in two other special cases of Pappus' problem did Descartes suggest a construction of a locus as curve (rather than as collection of points); these cases will be discussed in the next chapter.

de lignes, qui estoient inconnuës. Oubien s'il ne s'en trouve pas tant, et que nonobstant on n'omette rien de ce qui est desiré en la question, cela tesmoigne qu'elle n'est pas entierement determinée. Et lors on peut prendre a discretion des lignes connuës, pour toutes les inconnuës ausqu'elles ne correspond aucune Equation."

[11]In a short section at the end of book II ([Descartes 1637] pp. 368–369; II-F in Table 20.1) Descartes discussed curves on non-plane surfaces; he did not, however, remark that such a surface corresponds to an equation in three unknowns.

[12]Cf. [Descartes 1637] p. 313, where Descartes described the procedure in relation to a Pappus problem: ". . . on peut prendre a discretion l'une des deux quantités inconnues x ou y, et chercher l'autre par cete Equation. . . . Mesme prenant successivement infinies diverses grandeurs pour la ligne y, on en trouvera aussy infinies pour la ligne x, et ainsi on aura une infinité de divers poins, . . . par le moyen desquels on descrira la ligne courbe demandée."

Chapter 23

Descartes' solution of Pappus' problem

23.1 The problem

Locus problems

Descartes first studied Pappus' problem during late 1631 and early 1632, on the instigation of Golius. In Chapter 19 I argued that the confrontation with the problem was decisive for the final stage of the development of his programmatic ideas on geometry. I now turn to his treatment of the problem in the *Geometry*, where he used it as the central example for illustrating his techniques and showing their power.

Pappus' problem was a locus problem, that is, an indeterminate problem whose infinitely many solutions form a one-dimensional locus.[1] Such loci are curves (or sometimes straight lines). Descartes' primary approach to locus problems was (cf. Section 22.3) to find a pointwise construction of the locus. He prescribed assuming arbitrary values for one of the two unknowns in the final indeterminate equation, and to determine the corresponding values for the other unknown by the methods suitable for determinate problems. The pairs of values thus constructed were coordinates of points on the locus. In principle, any number of such points could be determined; hence, the result of this procedure was a pointwise construction of the required locus.[2]

The problem

I recall that Pappus' problem is as follows (cf. Problem 19.1):[3]

[1] Cf. Descartes' own characterization of locus problems, [Descartes 1637] pp. 334–335: "Car ces lieux ne sont autre chose, sinon que lors qu'il est question de trouver quelque point auquel il manque une condition pour estre entierement determiné."

[2] Cf. Table 4.2 item 3.2 and Notes 10 and 12 of Chapter 22.

[3] Cf. also Notes 3 and 5 of Chapter 19.

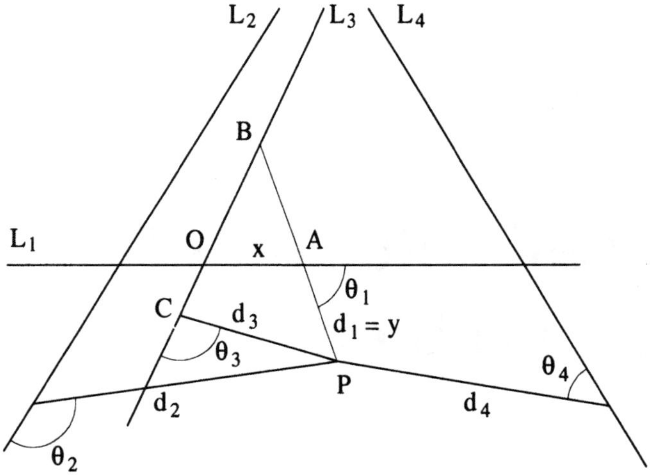

Figure 23.1: Pappus' problem

Problem 23.1 (Pappus' problem)[4]
Given n straight lines L_i in the plane (see Figure 23.1), n angles θ_i, and a line segment a. For any point P in the plane, the oblique distances d_i to the lines L_i are defined as the (positive) lengths of segments that are drawn from P toward L_i making the angle θ_i with L_i. It is required to find the locus of points P for which a certain ratio, involving the d_i and depending on the number of lines, is constant. The relevant ratios are:

For 3 lines:	d_1^2	:	$d_2 d_3$	(23.1)
For 4 lines:	$d_1 d_2$	:	$d_3 d_4$	(23.2)
For 5 lines:	$d_1 d_2 d_3$	:	$a d_4 d_5$	(23.3)
For 6 lines:	$d_1 d_2 d_3$	:	$d_4 d_5 d_6$	(23.4)
In general for an even number $2k$ of lines:	$d_1 \ldots d_k$	:	$d_{k+1} \ldots d_{2k}$,	(23.5)
and for an uneven number $2k+1$:	$d_1 \ldots d_{k+1}$	:	$a d_{k+2} \ldots d_{2k+1}$.	(23.6)

[4][Pappus Collection] pp. 507–510; cf. [Pappus 1876–1878] vol. 2, pp. 676–681 and [Pappus 1986] vol. 1 pp. 118–123.

23.2 The general solution: equations and constructions

Equations

Descartes' solution of the problem as developed in the *Geometry* started in the first book[5] with the general derivation of the equation of the locus. I have briefly sketched his procedure in Section 19.2; I now analyze it in more detail. Descartes introduced a coordinate system with its origin O at the intersection of L_1 and one of the other lines (L_3 in the figure), its X-axis along L_1 and its ordinate angle equal to θ_1. With respect to that system d_1 is equal to y. He showed that for any point P with coordinates x and y, each d_i could be written as

$$d_i = \alpha_i x + \beta_i y + \gamma_i \,, \tag{23.7}$$

in which the α_i, β_i, and γ_i were constants expressed in terms of ratios of line segments determined by the θ_i and the segments along the lines L_i between their points of mutual intersection;[6] because the θ_i and the positions of the lines L_i in the plane were given, the α_i, β_i, and γ_i were known.[7] Descartes was aware that if all the lines L_i are parallel, the x did not occur in the equation, because $d_1 = y$, $d_i = \beta_i y + \gamma_i$ for all $i > 1$.

The requirement that the given ratio (cf. Equations 23.5 and 23.6) be constant leads to the equation:

$$y(\alpha_2 x + \beta_2 y + \gamma_2)\cdots = \delta\overline{a}(\alpha_l x + \beta_l y + \gamma_l)(\alpha_{l+1}x + \beta_{l+1}y + \gamma_{l+1})\cdots \,; \tag{23.8}$$

with $l = k+1$ or $k+2$ depending on the number of given lines, $\overline{a} = a$ for an uneven number of lines and $= 1$ otherwise, and δ is the given constant value of the ratio. These are polynomial equations[8] in x and y (or in y alone if the

[5] [Descartes 1637] pp. 310–314.

[6] Thus to express d_3 in x and y Descartes considered the triangles OAB and CPB (A and B are the intersections of the extension of d_1, with L_1 and L_3 respectively, C is the intersection of d_3 with L_3); he noted that although P was unknown, the angles of these triangles, and thereby the ratios of their sides, were known. Thus if $AB : OA = \lambda$ and $CP : BP = \mu$, λ and μ are known and $d_3 = \mu BP = \mu(y + AB) = \mu(y + \lambda x) = \mu y + \mu\lambda x$. Descartes wrote the ratios not as single letters but as ratios between constant line segments; thus for the μ and λ above he wrote $c : z$ and $b : z$, respectively. Note that, contrary to Descartes' usual practice, z denoted an indeterminate, not an unknown.

[7] [Descartes 1637] p. 312; see below, Equations 23.13 and 23.14 for the case of the three- and four-line locus.

[8] If the problem is taken in its strict classical sense, the d_i, as well as the line segment a and the ratio δ, should be interpreted as positive, whence the equation should be

$$|y(\alpha_2 x + \beta_2 y + \gamma_2)\ldots| = \delta|\overline{a}(\alpha_l x + \beta_l y + \gamma_l)(\alpha_{l+1}x + \beta_{l+1}y + \gamma_{l+1})\ldots| \,, \tag{23.9}$$

which is equivalent to

$$y(\alpha_2 x + \beta_2 y + \gamma_2)\cdots = \pm\delta\overline{a}(\alpha_l x + \beta_l y + \gamma_l)(\alpha_{l+1}x + \beta_{l+1}y + \gamma_{l+1})\cdots \,. \tag{23.10}$$

The solution of one Pappus problem, therefore, consists of two curves. For a given set of n straight lines the collection of Pappus loci with respect to these lines and arbitrary constant values for the ratio thus constitutes a one-parameter family of curves represented by the equation

$$y(\alpha_2 x + \beta_2 y + \gamma_2)\cdots = \delta\overline{a}(\alpha_l x + \beta_l y + \gamma_l)(\alpha_{l+1}x + \beta_{l+1}y + \gamma_{l+1})\cdots \,, \tag{23.11}$$

Number of given lines	Degree of the equation	Highest power of x in the equation	Case of parallel given lines; degree of the equation in y
3	2	2	2
4	2	2	2
5	3	2	3
6	3	3	3
⋮	⋮	⋮	⋮

Table 23.1: Pappus' problem — the degrees of the equations

given lines are parallel). As all d_i are linear in x and/or y, the degrees of these equations depend on the number of given lines. Descartes did not explicitly note these dependencies, but it appears from his further statements about the constructibility of points on the loci that he was aware of them. They are listed in Table 23.1.

Constructions On the basis of the numbers in Table 23.1 Descartes was able to make general statements about the means of construction (plane, solid, or higher-order) necessary in the pointwise construction of the loci. Points on the loci could be constructed (cf. Section 22.3) by giving arbitrary values to y and constructing the corresponding xs as root(s) of the resulting equations. Choosing fixed values for y had the advantage that in the case of five, seven, nine, etc., lines, the resulting equations had degrees two, three, four, etc., whereas for fixed values of x the degrees were higher, namely, three, four, five, etc. The special cases in which all given lines were parallel led, as is easily seen, to equations in one unknown only, namely y. The resulting loci consisted of straight lines parallel to the given ones; the positions of these lines were determined by the roots of this equation in y.

The degrees of the final equations in x, or, for parallel lines, in y, are listed in the last two columns of Table 23.1. These degrees determined by what means the roots could be constructed. Descartes explained the relation between degree and constructibility in the third book of the *Geometry* (cf. Chapter 26), but anticipating his results there, he classified in the first book the cases of Pappus'

where δ now ranges over all (positive and negative) values. Usually Descartes started by considering one point on the locus and adjusted the coordinate system such that its x and y coordinates were positive. He then read off the values of the coefficients α_i, β_i, γ_i from the figure and tacitly assumed that the expressions thus gained applied generally. Moreover, he usually took the constant δ to be 1. The effect of these choices was that in dealing with a Pappus problem he considered one solution curve only. Yet the figures he provided suggest that he was well aware of the other solutions and realized that an obvious adjustment of the equation would produce them.

Number of given lines:	Degree of the equation in one unknown:	Points on the locus constructible by:
3, 4, 5 but not 5 parallel	2	plane means (circles and straight lines)
5 parallel, 6, 7, 8, 9 but not 9 parallel	3 or 4	solid means (conic sections)
9 parallel, 10, 11, 12, 13 but not 13 parallel	5 or 6	circles and a curve "only one degree more composite than the conics"
	etc.	

Table 23.2: Pappus' loci — pointwise constructibility

problem according to the means necessary for their pointwise construction. I summarize his classification[9] in Table 23.2. Descartes postponed the explanation of the expression "a curve only one degree more composite than the conics:"[10] he had the "Cartesian parabola" in mind, cf. Section 26.3.

These results concluded the first book of the *Geometry*. They convincingly illustrated the power of Descartes' method by surveying the various cases of a difficult problem, classifying these, and determining the status as to constructibility of each class.

23.3 The three- and four-line Pappus problem

The problem

However impressive, the result reached at the end of Book I was a classification only, it did not provide the actual constructions. In Book II Descartes dealt in much more detail with one Pappus problem, namely, the problem in three or four lines:

Problem 23.2 (Pappus' problem in three and four lines)[11]
Given four straight lines L_i in the plane (see Figures 23.1 and 23.2, the problem "in three lines" arises if two of the given lines coincide) and four angles θ_i. For any point P in the plane, the oblique distances d_i to the lines L_i are defined as in Problem 23.1. It is required to find the locus of points P for which

$$d_1 d_2 = d_3 d_4 \ . \tag{23.12}$$

[9][Descartes 1637] pp. 313–314.
[10][Descartes 1637] p. 314: ". . . une ligne, qui n'est que d'un degré plus composée que les sections coniques, en la façon que i'expliqueray cy aprés."
[11][Descartes 1637] pp. 324–334

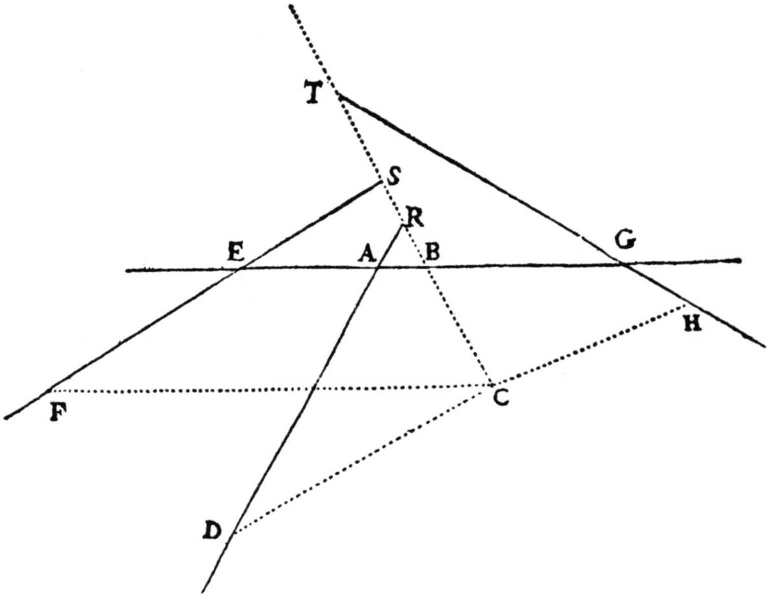

Figure 23.2: Pappus' problem in four lines (*Geometry* p. 309)

Equation 23.12 implies that Descartes took the given ratio δ to be equal to 1; he did not comment on this point, in fact his analysis can be easily adjusted to other values of δ.

In dealing with the three- and four-line locus Descartes did not pursue the approach by pointwise constructions; rather he provided constructions of the loci based on Apollonius' theory of conic sections. Descartes dealt with almost[12] all variants of the three- and four-line locus, and he presented his results in the classical form of constructions with proofs. His presentation was rather involved because of the many case distinctions he made. Descartes did not explain how he had arrived at the construction, but from the construction and the proof it is clear that he had used some indeterminate coefficient procedure.

I illustrate his analysis and construction by following one case, the one in which the locus is an ellipse. Afterward I briefly reconstruct the indeterminate coefficients method by which he probably found it.

The equation Descartes had explained the method for deriving the equation of the locus in the first book (cf. Analysis 19.2). Now he introduced letters for the unknowns and the various given parameters in the four-line Pappus problem as follows[13]

[12]Cf. Note 18.

[13]Contrary to the procedure explained in Section 1.7, I have not changed Descartes' lettering, partly because there are so many letters that some complexity cannot be avoided, partly to facilitate comparison with the original, which, because the *Geometry* is readily available, more readers may want to do than in the case of my other sources. I retain the notation L_i

(see Figure 23.2):

$$\begin{aligned} x &= AB \\ y &= BC \\ z:b &= AB:BR \\ z:c &= CR:CD \\ k &= AE \\ z:d &= BE:BS \\ z:e &= CS:CF \\ l &= AG \\ z:f &= BG:BT\,. \end{aligned} \tag{23.13}$$

Note that z was not an unknown but an indeterminate serving as common term in all the given ratios. He then expressed[14] the segments CB, CF, CD, and CH in terms of the known and unknown line segments introduced in Equation 23.13:

$$\begin{aligned} CB &= y \\ CF &= \frac{ezy + dek + dex}{z^2} \\ CD &= \frac{cyz + bcx}{z^2} \\ CH &= \frac{gzy + fgl - fgx}{z^2}\,. \end{aligned} \tag{23.14}$$

From these results, which were already given in Book I, the equation of the curve was readily calculated by inserting the values above in the defining property $CB \times CF = CD \times CH$ of the locus:[15]

$$y^2 = \frac{(cfglz - dekz^2)y - (dez^2 + cfgz - bcgz)xy + bcfglx - bcfgx^2}{ez^3 - cgz^2}\,. \tag{23.15}$$

Descartes then introduced further letters for abbreviation:[16]

$$\begin{aligned} 2m &= \frac{cfglz - dekz^2}{ez^3 - cgz^2} \\ \frac{2n}{z} &= \frac{dez^2 + cfgz - bcgz}{ez^3 - cgz^2} \end{aligned} \tag{23.16}$$

---

for the given lines and d_i for the distances; later on I introduce new coordinates u and v and letters r, t, and s for certain terms and line segments; they do not overlap with Descartes' lettering.

[14]Cf. Note 6.

[15][Descartes 1637] p. 325. Descartes did not separately discuss the case in which $ez^3 - cgz^2$ is zero, which leads to an equation without y^2-term; cf. Note 18 below.

[16]I have added a minus sign in the left-hand term of the last equation, where the text has $\frac{p}{m}$ ([Descartes 1637] p. 326). However, as Tannery has remarked (cf. [Descartes 1964–1974] vol. 6 p. 399), Descartes calculated further as if the left-hand term was $\frac{-p}{m}$. In their translation Smith and Latham stick to the + sign, whereby their formulae don't agree with those of the original.

$$\begin{aligned} o &= \frac{-2mn}{z} + \frac{bcfgl}{ez^3 - cgz^2} \\ \frac{-p}{m} &= \frac{n^2}{z^2} - \frac{bcfg}{ez^3 - cgz^2} . \end{aligned} \tag{23.17}$$

Note that Descartes did not introduce a unit length (cf. Section 21.3); as a result all his equations were homogeneous. Inserting the abbreviations and solving with respect to y, Equation 23.15 became[17]

$$y = m - \frac{n}{z}x + \sqrt{m^2 + ox - \frac{p}{m}x^2} , \tag{23.18}$$

which constituted the end result of the analysis of the three- and four-line problem.

The construction Descartes then turned to the construction of the locus on the basis of this equation. He did so by constructing, within the given configuration of lines, a conic section whose position and parameters depended on the coefficients of Equation 23.18. For the actual construction of this conic section he refered to the classical constructions of conic sections with given center, diameter direction, ordinate angle, and parameters as explained in Apollonius' *Conics*. He then proved that the constructed conic section was the required locus by showing that its equation coincided with the equation of the locus (i.e., equation 23.18). The total argument (construction and proof) implied an almost complete[18] proof that all quadratic equations in two unknowns represent conic sections and that therefore all three- and four-line loci are conic sections. It also implied a classification of the different cases (straight line, parabola, hyperbola, ellipse, circle).

Reading Descartes' argument is complicated by the fact that his terminology was based on the assumption that all letters in formulas denoted positive magnitudes and that therefore it was only the sign of a term in an equation that determined whether it should be added or subtracted. For instance in step 1 of Construction 23.3 below Descartes took $BK = m$ along BC downward from B "because here there is $+m$," and he stated that K should be taken upward from B "if there had been $-m$." This formulation ignored the possibility that m itself could be negative, whereas the definition of m in Equation 23.16 not at all excluded that possibility. Descartes was surely aware that terms such as m could turn out to be negative, but his terminology was not developed far enough to distinguish between the sign preceding a term and the positivity (or negativity) of that term.

[17] [Descartes 1637] p. 326.

[18] Actually, as he noted himself in a letter to Debeaune of 20 II 1639 ([Descartes 1964–1974] vol. 2 p. 511), he had overlooked the case in which the coefficient of y^2 in the equation of the curve is zero; cf. Note 15 above. In the letter he stated (correctly) that in that case the locus is a hyperbola; he also claimed that one of its asymptotes was parallel to the line AB, which is incorrect; it should be BC.

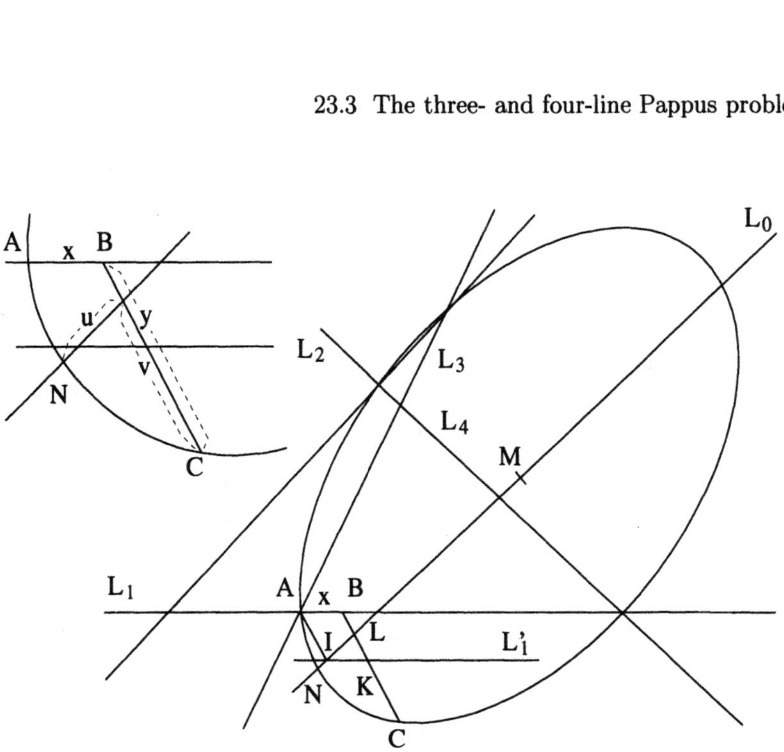

Figure 23.3: Pappus' problem in four lines — construction in the case of an ellipse

In rendering his construction and proof I paraphrase only one of the cases he distinguished, namely, the one in which the locus was an ellipse. It was as follows:

Construction 23.3 (Pappus' problem in three or four lines — case ellipse)[19]

Given and required: see Problem 23.2; the analysis of the problem has led to Equation 23.18.

Construction:

1. (Construction of the point (I) for which $x = 0$ and $y = m$.) Draw a line $L_1' \parallel L_1$ (see Figure 23.3[20]) intersecting BC in K and such that $BK = m$ (K taken below B because of + sign of m, see above); take point I on it such that $IK = x$.

(Descartes took the line BC as generic ordinate. The point I is independent of the value of x and therefore well defined; $\angle BAI$ is equal to the given angle θ_1 between d_1 and L_1, and $AI = m$. It

[19] [Descartes 1637] pp. 327–332.

[20] The figure is adapted from Descartes' figure in [Descartes 1637] p. 327. The locus is there drawn as a circle; I have stressed the generality of the case by drawing an explicit ellipse. I have removed the letters that don't occur in the argument and I have added the L_i to indicate the given and constructed straight lines, the x and y that Descartes did not incorporate in the figure, and the u and v that I use in rendering the argument.

is not clear why Descartes didn't introduce the point I directly by taking $AI = m$ on a line through A parallel to the ordinate direction BC.)

2. (Construction of the straight line along which the diameter of the ellipse is situated.) Draw a straight line L_0 through I, intersecting BK at L such that $IK : KL = z : n$.

(L_0 passes through the center of the ellipse hence its part within the ellipse is a diameter; the ordinates remain parallel to BC, the ordinate angle, therefore, is equal to $\angle AIL$. With $BC = y$ we have $LC = y - BK + KL = y - m + \frac{nx}{z}$, and therefore (cf. Equation 23.18)

$$LC = \sqrt{m^2 + ox - \frac{p}{m}x^2}\,. \tag{23.19}$$

LC is the ordinate of the ellipse with respect to the diameter along L_0 and the ordinate angle $\angle AIL$.)

2a. (Classification) Descartes noted at this point that if the square root in Equation 23.19 was zero, the locus coincided with the straight line L_0, and that if the root could be extracted (meaning $m^2 + ox - \frac{p}{m}x^2 = (\alpha x + \beta)^2$ for some α and β), the locus was "another straight line not harder to find than IL."[21] He then claimed that in all other cases the locus would be a conic section, namely:[22] a parabola if the term $\frac{p}{m}x^2$ was zero, a hyperbola if that term was preceded by the sign $+$, an ellipse if preceded by a $-$, and, in particular, a circle if $\angle ILC$ was a right angle and $a^2m = pz^2$. Descartes' construction and his proof for the ellipse was as follows:

3. (Construction of the center.) Introduce a new parameter a defined by $KL : IL = n : a$; then, because $KL = \frac{nx}{z}$, the abscissa IL along L_0 measured from I is $\frac{a}{z}x$. Take M on L_0 with $IM = \frac{aom}{2pz}$; M is the center of the ellipse.

4. (Construction of the parameters.) Take $t = \frac{ma}{pz}\sqrt{o^2 + 4pm}$ as *latus transversum* and $r = \frac{z}{a}\sqrt{o^2 + 4pm}$ as *latus rectum* of the ellipse.[23]

5. (Construction of the ellipse.) Use the constructions from Apollonius' *Conics*[24] to construct an ellipse with center M, diameter along L_0, ordinate angle equal to $\angle AIL$, *latus rectum* r, and *latus transversum* t. This ellipse is the required locus.

(Apollonius' construction, to which Descartes referred explicitly, proceeds by locating a cone in space that intersects the plane according

---

[21] [Descartes 1637] p. 328: "une autre ligne droite qui ne seroit pas plus malaysée a trouver qu'IL." He did not work out this (degenerate) case in more detail and did not note that the solution would actually consist of two straight lines.

[22] [Descartes 1637], cf. the remark on Descartes' terminology above.

[23] The letters r and t are mine and I have simplified the expressions; Descartes gave them as $\sqrt{\frac{a^2o^2m^2}{ppz^2} + \frac{4a^2m^3}{pz^2}}$ and $\sqrt{\frac{o^2z^2}{a^2} + \frac{4mpz^2}{a^2}}$, respectively.

[24] Propositons I-52–60 of [Apollonius Conics] contain the constructions of the conic sections; Props. 56–58 concern the ellipse.

the required ellipse.)

Proof:[25]

6. Take from M towards I a distance along L_0 equal to half the *latus transversum*, call its endpoint N; so $NM = \frac{1}{2}t$. N is the vertex of the ellipse corresponding to the diameter along L_0. Consider point C as on the ellipse. The Apollonian theory of conics yields the relation

$$v^2 = ru - \frac{r}{t}u^2 , \tag{23.20}$$

in which u is the abscissa NL measured from the vertex N, v the ordinate LC, r the *latus rectum*, and t the *latus transversum* (cf. Figure 23.3, detail).

7. Now insert for u the values used in the construction, taking $x = IK$, $y = BC$:

$$\begin{aligned} u &= NL = (NM - IM + IL) = \\ &= \frac{1}{2}t - \frac{aom}{2pz} + \frac{a}{z}x . \end{aligned} \tag{23.21}$$

Inserting this result in Equation 23.20 leads, after some calculation, to

$$v^2 = m^2 + ox - \frac{p}{m}x^2 . \tag{23.22}$$

Hence,

$$LC = v = \sqrt{m^2 + ox - \frac{p}{m}x^2} . \tag{23.23}$$

8. As $LC = y - m + \frac{nx}{z}$ (see **2**) Equation 23.23 yields

$$y = m - \frac{n}{z}x + \sqrt{m^2 + ox - \frac{p}{m}x^2} . \tag{23.24}$$

Hence the points on the ellipse satisfy the equation of the locus derived in the analysis (i.e., Equation 23.18), so the locus is an ellipse.

Derivation of the solution

The formulas and the argument of Descartes' proof strongly suggest that he found the construction as follows by means of a indeterminate coefficients procedure. From the equation of the curve

$$y = m - \frac{n}{z}x + \sqrt{m^2 + ox - \frac{p}{m}x^2} , \tag{23.25}$$

Descartes could recognize the line defined by

$$y = m - \frac{n}{z}x \tag{23.26}$$

[25] [Descartes 1637] pp. 332–333.

as the diameter of the curve; items **1** and **2** of his construction locate this diameter (L_0) with respect to the given straight lines and fix the point I corresponding to $x = 0$. The abscissae as measured from I along the diameter then are $\frac{a}{z}x$ with a as introduced in **3**. Calling u the abscissa as measured from a vertex of a conic along the diameter, and v the corresponding ordinate, r the *latus rectum*, and t the *latus transversum*, the general equation of the conic is

$$v^2 = ru \pm \frac{r}{t}u^2 . \tag{23.27}$$

Now $u = (a/z)x - s$ for some line segment s, so the right-hand side of Equation 23.27 can be considered as a second-degree polynomial in x. But we also have $v = y - m + (n/z)x$, hence it follows from Equation 23.25 that

$$v^2 = m^2 + ox - \frac{p}{m}x^2 . \tag{23.28}$$

Equating the coefficients of the powers of x on the right-hand sides of Equations 23.27 and 23.28 provides three equations from which r, t, and s can be determined; the location of M also follows immediately. The values found are precisely the ones Descartes used in his construction and proof.

Significance Descartes realized that his solution of the general three- and four-line locus problem had a significance beyond the special sphere of the Pappus problems. He wrote at the end of his solution:

> Finally, because all equations of degree not higher than the second are included in the discussion just given, not only is the problem of the ancients relating to three or four lines completely solved, but also the whole problem of what they called the composition of solid loci, and consequently that of plane loci, since they are included among the solid loci. . . . The ancients attempted nothing beyond the composition of solid loci, and it would appear that the sole aim of Apollonius in his treatise on the conic sections was the solution of problems of solid loci.[26]

The reference is to the introduction of the third book of Apollonius' *Conics* where "many surprising theorems" are announced, "that are useful for the syntheses of the solid loci and for the diorisms." Solid loci[27] were curves obtained

[26] [Descartes 1637] p. 334–335: "Au reste a cause que les equations, qui ne montent que iusques au quarré, sont toutes comprises en ce que ie viens d'expliquer; non seulement le problesme des anciens en 3 et 4 lignes est icy entierement achevé; mais aussy tout ce qui appartient à ce qu'ils nommoient la composition des lieux solides; et par consequent aussy a celle des lieux plans, a cause qu'ils sont compris dans les solides. . . . Mais le plus haut but qu'ayent eu les anciens en cete matiere a esté de parvenir a la composition des lieux solides: Et il semble que tout ce qu'Apollonius a escrit des sections coniques n'a esté qu'à dessein de la chercher."

[27] Cf. A.Jones' essay "The loci of Aristaeus, Euclid, and Eratosthenes" in his Pappus edition, [Pappus 1986] vol. 2 pp. 573–599; the two short quotations above are from his translation *ibid.* p. 585.

by the intersection of spheres, cylinders, or cones with planes, so they were conic sections. The synthesis of these loci was necessary in the solution of solid problems (cf. Section 5.5). Once the analysis of such a problem had revealed that the required point was on two solid loci, each defined by a certain property, the synthesis (construction) of the problem required that these loci were in fact constructed. This meant that the nature of the locus (ellipse, hyperbola, parabola) had to be determined and that a vertex had to be given in position, together with the direction of the corresponding diameter, while the ordinate angle, the *latus rectum*, and the *latus transversum* had to be given in magnitude. Given these elements the loci could indeed be constructed by Propositions I-52–60 of Apollonius' *Conics*. Thus the synthesis of solid loci consisted in the determination, given the locus-property, of the nature of the conic, its diameter, ordinate angle, vertex (or center), *latus rectum*, and *latus transversum*. This was indeed precisely what Descartes did.

Constructibility

When presenting his general results on Pappus' problem in Book I, Descartes had concentrated on the constructibility of points on the locus by plane, or solid, or higher-order means (cf. Table 23.2). However, such pointwise constructions beg the question in what sense they provide the whole locus. In his treatment of Pappus' problem in three and four lines, discussed above, Descartes achieved the required locus by Apollonian constructions, which did provide the whole conic section, namely, as the intersection figure of a cone and a plane. Thus in a sense his solution in this case was stronger than in the general case. On the other hand, Apollonius' constructions presupposed the possibility of locating a cone in a prescribed position with respect to a plane. This is not a method of construction that immediately presents itself to the mind as clear and distinct. The more evident alternative was to generate a curve by motions in the plane. As we will see in the next section, Descartes solved some instances of the five-line locus by specifying such a generation of the locus. It is remarkable that in the case of the three- and four-line locus he did not do so. He may have considered it superfluous to work out a complete method of tracing conic sections by motion, knowing several instances of such procedures; elsewhere in the *Geometry* he gave a tracing procedure for the hyperbola (cf. Section 19.4, Problem 19.5) and referred to the familiar one (by means of strings) for the ellipse (cf. Section 24.4). However this may have been, the solution of the three- and four-line locus raised the question of the relative acceptability of the various ways of generating curves in geometry. We will see in Chapter 24 that Descartes devoted a considerable part of his book to this issue.

23.4 Pappus' problem "in five lines"

The problem and the equation

After dealing with the general case of the three- and four-line problem Descartes gave the solution of two special cases of the five-line problem, namely, what he called the "simplest case in five lines" and a variant of that case. The former was the special problem, which, if my conjecture in Chapter 19 is valid,

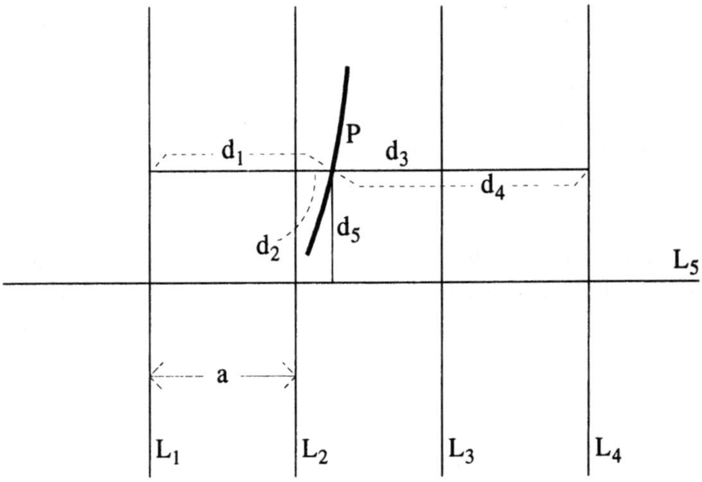

Figure 23.4: Pappus' problem in five lines

led Descartes to the turning ruler and moving curve procedure and inspired several new ideas in the development of his doctrine of construction. The locus in this case was the "Cartesian parabola" which played a central role in the *Geometry*. For convenience I repeat the formulation of the problem as given in Chapter 19 (Problem 19.3):

Problem 23.4 (Pappus' problem in five lines)[28]
Given (see Figure 23.4) four parallel, equidistant lines $L_1, \cdots, L_4$ (distance a), and one line L_5 perpendicular to them. It is required to find the locus of points whose perpendicular distances d_i to L_i satisfy

$$ad_3d_5 = d_1d_2d_4 \,. \tag{23.29}$$

In the *Geometry* Descartes first derived the equation of the locus. Taking $d_3 = y$, $d_5 = x$, and hence $d_1 = 2a - y$, $d_2 = a - y$, $d_4 = a + y$, he found:

$$axy = (2a - y)(a - y)(a + y) = y^3 - 2ay^2 - a^2y + 2a^3 \,. \tag{23.30}$$

The construction Descartes then provided the construction of the curve by the turning ruler and moving curve procedure, which I discussed above in Section 19.3 (Construction 19.4):

[28][Descartes 1637] pp. 335–338.

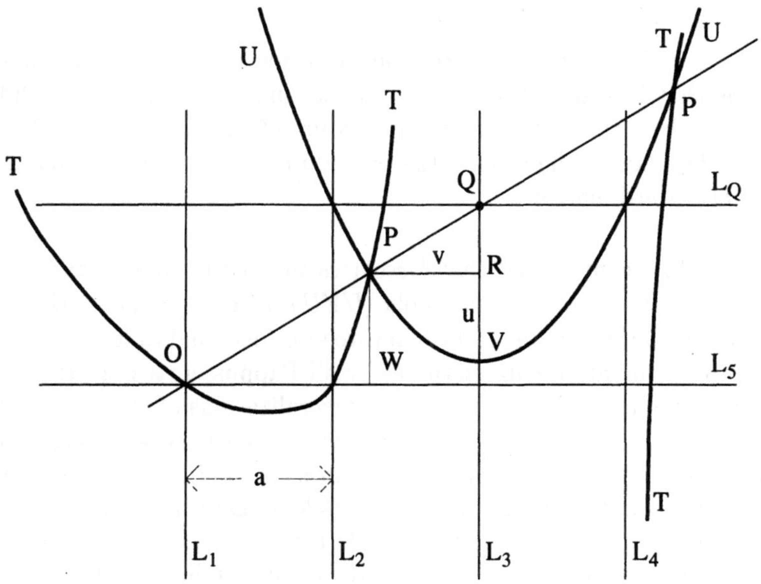

Figure 23.5: Tracing the five-line locus

Construction 23.5 (Five-line locus — Descartes)[29]
Given and required: see Problem 23.4.

Construction:
1. Consider (see Figure 23.5) a parabola UVU with vertical axis along L_3 and *latus rectum* equal to a (which means that its equation in rectangular coordinates u and v as indicated in the figure is $au = v^2$). The parabola can move up and down while keeping its axis along L_3. Moving with it is a point Q on the axis inside the parabola with distance a to the vertex V.
2. Consider also a straight line OQ that can turn around the intersection O of L_1 and L_5 while Q moves along L_3.
3. During the combined motion the points P of intersection of the parabola and the straight line move over the plane; they trace a new curve $TOPT$ TPT (consisting of two branches); this curve is the required five-line locus.
[**Proof**: During the process the coordinates u and v satisfy $au = v^2$. Now $a + v = d_4$ and $a - v = d_2$ from which it follows that $v^2 = a^2 - d_2d_4$ (1). Moreover, because the triangles PRQ and OWP along OQ are similar and $QR = QV - VR = a - u$, it follows that $(a - u) : d_3 = d_5 : d_1$, whence $au = a^2 - ad_3d_5/d_1$ (2). Equating the expressions (1) and (2) for au and v^2, respectively, yields

29 [Descartes 1637] p. 337.

$ad_3d_5 = d_1d_2d_4$.]

It should be noted that this construction does not follow from the analysis. The equation that Descartes had derived for the curve, Equation 23.30, did not suggest in any obvious way the tracing procedure of the construction. Nor did Descartes explain how he had found the procedure (or even that finding it was not a straightforward matter).

Origin and generalizations

In Chapter 19 I conjectured that Descartes found the construction by the turning ruler (OP) and moving parabola (UVU), discussed above, through a certain reduction of the five-line locus to a three-line one, and that for some time he hoped to find similar tracing methods for all Pappus curves. In the *Geometry* he made some remarks about possible generalizations of the construction, which corroborate my conjecture because they concern the generalization of the tracing procedure to cases in which the given lines are in different configurations than the one given in the special five-lines problem. Descartes stated that the procedure could be modified to apply also (1) if the distances d_i were not perpendicular, (2) if the transversal line L_5 were not perpendicular either, and (3) if the four parallel lines were no longer equidistant.[30] Remarks (1) and (2) are readily explained if we assume that he was thinking of the derivation of the construction I conjectured in Section 19.3 (Equations 19.11–19.14): Taking oblique rather than perpendicular distances only affects the value of a in the equation $ad_3d_5 = d_1d_2d_4$. The derivation of the construction, however, does not depend on the particular value of a, so statement (1) follows as an obvious generalization. Similarly, it is easily seen that the derivation can be adjusted to the case of non-equidistant parallels, whence statement (3). If, as in statement (2) the transversal is not perpendicular (cf. Figure 23.6), one can take the $d_1, \cdots, d_4$ parallel to it, by which the same proportionalities as in Equations 19.10 and 19.12 apply and analogously the problem is reduced to a three-line locus with respect to two vertical lines and one oblique transversal. It is not difficult to derive that this three-line locus is a parabola with vertical axis and passing through the two points of intersection of L_Q, L_2 and L_4. We can, therefore, also understand statement (2) as an obvious corollary to the construction as found according to my conjecture.

In a further comment on the five-line locus in the *Geometry*, Descartes claimed (4) that the tracing procedure would apply also in some instances of the five-line problem in which the four lines are not parallel.[31] In order to apply the procedure as explained in Chapter 19 the lines L_2 and L_4 have to be parallel (otherwise the moving three-line locus would change its shape) and L_1 and L_3 have to be parallel (otherwise the basic proportionality introduced by the ruler no longer applies), but the two pairs of parallels need not have the

[30] [Descartes 1637] pp. 338–339: "Or encore que les paralleles données AB, IH, ED, et GF ne fussent point esgalement distantes, et que GA ne les couppast point a angles droits, ny ausy les lignes tirées du point C vers elles, ce point C ne laisseroit pas de se trouver tousiours en une ligne courbe, qui seroit de cete mesme nature. Et il s'y peut aussy trouver quelquefois, encore qu'aucune des lignes données ne soient paralleles."

[31] Cf. the last sentence of the quotation in Note 30.

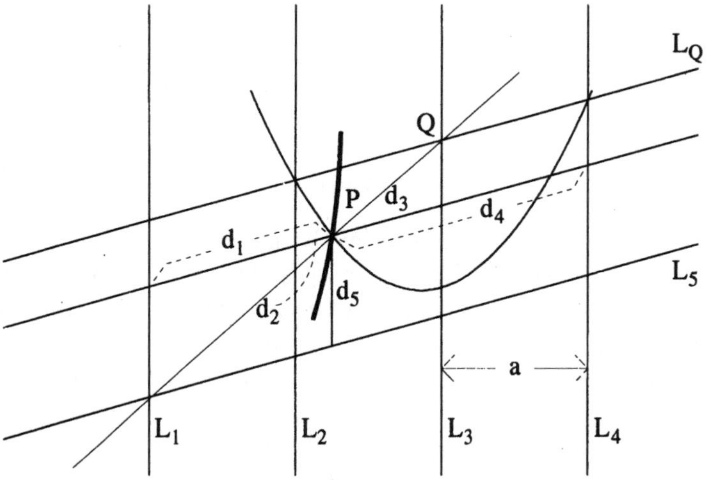

Figure 23.6: Five-line locus with oblique transversal

same direction. Descartes may have seen this and he may have had this case in mind when writing down his comment. For arbitrary positions of the given five lines, however, the turning ruler and moving curve procedure does not apply. Perhaps Descartes' statement is to be considered as an echo of the hopes he may have had about the procedure, namely, that with suitable adjustments it would cover any position of the given lines. That hope may have been based on generalizations as the one we have in the present comment, untenable in its generality but suggestive and hopeful.

23.5 The "simplest" five-line locus

Locus depends on order of the d_i

In the *Geometry* Descartes repeatedly asserted that the case of the five-line locus which yielded the Cartesian parabola, was the simplest case.[32] The statement is usually interpreted to mean that the configuration of four equidistant parallels and one perpendicular transversal is the simplest of five lines (after the one of five equidistant parallel lines).[33] However, this configuration of lines

[32] Cf., e.g., the margin title of the section on the five-line locus, [Descartes 1637] p. 335: "Quelle est la premiere et la plus simple de toutes les lignes courbes qui servent en la question des anciens quand elle est proposée en cinq lignes."

[33] This is how Newton interpreted Descartes' claim when he criticised it as one of the errors of the *Geometry*, cf. [Newton 1967–1981] vol. 4 pp. 336–345. He argued that by analogy the parabola or the hyperbola, originating as loci with respect to two parallel lines and one perpendicular transversal (cf. Figure 19.3), would be the simplest second-degree curve, simpler,

gives rise to several types of Pappus curves, only one of which was singled out as the simplest by Descartes. Thus the usual interpretation is incomplete because it leaves Descartes' further choice unexplained.

If we consider four given parallel lines $L_1, \cdots, L_4$ and one perpendicular transversal L_5 and denote the corresponding distances in the way introduced above, the general Pappus problem with respect to these lines is

$$ad_i d_j = d_k d_l d_m \,, \tag{23.31}$$

in which (i, j, k, l, m) is any permutation of the numbers $1, \cdots, 5$. Descartes' curve arises in the particular case $ad_3d_5 = d_1d_2d_4$. Descartes' choice is not the most obvious one among the possible permutations, nor are the curves arising in the other permutations in any obvious sense similar to the one he chose. It is therefore of interest to study the various loci defined by Equation 23.31 in order to understand, if possible, Descartes' choice and his assertion that his curve in particular was the simplest. I have analyzed this question elsewhere; in the following I summarize the results of that analysis, referring for a more-detailed explanation and for the calculations to my earlier publication.[34]

Variants of the five-line problem

The possible types of the five-line problem can be divided into two categories depending on the position of the factor d_5 in the equation. Denoting d_5 by y (cf. Figure 23.7 and 23.8; note that Descartes denoted d_5 by x), the two categories are:

$$\text{Category I:} \quad ayd_i = d_j d_k d_l \,, \tag{23.32}$$

$$\text{Category II:} \quad ad_i d_j = yd_k d_l \,. \tag{23.33}$$

Descartes' case belongs to category I.

For category I the possible permutations yield, if we disregard symmetries,[35] essentially two loci; they are drawn in Figure 23.7. The one to the left is the curve which Descartes proposed (i.e., the Cartesian parabola); the other is different in that one of its branches has two local extremes.[36] For the loci of Category II there are (again disregarding symmetries) four different types (see Figure 23.8).[37]

Simplicity

A further analysis,[38] shows that in each of these cases an argument similar to

therefore, than the circle, which he considered absurd.

[34][Bos 1992] pp. 78–90.

[35]That is: the symmetry with respect to the X-axis and the symmetry with respect to a vertical (the latter is obtained by numbering the lines from right to left); thus $ayd_1 = d_2d_3d_4$, $-ayd_1 = d_2d_3d_4$ and $ayd_4 = d_1d_2d_3$ are considered the same.

[36]In Newton's classification of cubics both curves belong to the class of "tridents" or "Cartesian parabolas" because both have an equation of the form $xy = Ax^3 + Bx^2 + Cx + D$, cf. [Newton 1967–1981] vol. 7 pp. 630–631, Figure 76. I use the term "Cartesian parabola" to refer exclusively to the type proposed by Descartes.

[37]The four types of Category II, coresponding to items a-d of the figure, are IIa: $ad_3d_4 = yd_1d_2$; IIb: $ad_2d_4 = yd_1d_3$; IIc: $ad_2d_3 = yd_1d_4$; IId: $ad_1d_4 = yd_2d_3$.

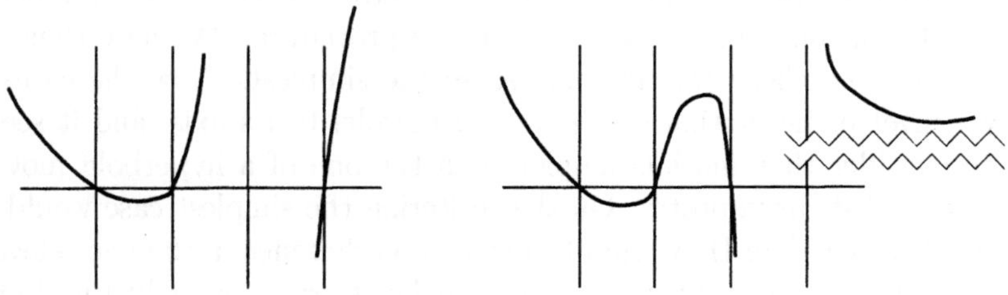

Figure 23.7: The five-line problem — loci of Category I

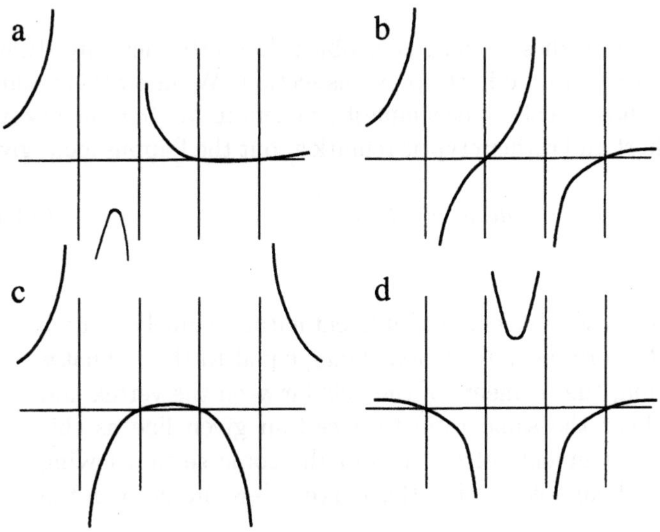

Figure 23.8: The five-line problem — loci of Category II

the one I conjectured in Chapter 19 leads to a turning ruler and moving curve construction of the locus. For Category I the moving curves are parabolas with axes parallel to L_1; their motion is also parallel to L_1. For Category II the curves are hyperbolas with one asymptote parallel to L_1; also their motion is parallel to L_1.

Thus, if Descartes had considered other versions of Equation 23.31 than the one he discussed in the *Geometry*, he would have encountered a number of variants of the turning ruler and moving curve procedure. We may then ask whether the one he chose was in some sense the simplest. Now the cases of Category I involve the motion of a parabola parallel to its axis, and it seems natural to consider that motion simpler than the one of a hyperbola moving parallel to one of its asymptotes. On that criterion the simplest case would be of Category I rather than II. Within Category I a difference in the two relevant turning ruler and moving curve procedures may be observed, namely, that in the case singled out by Descartes the point (Q in Figure 23.5), which connects the motion of the ruler to the motion of the parabola, is on the axis of the parabola, whereas in the other case it lies at some distance from the axis,[39] which gives, one might say, a certain imbalance to the corresponding tracing motion.

Thus there is a sense in which the five line problem Descartes chose to expose in the *Geometry* is indeed the simplest one. And even if we do not suppose that Descartes has surveyed all possibilities, it may well be that he saw some of them and chose the simplest of those.

23.6 Another five-line locus

The text At some stage of his studies of Pappus' problem Descartes actually studied loci of the Category II introduced in the previous section. We know this because after his solution of what he called the simplest problem in five lines he inserted in the *Geometry* a short and rather cryptic remark about the Pappus locus given by[40]

$$ad_id_j = yd_kd_l. \tag{23.34}$$

He wrote:

> The required point lies on a curve of different nature, namely, a curve such that, all the ordinates to its axis being equal to the ordinates of a conic section, the segments of the axis between the vertex and the ordinates bear the same ratio to a certain given line as this line bears to the segments of the axis of the conic section having equal ordinates. I cannot say that this curve is less simple than the

[38] For details see [Bos 1992] pp. 83–87.

[39] This argument is explained in more detail in [Bos 1992].

[40] Descartes formulated the problem in words, [Descartes 1637] p. 339: "Mais si lorsqu'il y en a 4 ainsi paralleles, et une cinquiesme qui les traverse: et que le parallelepipede de trois des lignes tirées du point cherché, l'une sur cete cinquiesme, et les 2 autres sur 2 de celles qui sont paralleles; soit esgal a celuy, des deux tirées sur les deux autres paralleles, et d'une autre ligne donnée."

preceding; which nevertheless I believed should be taken as the first, since its description and calculation are somehow easier."[41]

Comments

Descartes left the permutation of the distances unspecified; thereby his formulation covers all problems of Category II. As I mentioned above, Descartes could have found a procedure for tracing the loci by a turning ruler and a moving hyperbola. However, although the passage does mention the reduction to a conic section, it does not at all suggest tracing by motion. There are several possible interpretations of this text,[42] none of which, however, explicitly relates to a procedure similar to the one of my conjecture. I like to see Descartes' remarks on this five-line locus as an echo of some intermediate, primarily algebraic, investigation of the problem, undertaken sometime after the completion of the "écrit" mentioned in Section 19.1, perhaps as a result of doubts about the generality of the earlier solution. The uncertainty, voiced in the final sentence of the passage, about whether, after all, this curve or the other is the simplest, seems to me an indication of such doubts.

23.7 Clarity and concealment

Power and virtuosity

Descartes' solution of Pappus' problem as presented in the *Geometry* was impressive indeed and well suited to convince his readers of the power of his new method and of his own virtuosity in handling it. The conclusions Descartes reached about the solvability of the problem in general for any number of lines showed that he had definitely extended his method to the realm beyond the classical solid problems. He did not fail to stress the classical standing of the problem and its relation to the ancient method of analysis. Classical analysis had been restricted by the difficulty of the composition of solid loci — a difficulty now solved as a corollary to Descartes' solution of Pappus' problem.

Secrecy

The discussion above of Descartes' solution of Pappus' problem also illustrates the curious mixture of clarity and concealment in the *Geometry*. On one level the book was indeed a treatise on method; it explained with great clarity a new method for finding the solution of geometrical problems. This clarity applied in particular to the analytical part of the method, the procedure leading from the problem to the equation. On a deeper level, however, Descartes' attitude to his readers might almost be called secretive. This secrecy concerned primarily the heuristics of the synthetic side of the method, the way

---

[41] [Descartes 1637] p. 339: "Ce point cherché est en une ligne courbe d'une autre nature, a sçavoir en une qui est telle, que toutes les lignes droites appliquées par ordre a son diametre estant esgales a celles d'une section conique, les segmens de ce diametre, qui sont entre le sommet et ces lignes, ont mesme proportion a une certaine ligne donnée, que cete ligne donnée a aux segmens du diametre de la section conique, ausquels les pareilles lignes sont appliquées par ordre. Et ie ne sçaurois veritablement dire que cete ligne soit moins simple que la precedente, laquelle iay creu toutefois devoir prendre pour la premiere, a cause que la description, et le calcul en sont en quelque façon plus faciles."

[42] Cf. [Bos 1992] pp. 89–90.

Descartes had found the constructions he proposed. He explained and proved the constructions of the four-line loci clearly enough, but he remained silent on how he determined the location and the parameters of the conic sections from their equation; apodictically, he only gave the values and proved them to be the correct ones. The method behind his findings, which, as noted above, was probably some kind of indeterminate coefficients technique, was not explained.[43] Similarly, although the tracing procedure of the five-line locus by turning ruler and moving parabola was exposed in detail, no clarification was offered of how Descartes had found it. We will find the same uncommunicative attitude later with respect to the standard constructions of solid and higher-order problems.

One important aspect of Pappus' problem that also remained concealed in the *Geometry* was its role in providing the ingredients of Descartes' classification of curves and his demarcation of geometry. In the next chapter I discuss what Descartes wrote on these issues.

[43] Apparently Descartes referred to this omission in his letter to Mersenne of March 31, 1638 ([Descartes 1964–1974] vol. 2 p. 83): "Mais le bon est, touchant cette question de Pappus, que je n'en ay mis que la construction et la demonstration entiere, sans en mettre toute l'analyse."

Chapter 24

Curves and the demarcation of geometry in the *Geometry*

24.1 The demarcation

"Geometrical" and "mechanical" curves

I now come to a crucial issue in Descartes' geometrical doctrine: the demarcation of geometry.[1] His program required a reinterpretation of geometrical exactness concerning constructions. Constructions were to be performed by means of curves; they had to be geometrically acceptable and as simple as possible. Consequently, his new doctrine had to provide clear answers to the following two questions:

> **A.** Which curves are acceptable as means of exact construction in geometry?
>
> and
>
> **B.** When is one curve simpler than another?

I will discuss the second question in Chapter 25. The first question concerned the demarcation between exact, geometrical procedures, on the one hand, and non-exact, non-geometrical ones on the other. Descartes gave such a demarcation in terms of the curves used in the procedures. He distinguished between "geometrical" and "mechanical" curves; the former were acceptable in geometry, in particular for use in constructions, the latter were not. In effect, his distinction was straightforward: "geometrical" curves were those that, with respect to rectilinear coordinates, had an algebraic equation; all others (in particular the spiral and the quadratrix) were "mechanical." This Cartesian demarcation of geometry had a great influence in the second half of the seventeenth century,

[1] I have dealt with many of the themes of the present chapter in [Bos 1981].

especially as the background for discussions on the exactness of various methods for tracing curves.[2]

The curves that Descartes allowed in geometry are now called "algebraic," the others "transcendental." Leibniz introduced these terms for Descartes' "geometrical" and "mechanical" as part of his own re-interpretation of mathematical exactness, thereby removing the implication that the latter class of curves should be excluded from true geometry.[3] For Descartes, however, the demarcation was not fundamentally algebraic. Behind his answer — algebraic vs. non-algebraic — lay an argument of considerable complexity. The present chapter deals with that argument.

Acceptable curves The principal text on the demarcation of geometry is at the beginning of Book II of the *Geometry*; its margin title is

> Which curved lines can be admitted in geometry.[4]

Descartes' criterion for accepting a curve concerned the manner of its tracing. I argued in Chapter 19 that the confrontation with Pappus' problem in 1632 strengthened his idea that curves were to be accepted in geometry on the basis of the methods used to trace them. It also gave him the conviction that acceptable curves were precisely those that had algebraic equations. However, an argument for this equivalence was lacking. Simply postulating that geometry should be restricted to algebraic curves was not convincing, for why should algebra provide a criterion for the demarcation of geometry? The question was too crucial for such a facile answer. The matter was further complicated by the fact that, besides tracing by motion, he had to consider two other ways of generating or representing curves, namely, pointwise construction and tracing procedures involving strings. For each of the three methods for generating curves Descartes had to formulate criteria that excluded the spiral, the quadratrix, and similar curves, while including the algebraic curves. He took these questions very seriously indeed, and in my opinion his arguments about them, although ultimately inconclusive, form the deepest and most impressive part of the intellectual effort that produced the *Geometry*.

Survey of the arguments Descartes did not collect all his arguments in one particular section of the *Geometry*, nor were all his arguments as explicit as one could wish. Before analyzing the various relevant passages of the *Geometry* it is useful to give a schematic survey of Descartes' reasoning; this is done in Table 24.1. The arrows in the table ($\cdots \mathbf{A} \rightarrow \mathbf{B} \rightarrow \mathbf{E} \rightarrow \mathbf{D} \rightarrow \mathbf{A} \cdots$) indicate how a proof that the algebraic curves are precisely those that are acceptable in geometry can be reconstructed from Descartes' statements: **A** Curves traced by coordinated continuous motions are acceptable in geometry; **B** curves that are acceptable

[2] I have dealt with some of these discussions in [Bos 1987], [Bos 1988] and [Bos 1996].

[3] Cf. [Breger 1986], pp. 122–123.

[4] [Descartes 1637] pp. 315: "Quelles sont les lignes courbes qu'on peut recevoir en Geometrie:" the pertaining text is on pp. 315–319.

	Acceptance		Equations	Rejection
Tracing by motion	**A.** Curves traced by coordinated continuous motions are acceptable in geometry. In particular: if lines or curves are moved continuously one by the other, then the curves traced by their intersections are acceptable in geometry.	→ →	**B.** Curves that are acceptable in geometry according to **A** have algebraic equations; several examples show how the equation can be deduced from the method of tracing.	**C.** If the separate motions involved in the tracing process are not coordinated, i.e., if they have no measurable relation, then the resulting curve has to be rejected.
	↑ ↑		↓ ↓	
Pointwise construction	**D.** If there is a pointwise construction of the curve for which in principle any of the curve's points may occur among the constructed ones (generic pointwise construction), then the curve can also be generated by coordinated continuous motion.	← ←	**E.** The algebraic equation implies a generic pointwise construction of the curve (presupposing a complete technique of constructing roots of polynomial equations).	**F.** If there are points on the curve that cannot occur among those provided by the pointwise construction, then the curve has to be rejected.
Tracing by procedures involving strings.	**G.** If strings are used in a curve tracing process only to ensure the validity of relations between straight lines, then the resulting curve is acceptable.		**H.** [Descartes made no remarks on the equations of these curves.]	**I.** If a curve tracing process involves strings that during the process change from straight into curved, then the curve has to be rejected.

Table 24.1: Descartes' arguments about the demarcation of geometry

in this sense have algebraic equations; **E** if a curve has an equation, it can be constructed by a generic pointwise construction, that is, a pointwise construction in which in principle any point of the curve may occur among those constructed; **D** if a curve admits a generic pointwise construction, it can also be traced by coordinated continuous motion. **A**, **B**, **E**, and **D**, in this order, combine to a proof that acceptable curves are precisely the ones that have algebraic equations.

I now analyze Descartes' arguments about these questions of demarcation in more detail; I follow the order indicated by the left column in Table 24.1.

24.2 Curves traced by motion

Acceptable tracing

For Descartes the primary criterion for accepting curves as geometrical was whether they could be traced by what in **A** in the table I have called "coordinated continuous motion." Descartes formulated his views on acceptable curves at the beginning of the second book of the *Geometry*. He started by criticizing the ancient geometers for not classifying problems and curves beyond solid problems and conic sections, respectively. Moreover, they had rejected other curves because they were "mechanical," that is, traced by a machine. But, Descartes wrote, circles and straight lines are also traced by machines and yet they were accepted in geometry. The precision with which a curve could be understood should be the criterion in geometry, not the precision with which it could be traced by hand or by instruments. Nor were the classical Euclidean postulates sufficient argument for excluding curves, because the ancient geometers themselves were prepared to go further, by postulating, for instance, that a curve could be generated by the intersection of a cone and a plane. Descartes' own criterion was, he claimed, easier than that:

> nothing else need be supposed than that two or several lines can be moved one by the other and that their intersections mark other lines.[5]

Perhaps, Descartes wrote, the ancients did not quite accept conic sections as geometrical but such a restriction had to be rejected, and

> . . . it seems very clear to me that if we make the usual assumption that geometry is precise and exact, while mechanics is not, and if we think of geometry as the science which furnishes a general knowledge of the measurement of all bodies, then we have no more right to exclude the more complex curves than the simpler ones, provided they can be conceived of as described by a continuous motion or by several successive motions, each motion being completely determined by those which precede; for in this way an exact knowledge of the measure of each is always obtainable.[6]

[5][Descartes 1637] p. 316: "Et il n'est besoin de rien supposer pour traçer toutes les lignes courbes, que je pretens icy d'introduire; sinon que deux ou plusieurs lignes puissent estre meuës l'une par l'autre, et que leurs intersections en marquent d'autres."

[6][Descartes 1637] p. 316: ". . . il est, ce me semble, tres clair, que prenant comme on

Criteria for acceptable motion

Descartes' arguments paraphrased and quoted above enable us to reconstruct his primary criteria for the tracing motions to produce acceptable curves. They were: (1) that the moving objects were themselves straight or curved lines, (2) that the tracing point was defined as the intersection of two such moving lines, (3) that the motions of the lines were continuous (which Descartes did not specify further), and (4) that they were strictly coordinated by one initial motion. I use the term "coordinated continuous motion" for curve tracing movement satisfying these criteria. I illustrate these criteria below by the example of the mesolabum; the "turning ruler and moving curve procedure" (Section 19.4) provides another instance of coordinated continuous motion.

Obviously the motion criteria summarized a manner of conceiving or imagining a motion, and a resulting curve, which Descartes accepted as sufficiently clear and distinct to count as certain. In this sense the criteria implemented Descartes' first rule of method from the *Discourse*:

> ... never to accept anything as true if I did not have evident knowledge of its truth: that is, carefully to avoid precipitate conclusions and preconceptions, and to include nothing more in my judgements than what presented itself to my mind so clearly and distinctly that I had no occasion to doubt it.[7]

It is also clear that in referring to methods of tracing curves, Descartes did not envisage the actual physical processes of tracing on paper or other surfaces, but the mental processes by which the mind contemplated the generation of curves by motion. The instruments such as those described in the *Geometry* were not meant to be made and used, but to be contemplated and thus to be helpful for the mind in intuiting the tracing process and in determining whether it was sufficiently clear and distinct to accept the resulting curve as geometrical.

The "mesolabum"

One of the instruments Descartes used in the *Geometry* to illustrate the kind of tracing he envisaged was the device, later called "Descartes' Mesolabum," for constructing mean proportionals;[8] I discussed it above in Chapter 16 (cf. Instrument 16.3). The instrument is often reproduced (not least for its pictorial appeal) in studies about Descartes' mathematics. The curves AB, AD, AF, and AH, traced by the Mesolabum (cf. Figure 24.1), satisfied the four criteria

---

fait pour Geometrique ce qui est precis et exact, et pour Mechanique ce qui ne l'est pas; et considerant la Geometrie comme une science, qui enseigne generalement a connoistre les mesures de tous les cors, on n'en doit pas plutost exclure les lignes les plus composées que les plus simples, pourvû qu'on les puisse imaginer estre descrites par un mouvement continu, ou par plusieurs qui s'entresuivent et dont les derniers soient entierement reglés par ceux qui les precedent. Car par ce moyen on peut tousiours avoir une connaissance exacte de leur mesure." Cf. [Molland 1976] for a detailed discussion of these arguments, in particular of whether they accurately reflected the ideas of the "ancient geometers."

[7] [Descartes 1637b] p. 18 ". . . de ne recevoir iamais aucune chose pour vraye, que ie ne la connusse evidemment estre telle: c'est a dire, d'eviter soigneusement la Precipitation, et la Prevention; et de ne comprendre rien de plus en mes iugemens, que ce qui se presenteroit si clairement et si distinctement a mon esprit, que ie n'eusse aucune occasion de le mettre en doute;" translation quoted from [Descartes 1985–1991] vol. 1 p. 120.

[8] [Descartes 1637] pp. 317–319, 370–371.

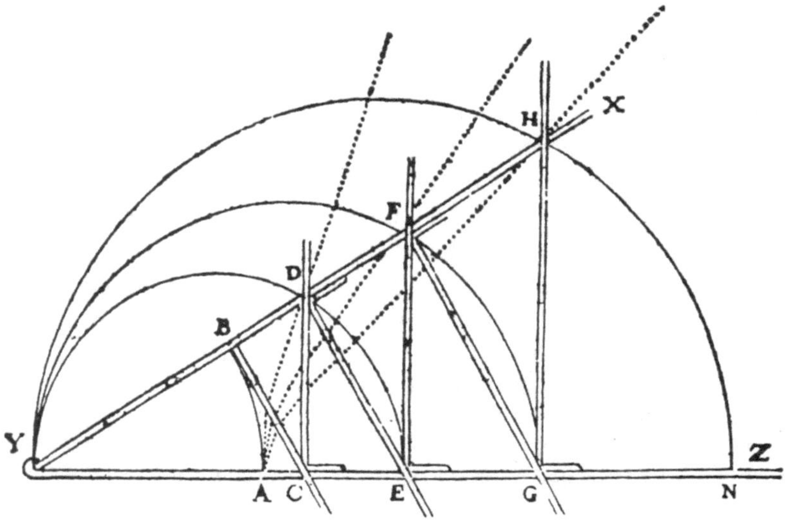

Figure 24.1: Descartes' Mesolabum (*Geometry* p. 318)

for acceptable motion formulated above. The moving parts were straight lines; the curves were traced by the points B, D, F, and H, which were points of intersection of the moving straight lines; the motions of the lines were continuous; and they were all coordinated by the initial movement of turning the ruler YX around the fixed point Y. Descartes noted that, although these curves were successively more complicated, their manner of tracing remained clear and distinct, and therefore they should be accepted in geometry:

> But I do not see what could prevent us from conceiving the description of the first [i.e., the curve traced by D] as clearly and distinctly as that of the circle, or at least as that of the conic sections, nor what could prevent us from conceiving the second one and the third one and all the others, which one can describe equally well as the first one; nor therefore what could prevent us from accepting all these curves in the same manner, to serve the speculations of geometry.[9]

Descartes then explained that the equations of these curves were the best means to classify them:

[9] [Descartes 1637] pp. 318–319: "Mais je ne voy pas ce qui peut empescher, qu'on ne concoive aussy nettement, et aussy distinctement la description de cete premiere, que du cercle, ou du moins que des sections coniques; ny ce qui peut empescher, qu'on ne concoive la seconde, et la troisiesme, et toutes les autres, qu'on peut descrire, aussy bien que la premiere; ny par consequent qu'on ne les recoive toutes en mesme façon, pour servir aux speculations de Geometrie."

> I could present here several other ways of tracing and conceiving curved lines which would be increasing in complexity by degrees to infinity. But for comprising together all those that are in nature and for distinguishing them by order in certain classes, I know no better way than to say that all the points of those which one may call geometrical, that is, of those that are subject to precise and exact measurement, necessarily have some relation to all the points of one straight line which can be expressed by an equation, the same for all points.[10]

This passage is of importance because it contains the argument (marked **B** in Table 24.1) that acceptable curves have algebraic equations. Geometrically acceptable curves were traced and conceived in the manner illustrated by the Mesolabum; they were "subject to precise and exact measurement" and hence they admitted an algebraic equation. It should be noted that at this point Descartes did not claim that, conversely, every algebraic equation (in two unknowns) described a geometrical curve.

Descartes did not further explain his statement that acceptable curves had algebraic equations. In several examples he calculated the equations of acceptably traced curves (for instance, the hyperbola traced by a turning ruler and moving straight line[11] and the Cartesian parabola traced by a turning ruler and moving parabola[12]) and these calculations illustrated that curves traced by acceptable motions indeed have algebraic equations.

Tracing non-geometrical curves

The "mechanical" curves such as the spiral and the quadratrix were defined by certain tracing procedures. Evidently Descartes had to argue that the motions involved in these procedures were not coordinated continuous motions in the sense explained above. We have seen (Definitions 3.2 and 3.3) how the spiral and the quadratrix were generated by combinations of rectilinear and circular motions. In these cases Descartes stated that the coordination of the two motions could not be measured and therefore could not be conceived with sufficient clarity and distinction. Thus he wrote about the spiral and the quadratrix that they

> . . . in truth only belong to mechanics, and are not at all among those which I think should be accepted here, because one imagines them described by two separate movements, between which there is no relation at all that one could measure exactly.[13]

---

[10] [Descartes 1637] p. 319: "Ie pourrois mettre icy plusieurs autres moyens pour tracer et conçevoir des lignes courbes, qui seroient de plus en plus composées par degrés a l'infini. Mais pour comprendre ensembles toutes celles, qui sont en la nature, et les distinguer par ordre en certaines genres; ie ne sçache rien de meilleur que de dire que tous les poins, de celles qu'on peut nommer Geometriques, c'est a dire qui tombent sous quelque mesure precise et exacte, ont necessairement quelque rapport a tous les poins d'une ligne droite, qui peut estre exprimé par quelque equation, en tous par une mesme."

[11] [Descartes 1637] pp. 319–322, cf. also Section 19.4 Problem 19.5.

[12] [Descartes 1637] p. 337, cf. Construction 23.5.

[13] [Descartes 1637] p. 317 ". . . la Spirale, la Quadratrice, et semblables, qui n'appartienent

The absence of a *measurable* relation ("raport") of the motions is the essential point here; both for the quadratrix and the spiral the two movements could in principle be coordinated in such a way that the one determined the other, namely, by a string mechanism such as the one I illustrate in Section 24.4 below (Instrument 24.1). When Descartes spoke about measuring the relation of the motions, he evidently thought of the ratio of their velocities or the ratio of the distances traversed by each in equal time intervals. In the case of the quadratrix, as in that of the spiral, the two velocities involved in tracing the curve had no exactly measurable "raport:" measuring their relation would involve comparing the lengths of straight and curved lines, in particular those of the diameter and the circumference of a circle. Descartes returned to this argument some pages later in connection with the tracing of curves by instruments involving strings (see Section 24.4). He argued that one should not accept lines in geometry which resemble strings

> . . . that is to say which are sometimes straight and sometimes curved, because the proportion between straight lines and curves is not known, and, I even believe, will never be known to man, and therefore one cannot conclude anything exact and certain on that basis.[14]

Thus the separation between geometrical and non-geometrical curves, fundamental in Descartes' vision of geometry, rested ultimately on his conviction that proportions between curved and straight lengths cannot be known exactly. This, in fact, was an old doctrine, going back to Aristotle.[15] The central role of the incomparability of straight and curved in Descartes' geometry was the reason why the first rectifications of algebraic (i.e., for Descartes: geometrical) curves[16] in the late 1650s were so revolutionary: they undermined a cornerstone of the edifice of Descartes' geometry.

24.3 Pointwise construction of curves

From the equation to a pointwise construction

In his solution of the general Pappus problem, Descartes had shown how the equation of a curve implied the possibility, in principle, of constructing arbitrarily many points on it (cf. Section 23.2). If the given polynomial equation was $F(x, y) = 0$, one could choose an arbitrary value x_1 for x and solve the resulting equation in one unknown $F(x_1, y) = 0$ geometrically, that is, one could

veritablement qu'aux Mechaniques, et ne sont point du nombre de celles que ie pense devoir icy estre receues, a cause qu'on les imagine descrite par deux mouvemens separés, et qui n'ont entre eux aucun raport qu'on puisse mesurer exactement . . ."

[14] [Descartes 1637] pp. 340–341: ". . . lignes qui semblent a des cordes, c'est a dire qui devienent tantost droites et tantost courbes, a cause que la proportion, qui est entre les droites et les courbes, n'estant pas connuë, et mesme ie croy ne le pouvant estre par les hommes, on ne pourroit rien conclure de là qui fust exact et assuré."

[15] Cf. [Heath 1949] pp. 140–142.

[16] These rectifications were achieved around 1658, independently by Van Heuraet, Neile, and Fermat; cf. [Baron 1969] pp. 223–228.

construct its root or roots $y_{1,1}, \cdots, y_{1,k}$. (Note that Descartes here assumed that the roots of any polynomial equation could be constructed geometrically, but that was indeed what he claimed to achieve in Book III of the *Geometry*, cf. Section 26.4 and 26.5.) Thereby the points on the curve with coordinates $(x_1, y_{1,i})$ could be constructed and the procedure could be repeated for other values x_2, x_3, $\cdots$ of x, yielding arbitrarily many points on the curve.[17] By this process of pointwise construction one could determine a net of points distributed along the curve with any required density. Moreover, because the x_i could be chosen arbitrarily, in principle any point of the curve could occur among those constructed. I use the term "generic pointwise construction" for pointwise constructions that have this property of randomness; Descartes used the property to distinguish these pointwise constructions from others which he deemed less acceptable.

Provided a complete method for constructing roots of algebraic equations was available (and Descartes claimed so), any algebraic curve could be constructed generically pointwise in this way. Thus his results implied the statement marked **E** in Table 24.1, namely that any curve which admits an algebraic equation can be achieved by a generic pointwise construction.

Objections against pointwise constructions

The obvious difference between tracing curves by motion and constructing them pointwise was that the former produced all points of the curve, while the latter only produced a (possibly large) number of such points. We have seen how Clavius struggled with this aspect when he argued that his pointwise construction of the quadratrix was "geometrical" (Section 9.3) and how Kepler criticized pointwise construction because the parts of the curve between the constructed points remained unconstructed (Section 11.4).

The question whether or not a procedure achieved all points of a curve was important in connection with the use of curves in constructions. This use was based on the assumption that if two constructing curves (or straight lines) were given, their point or points of intersection were given as well. For curves traced by motion this assumption was warranted by the continuity of the motion. But if, as in the case of pointwise construction, only a number (however large) of points of a curve were given, it was not at all obvious that its points of intersection with any other curve were also given. Consider, for instance, two ellipses C and D, both generically constructible pointwise (cf. Figure 24.2). For any arbitrarily chosen abscissa x, the corresponding points P on the ellipses could be exactly constructed. However, a point I of intersection could only occur among the so-constructed points if by accident its abscissa x_I was chosen, and one could never be certain whether or not that choice had been made.

Yet Descartes argued that if a curve could be produced by a generic pointwise construction, it could also be traced by coordinated continuous motion (item **D** in Table 24.1). He did so while explaining the difference between generic pointwise constructions and pointwise constructions employed for "mechanical" curves such as the quadratrix.

[17]The notation here is mine, cf. Note 3 of Chapter 19.

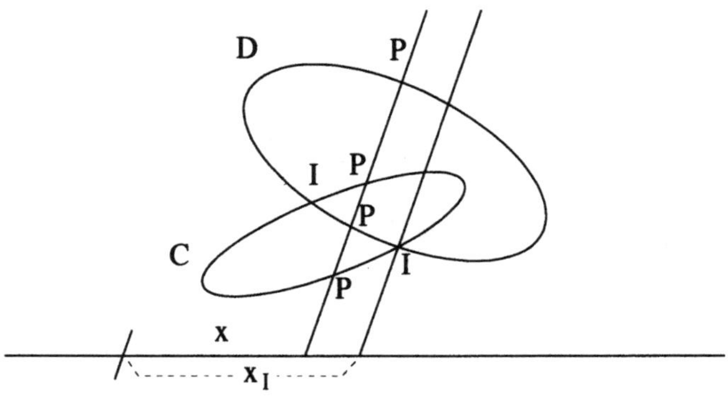

Figure 24.2: Intersection of pointwise constructible curves

Pointwise construction of non-geometrical curves

Descartes knew that curves such as the quadratrix[18] and the spiral can be constructed pointwise (cf. Section 16.5 in particular Note 29). So he had to explain in what sense these procedures differed from the ones that yielded "geometrical" curves. He did so in a section of Book II whose margin title was

> Which are the curved lines that one describes by finding many of their points and that can be accepted in geometry.[19]

Descartes wrote:

> It is worthy of note that there is a great difference between this method of finding different points to trace a curved line [sc. the pointwise construction of Pappus curves], and that used for the spiral and similar curves. For with the latter one does not find indifferently all points of the required curve, but only those points which can be determined by a simpler measure than is required for the generation of the curve itself. Therefore, strictly speaking, one does not find any one of its points, that is, not one of those which are so properly points of the curve that they cannot be found except by means of it. On the other hand there is no point on the curves which are of

[18]On Descartes' knowledge and opinion on the quadratrix see also [Gäbe 1972] in particular pp. 123–128.

[19][Descartes 1637] p. 340: "Quelles sont les lignes courbes qu'on descrit en trouvant plusieurs de leurs poins, qui peuvent estre receues en Geometrie."

> use for the proposed problem [the Pappus problem] that could not occur among those which are determined by the method explained above. And because this method of tracing a curved line by finding a number of its points taken at random is only applicable to curves that can also be described by a regular and continuous motion, one may not exclude it entirely from geometry.[20]

Earlier, in 1629, Descartes had expressed the same ideas in a letter to Mersenne, prompted by a report of the latter on a procedure for the general multisection of angles that had been explained to him.[21] From Descartes' remarks (Mersenne's letter is lost) it appears that the procedure involved a cylinder, a string, and a spiral curve. Probably, then, the procedure was a variant of the multisection of the angle by means of the spiral that Pappus had described (cf. Construction 3.4). At least Descartes came to that conclusion and suggested a method for tracing a spiral by means of a string wound off from a cylinder; I return to that method in the next section. Descartes further commented that the procedure Mersenne described, although very precise in practice, was not geometrical. It implicitly used the spiral, which, like the quadratrix, could neither be traced by sufficiently interdependent motions nor generated by sufficiently general pointwise constructions. For the pointwise construction of the quadratrix Descartes referred to Clavius' Euclid edition, which shows that in his comments above he had Clavius' pointwise construction of the quadratrix in mind (cf. Construction 9.1). Indeed Clavius' construction did not find "indifferently all points" but only points corresponding to divisions of the right angle that could be constructed by Euclidean means. Descartes made a sharp distinction between pointwise constructions such as these and the generic ones. The latter, unlike the former, could in principle yield any point on the curve. The distinction is indeed one between the particular and the generic; Descartes' remark that the constructed points on the quadratrix do not belong to "those which are so properly points of the curve that they cannot be found except by means of it" is a very apt characterization of the fact that they are not generic points of the curve.

Generic pointwise construction acceptable in geometry

The last sentence in the quotation above shows that Descartes saw an analogy

[20][Descartes 1637] pp. 339–340: "Mesme il est a propos de remarquer, qu'il y a grande difference entre cete façon de trouver plusieurs poins pour tracer une ligne courbe, et celle dont on se sert pour la spirale, et ses semblables. Car par cete derniere on ne trouve pas indifferemment tous les poins de la ligne qu'on cherche, mais seulement ceux qui peuvent estre déterminés par quelque mesure plus simple, que celle qui est requise pour la composer, et ainsi a proprement parler on ne trouve pas un de ses poins. C'est a dire pas un de ceux qui luy sont tellement propres, qu'ils ne puissent estre trouvés que par elle: Au lieu qu'il n'y a aucun point dans les lignes qui servent en la question proposée, qui ne se puisse rencontrer entre ceux qui se determinent par la façon tantost expliquée. Et pourceque cete façon de tracer une ligne courbe, en trouvant indifferemment plusieurs de ses poins, ne s'estend qu'a celles qui peuvent aussy estre descrites par un mouvement regulier et continu, on ne la doit pas entierement reietter de la Geometrie."

[21]The relevant passages are in Descartes' letters to Mersenne of 08/10/1629 and 13/11/1629, [Descartes 1964–1974] vol. 1 pp. 22–32 (esp. pp. 25–26) and pp. 69–75 (esp. pp. 70–71), re-

between generic pointwise construction of curves and their tracing by coordinated continuous motion. In curve tracing the continuity of the motion ensured that each point of the curve was generated; in generic pointwise construction, at least in principle, each point could be constructed. He considered the analogy significant enough to base on it the conclusion, marked **D** in Table 24.1, that curves generated by generic pointwise constructions could also be traced by coordinated continuous motion ("this method of tracing . . . is only applicable to curves that can also be described by a regular and continuous motion"). He gave no further argument.

At the beginning of Book II Descartes formulated the conclusion that generic pointwise constructions of curves were acceptable in geometry on a par with tracing by coordinated continuous motion. Further on in the *Geometry* Descartes repeatedly introduced curves by a generic pointwise construction, even without considering their equations. Thus the ovals[22] (treated in the section marked II-E in Table 20.1) were introduced by (generic) pointwise constructions; equations were not mentioned.

The arguments

Such were Descartes' intricate arguments about accepting or rejecting curves as geometrical. I have located in particular the round of arguments (Table 24.1 **A** → **B** → **E** → **D** → **A**) which combine to prove that the algebraic curves are precisely those that are acceptable in geometry. Descartes took these questions very seriously, and one may well admire the depth he achieved in treating this matter. Especially the arguments about pointwise construction, with their intuition of the difference between particular and generic points of a curve, are impressive.

24.4 Curves traced by means of strings

Acceptable use of strings

Immediately after the arguments on pointwise construction Descartes dealt with another method for generating curves, namely, tracing procedures involving strings. He used this mode of generating curves later in the section on the "ovals," as an alternative to generic pointwise construction.[23] But he probably realized that strings could also be used to combine rectilinear and circular motions so as to trace curves like the quadratrix or the spiral, which he wanted

---

spectively.

[22]The ovals that Descartes discussed on pp. 352–368 of the *Geometry* were curves whose surfaces of revolution provided shapes for lenses with the property that light rays coming from one point converged, after entering the lens, to another point (and variants of this property). Descartes explained how these ovals could be constructed when the positions of the light source and the converging point, and the refracting index of the lens material, were given. In the *Dioptrics* — the first "essay" in [Descartes 1637b] — he had discussed combinations of lenses with spherical, ellipsoid and/or hyperboloid surfaces that effectuate the same optical effects. He noted there (p. 110) the possibility of achieving these effects by single lenses of more complicated shape, but postponed their treatment to the *Geometry*; these were the ovals. Descartes considered them less suitable for the actual production of lenses. For further information on the Cartesian ovals see [Garibaldi 1991] and [Hara 1985].

[23][Descartes 1637] pp. 356–357.

to exclude. Hence curve tracing with strings required a separate clarification of the demarcation between geometrical and non-geometrical curves.

Descartes discussed this matter in a section with the margin title

> And which curves that one describes by means of a string can be accepted [in geometry].[24]

He wrote there:

> Nor should we reject the method in which a string or a loop of thread is used to determine the equality of or the difference between two or more straight lines which can be drawn from each point of the required curve to certain other points or towards certain other lines at certain angles. We have used this method in the Dioptrique to explain the ellipse and the hyperbola. It is true, though, that one cannot accept in geometry any lines which are like strings, that is to say which are sometimes straight and sometimes curved, because the proportion between straight lines and curves is not known, and, I even believe, will never be known to man, and therefore one cannot conclude anything exact and certain on that basis. Nevertheless, because in these constructions one uses strings only to determine straight lines whose lengths are perfectly known, this should not be a reason for rejecting them.[25]

Descartes referred to the familiar construction of the ellipse by means of a string fixed in the two foci, and a variant of this method, generating the hyperbola; he discussed these in more detail in his *Dioptrics*.[26] He considered these tracing methods acceptable because the strings were used merely to ensure a certain relation between straight lines (for instance that their sum was constant).

The spiral traced by using a string

In the previous section I mentioned a method for the general angular section about which Mersenne corresponded with Descartes in 1629. Descartes did not describe the procedure in detail, but in 1650 Christiaan Huygens sketched in his notebook an instrument that exactly corresponds to the remarks in the letter to Mersenne (cf. Note 21). It may very well be that he had heard about it

---

[24] [Descartes 1637] p. 340: "Quelles sont aussy celles qu'on descrit avec une chorde, qui peuvent y estre receues."

[25] [Descartes 1637] pp. 340–341: "Et on n'en doit pas reietter non plus, celle ou on se sert d'un fil, ou d'une chorde repliée, pour determiner l'egalité ou la difference de deux ou plusieurs lignes droites qui peuvent estre tirées de chasque point de la courbe qu'on cherche, a certains autres poins, ou sur certaines autres lignes a certains angles. Ainsi que nous avons fait en la Dioptrique pour expliquer l'Ellipse et l'Hyperbole. Car encore qu'on n'y puisse reçevoir aucunes lignes qui semblent a des chordes, c'est a dire qui devienent tantost droites et tantost courbes, a cause que la proportion, qui est entre les droites et les courbes, n'estant pas connuë, et mesme ie croy ne le pouvant estre par les hommes, on ne pourroit rien conclure de là qui fust exact et assuré. Toutefois a cause qu'on ne se sert de chordes en ces constructions, que pour déterminer des lignes droites, dont on connoist parfaitement la longeur, cela ne doit point faire qu'on les reiette." I quoted part of this passage above, see Note 14.

[26] [Descartes 1637b] pp. 90 and 102.

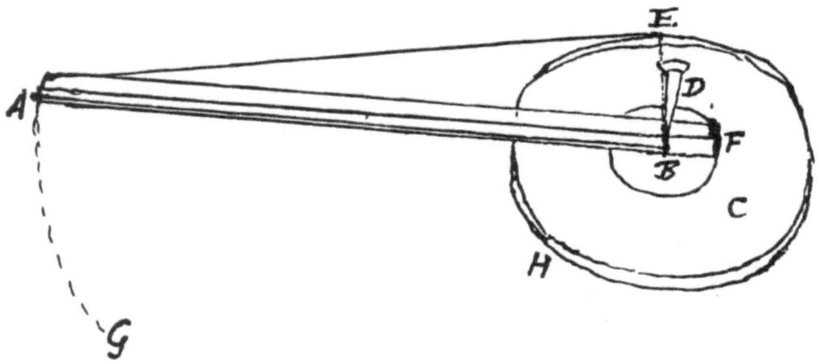

Figure 24.3: Huygens' spiral tracing machine, from a manuscript of 1650

from Descartes himself, who was a frequent guest in the Huygens mansion when Christiaan grew up.[27] I discuss it here as an example of a procedure in which strings change from curved to straight during the motion.

Instrument 24.1 (Spiral tracing instrument — Huygens)[28]

> **1.** C (see Figure 24.3) is a flat circular disk fixed upon the paper with its center at B; FA is a ruler which can turn around B; a string $H-E-A-D$ is fixed to the disk at H, slung around the disk and guided via A to B; at the string's end at B a tracing pin D is attached.
> **2.** If the ruler is turned uniformly counter-clockwise; the string winds up around the disk and the tracing pin D is drawn uniformly along the ruler in the direction of A.
> **3.** During that motion, the pin D traces the spiral.
> [**Proof** The ruler FA turns uniformly around B; at the same time the tracing pin D moves uniformly along FA, hence (cf. Definition 3.2) D traces a spiral.]

[27] It may be noted in this connection that Besson's treatise on instruments and machines ([Besson 1582] pp. 5v–6r) contains a machine for tracing the Archimedean spiral in which the two movements are interrelated by a screw mechanism; the text claims this arrangement to be better than the use of strings.

[28] [Huygens 1888–1950] vol. 11, p. 216.

Here clearly parts of the string change from straight to curved during the motion. Descartes considered that aspect of the procedure as geometrically unacceptable because, as we have seen above (Section 24.2), he was convinced that the ratio of straight and curved lines could never be known.

24.5 The demarcation of geometry before 1637

c. 1619 – c. 1630

From his letter to Beeckman and the other mathematical documents discussed in Chapter 16 we know that already in 1619 Descartes classified geometrical constructions according to the curves used in them. Beyond straight lines and circles he envisaged the use of curves "which arise from one single motion and which therefore can be traced by the new compasses, which I consider to be no less certain and geometrical than the usual compass by which circles are traced."[29] He realized that the class of such curves contained more than only the conic sections; but he considered the quadratrix and the linea proportionum to be outside that class. At the time he did not reject these curves from geometry; indeed he claimed that by their use "hardly anything would remain to be found in geometry."[30] However, he considered these curves as "only imaginary" and intimated that for any problem one could imagine a curve of that kind by which it could be solved. As I explained in Section 16.5 the quadratrix could indeed be seen as specially made for angular sections. In the letter to Beeckman Descartes set himself the task to demarcate precisely between, on the one hand, problems that can only be solved by such "imaginary" curves and, on the other hand, problems solvable by curves arising from one single motion. He also intended to classify the latter problems according to the curves used in their construction. As we have seen, that task he successfully completed in the *Geometry*.

The *Rules* of 1628 did not contain further arguments on the construction of problems by curves, but we know from the letters to Mersenne discussed above[31] that by 1629 Descartes was convinced that constructions by means of the spiral or the quadratrix were not geometrical because they could not be traced by sufficiently coordinated motions or by generic pointwise constructions.

The letter to Golius, 1632

Golius' challenge in 1631 to solve Pappus' problem made Descartes rethink the issue. We have seen in Section 23.1 that he wrote down a sketch of a classification of curves in a letter to Golius. The sketch was an addition to the "écrit" he had sent earlier (and which is lost). After the survey above of Descartes' ideas on the demarcation of geometry and the classification of curves in the *Geometry* it is useful to return to the 1632 document and analyze it in order to see whether it provides clues for the understanding of the development of Descartes' thought on the matter before 1637.

The text of the addition may be paraphrased as follows (see also Table 24.2):

[29] See Note 6 of Chapter 16.
[30] *Ibid.*
[31] See Note 21 and Instrument 24.1.

Number (n) of given lines in the Pappus problem	Number (k) of "simple relations"	Degree (d) of the equation
3, 4	2 or 3	≤ 2
5, 6, 7, 8	3 or 4	≤ 4
9, 10, 11, 12	5 or 6	≤ 6
etc.		

Table 24.2: Concerning Descartes' classification of curves, 1632

For any position of the given lines, the solution of a Pappus problem is a curve that can be traced by "one single continuous motion completely determined by a number of simple relations."[32] "Simple relations" are "those which involve only one geometrical proportion."[33] The number (k) of these simple relations depends on the number (n) of lines in the problem: if $n \leq 4$, then $k = 2$ or $k = 3$; if $n \leq 8$, then $k = 3$ or $k = 4$; if $n \leq 12$, then $k = 5$ or $k = 6$; etc. Conversely, any curve traced by a single continuous motion determined by k simple relations is a solution of a Pappus problem in n lines where, if $k \leq 2$, then $n \leq 4$; if $k \leq 4$, then $n \leq 8$; etc. The requirement of tracing by one continuous motion excludes curves like the spiral and the quadratrix because their description involves unrelated motions. The requirement that the motion should be determined by simple relations excludes a further class of curves, as yet unnamed. Finally, the number of simple relations in the tracing motion induces a classification of the remaining, not excluded curves; the first class consists of the conics, the second one contains, apart from some special curves mentioned in the écrit,[34] many others as well, too many to enumerate.[35]

[32]Cf. Note 35 below: ". . . unico motu continuo, et omni ex parte determinato ab aliquot simplicibus relationibus."

[33]Cf. Note 35 below: ". . . illas . . . quarum singulae non nisi singulas proportiones Geometricas involvunt."

[34]Or so I interpret the "supra" in the text, see Note 35.

[35]Descartes to Golius, January 1632, [Descartes 1964–1974] vol. 1 pp. 233–234. The classification is in Latin (the rest of the letter in French); its full text is: "Datis quotcunque rectis lineis, puncta omnia ad illas iuxta tenorem quaestionis relata, contingent unam ex lineis quae describi possunt unico motu continuo, et omni ex parte determinato ab aliquot simplicibus relationibus; nempe, à duobus vel tribus ad summum, si rectae positione datae non sint plures quam quatuor; à tribus vel quatuor relationibus ad summum, si rectae positione datae non sint plures quam octo; à quinque vel sex, si datae rectae non sint plures quam duodecim, atque ita in infinitum. Et vice versâ nulla talis linea potest describi, quin possit inveniri positio aliquot rectarum, ad quas referantur infinita puncta, iuxta tenorem quaestionis, quae illam contingent. Quae quidem rectae non erunt plures quam quatuor, si curva descripta non pendeat à pluribus quam duobus simplicibus relationibus; nec plures quam octo, si curva non pendeat à pluribus quam quatuor relationibus; et sic consequenter. Hîc autem simplices relationes illas appello, quarum singulae non nisi singulas proportiones Geometricas involvunt. Atque haec linearum

Interpretation

Unfortunately, the key terms "single continuous motion," "simple relation," and "geometrical proportion" in this text defy precise interpretation. It is tempting to suppose that these terms, and the numbers Descartes mentioned, correlated to algebraic properties of the equations of the Pappus curves, in particular their degree or the number of linear factors on their left- and right-hand sides. However, as Table 24.2 shows, Descartes' number of simple relations did not fully correspond to the degree, nor could the discrepancy easily be attributed to a mistake or oversight. Also, if Descartes had the degree of the equation in mind when writing about the number of simple relations, it is difficult to understand why he grouped the given numbers of lines in fours ($n \leq 4$, $n \leq 8$, $n \leq 12$, etc.) rather than in twos. Hence Descartes' "simple relations," involving, as he wrote "only one geometrical proportion," cannot refer to the linear (first-degree) terms of the equation.

Although it seems impossible to reconstruct what precisely Descartes meant by "a single continuous motion determined by a number of simple relations," the structure of the text allows some more global conclusions. When he wrote the letter, Descartes believed that any Pappus curve was traceable by such a single continuous motion determined by a number of simple relations, and conversely that any curve so traceable was the solution of some Pappus problem. I have argued earlier (Section 19.2) that he knew how to derive the equations of Pappus curves. It seems also that he had already come to the conclusion that conversely all algebraic curves were Pappus curves.[36] Furthermore, he envisaged a classification of these curves according to the number of given lines, which number he thought to correspond somehow to the number of "simple relations" determining the tracing motion. The last few sentences of the text show that Descartes believed that the requirement of a *single* continuous motion excluded curves as the quadratrix and the spiral, and that the requirement that the tracing motion be determined by simple relations excluded a further class consisting of curves that nobody yet had named. It is difficult to imagine which curves Descartes had in mind here; it may well be that his remark was prompted by the assumption that the two criteria for curves to be acceptable induced two classes of unacceptable curves.

---

quaesitarum definitio est, ni fallor, adaequata et sufficiens. Per hoc enim quod dicam illas unico motu continuo describi, excludo Quadratricem et Spirales, aliasque eiusmodi, quae non nisi per duos aut plures motus, ab invicem non dependentes, describuntur. Et per hoc quod dicam illum motum ab aliquot simplicibus relationibus debere determinari, alias innumeras excludo, quibus nulla nomina, quod sciam, sint imposita. Denique per numerum relationum singula genera definio; atque ita primum genus solas Conicas Sectiones comprehendit, secundum verò praeter illas quas supra explicui, continet alias quam plurimas, quas longum esset recensere."

[36]The words "et vice versa . . . contingent" (cf. Note 35) point in that direction; cf. Section 19.5, in particular Note 17. The result was never challenged during the early modern period; for a proof that it is incorrect see [Bos 1981] pp. 332–338.

24.6 The development of Descartes' ideas on demarcation between 1632 and 1637

Comparison In Table 24.3 I have collected the ideas from the 1632 text and added the corresponding ideas from the *Geometry* as discussed above. Both the 1632 text and the *Geometry* contained the concept of curve tracing by acceptable kinds of motion and the idea that curves not so traceable, such as the spiral and the quadratrix, had to be excluded. But in 1637 the nature of the motions was less precisely defined, in particular the "simple relations" were no longer mentioned.

Pappus curves and the way they could be traced were essential in the 1632 text. They were less so in the *Geometry*; Descartes showed that Pappus curves were algebraical by showing how their equations could be derived, and he claimed that any algebraic equation could occur as the equation of some Pappus curve. But these statements did not serve as arguments in support of the demarcation, nor in support of the equivalence of geometrical and algebraic curves.

In the 1632 text Descartes distinguished two kinds of excluded curves. The distinction is not mentioned in the *Geometry*; in 1637 Descartes offered no further structure within the class of curves that were excluded; they were all "mechanical."

Reconstruction The most striking difference between the two stages in Descartes' thinking about classification of curves and the demarcation of geometry is that in 1632 Descartes connected curve tracing directly to Pappus' problem, whereas in the *Geometry* the link between the two was very indirect. The comparison suggests the following reconstruction of the development of Descartes' ideas and insights about Pappus' problem and the demarcation of geometry: When first studying the problem in early 1632, he thought that for any Pappus curve he would be able to derive an explicit specification of the motion by which it could be traced. On the basis of that conviction he elaborated and refined his earlier programmatic and structural ideas about geometry, in particular the tenet that curves are to be accepted in or rejected from geometry on the basis of the nature of the motions that trace them. He formulated these ideas in the addition to the "écrit." Later he realized that his motion solution of Pappus' problem was incomplete. However, he retained the programmatic and structural ideas about geometry and he tried to save the arguments for it. The result was that in the *Geometry* Pappus' problem was no longer linked to curve tracing and that Descartes had to give up classifying acceptable curves on the basis of properties of the tracing motion. In general he was forced to base his methodological arguments on the equations of curves rather than on the procedures for tracing them. Yet several elements of his original arguments about explicit curve tracing were retained in the *Geometry*, where in fact they no longer served much purpose.

Curves	1632	1637
Curves traced by acceptable motion	Acceptable motion is characterized as "one single continuous motion completely determined by a number of simple relations:" curves not so traceable are excluded.	Acceptable motion is characterized as "a continuous motion or by several successive motions, each motion being completely determined by those which precede:" curves not so traceable are excluded.
Pappus curves	The class of Pappus curves coincides with the class of curves that can be traced by acceptable motion; it is an extensive class.	The class of Pappus curves coincides with the class of algebraic curves.
Algebraic curves	[Not mentioned explicitly.]	All algebraic curves are Pappus curves; because they allow generic pointwise construction they are also traceable by acceptable motion.
Excluded curves	Some curves, such as the spiral and the quadratrix, are excluded by the requirement of a *single* continuous motion; others are excluded by the requirement of motion determined by simple relations; such curves are as yet unnamed.	The excluded curves are called "mechanical:" examples: the quadratrix and the spiral.

Table 24.3: The ideas on classification and demarcation

In the next chapter we will see that also with respect to the classification of "geometrical" curves as to simplicity Descartes developed ideas that he probably acquired while studying Pappus' problem but which in 1637 he formulated in algebraic terms independently of that context.

Chapter 25

Simplicity and the classification of curves

25.1 The classification

The classes

In the previous chapter I analyzed Descartes' answer to the first of two crucial questions that his new doctrine of geometry had to address. These questions were:

> **A.** Which curves are acceptable as means of exact construction in geometry?
>
> and
>
> **B.** When is one curve simpler than another?

I now turn to the second question.

Descartes divided the "geometrical" curves in successive classes ("genres"). The relevant text is at the beginning of Book II (IIA in Table 20.1), after the passage, (quoted in Section 24.2), in which Descartes explained that the equations of curves were the best means for classifying them. Descartes then wrote:

> If this equation contains no term of higher degree than the rectangle of two unknown quantities, or the square of one, the curve belongs to the first and simplest class.[1]

Descartes took this class to contain circles, parabolas, hyperbolas, and ellipses, he did not mention straight lines.

> But when the equation contains one or more terms of the third or fourth dimension in one or both of the unknown quantities (for it

[1] [Descartes 1637] p. 319: "Et que lorsque cete equation ne monte que iusques au rectangle de deux quantités indeterminées, oubien au quarré d'une mesme, la ligne courbe est du premier et plus simple genre . . ."

> requires two unknown quantities to express the relation of one point to another) the curve belongs to the second class; and if the equation contains a term of the fifth or sixth degree in either or both of the unknown quantities the curve belongs to the third class, and so on indefinitely.[2]

Thus curves of degrees one and two formed the first class, those of third- and fourth-degree the second, etc. The same classification was implicit in Descartes' letter to Golius of 1632, where, as we have seen (Section 24.5) he grouped the Pappus loci according to whether the number of given lines was ≤ 4, ≤ 8, ≤ 12, etc., which corresponded to degrees ≤ 2, ≤ 4, ≤ 6, etc.

Grouping by pairs of degrees

Descartes gave one explicit argument for this classification of curves by pairs of degrees, namely, that it was based on the fact that fourth-degree problems ("difficultés") could always be reduced to third-degree ones, and sixth-degree problems to fifth-degree ones.[3] It is difficult to interpret this statement. First of all it is incorrect. Descartes obviously referred to the fact that fourth-degree equations (in one unknown) can be reduced to third-degree ones by procedures such as Ferrari's[4] or Viète's,[5] whereas third-degree equations withstood all attempts to reduce them to second-degree ones. He then generalized this pattern to higher degrees, although no procedure for reducing sixth-degree equations to fifth-degree ones was known at the time, and, as we now know, there are no such procedures. Yet the generalization apparently seemed plausible enough to Descartes to accept it without further proof.

The second difficulty with Descartes' statement is that he did not make clear why a pattern observed in a classification of equations in one unknown (equations related to problems) should apply for equations in two unknowns (equations of curves). Possibly he based this step on the pattern akin to the one he gave for the pointwise construction of Pappus curves (cf. Table 23.2 and Section 24.3): To construct a Pappus curve with $(2n-1)$th or $2n$th-degree equation $F(x, y) = 0$, Descartes prescribed choosing a number of ordinate values y_i and to construct the roots of the corresponding equations $F(x, y_i) = 0$ in one unknown; these equations are at most of degree $2n-1$ or $2n$. According to Descartes' canon of construction, to be discussed in the next chapter (Sections 26.4

[2][Descartes 1637] p. 319: "Mais que lorsque l'equation monte iusques a la trois ou quatriesme dimension des deux, ou de l'une des deux quantités indeterminées, car il en faut deux pour expliquer icy le rapport d'un point a un autre, elle est du second: et que lorsque l'equation monte iusques a la 5 ou sixiesme dimension, elle est du troisiesme; et ainsi des autres a l'infini."

[3][Descartes 1637] p. 323: "Au reste ie mets les lignes courbes qui font monter cete equation [sc. the equation of the curve] iusques au quarré de quarré, au mesme genre que celles qui ne la font monter que iusques au cube. Et celles dont l'equation monte au quarré de cube, au mesme genre que celles dont elle ne monte qu'au sursolide. Et ainsi des autres. Dont la raison est qu'il y a reigle generale pour reduire au cube toutes les difficultés qui vont au quarré de quarré, et au sursolide toutes celles qui vont au quarré de cube, de façon qu'on ne les doit point estimer plus composées."

[4]See Note 18 of Chapter 10.

[5]Cf. Note 16 of Chapter 10.

and 26.5), the roots of $(2n-1)$th- and $2n$th-degree equations are constructed by one and the same standard construction. Thus with respect to pointwise construction, curves of degrees $2n-1$ and $2n$ were of the same complexity and could therefore be considered as belonging to the same class. It should be noted that Descartes' canon of construction, and hence the argument above, did not require a reduction of $2n$-degree equations (in one unknown) to $2n-1$-degree ones, which Descartes, as we saw above, claimed to be possible. Thus, if he had the argument suggested above in mind when he decided to classify curves by pairs of degrees, his incorrect statement about the possibility of such a reduction was unnecessary.

There is, however, no direct textual evidence that Descartes adopted the argument via pointwise construction as basis for his classification of curves. It may well be that he simply grouped curves according to pairs of degrees because he had already done so with problems.

It should be noted that Descartes did not consider curves belonging to the same class as equally simple. As we will see (Sections 26.3 and 26.4) he used a third-degree curve, namely, the Cartesian parabola, for the construction of fifth- and sixth-degree equations and claimed that equations of degree seven and eight could be solved by using a fourth-degree curve. Had he considered third- and fourth-degree curves equally simple, these choices of procedures would have had little sense. Thus Descartes did not essentially use his classification of curves by pairs of degrees; rather he used a classification by single degrees. The terminology of "genres" was confusing and hardly functional.

25.2 Simplicity, tracing, and degree

Choosing the simplest constructing curve

Descartes' primary criterion for acceptability of curves was geometrical, it related to the motions by which they were traced. His criterion for classification, however, was algebraic, namely, the degree of the curve. The different nature of these criteria raises the question how he saw the relation between the simplicity of curves, their degrees, and the tracing procedures by which they could be generated. In the opening sentences of the third book Descartes referred to the simplicity of tracing:

> While it is true that every curve which can be described by a continuous motion should be accepted in geometry, this does not mean that we should use at random the first one that we meet in the construction of a given problem. We should always choose with care the simplest curve that can be used in the solution of a problem.[6]

Yet in the subsequent sentences he rejected tracing as a criterion for simplicity:

[6] [Descartes 1637] pp. 369–370: "Encore que toutes les lignes courbes, qui peuvent estre descrites par quelque mouvement regulier, doivent estre receuës en la Geometrie, ce n'est pas a dire qu'il soit permis de se servir indifferemment de la premiere qui se rencontre, pour la construction de chasque problesme: mais il faut avoir soin de choisir tousiours la plus simple, par laquelle il soit possible de le resoudre."

> But it should be noted that the simplest means not merely the one most easily described, nor the one that leads to the easiest demonstration or construction of the problem, but rather the one of the simplest class that can be used to determine the required quantity.[7]

In connection with this statement Descartes returned to the Mesolabum. At the beginning of Book II he had introduced this instrument to explain that the curves it traced, although successively more complex, were all equally acceptable in geometry (cf. Instrument 16.3, Figures 16.4 and 24.1, and Section 24.2). Now he noted that, although these curves doubtlessly provided the easiest constructions of mean proportionals and an immediate proof of the correctness of that construction, nevertheless, they were not the appropriate constructing curves, because they were not of lowest possible class. With obvious reference to Pappus' dictum on the "sin" committed by geometers when failing to use the simplest possible means of construction (cf. Section 3.4), Descartes stated that using these curves was an error ("faute") in geometry:

> But the curve *AD* is of the second class,[8] while it is possible to find two mean proportionals by the use of the conic sections, which are curves of the first class. Again, four or six mean proportionals can be found by curves of lower classes than *AF* and *AH* respectively. It would therefore be a geometric error to use these curves. On the other hand, it would be a blunder to try vainly to construct a problem by means of a class of lines simpler than its nature allows.[9]

A reversal It is important to pause and note the strangeness of these passages. Descartes first encouraged the reader to follow an obvious line of thought: to prefer the curves traced by the Mesolabum over any other curves for constructing mean proportionals, because they were most simply traced and they provided direct proof of the correctness of the construction. And then, by a sudden reversal of direction, he blocked this line of thought and ordered, without argument, that nevertheless one should employ curves of lowest possible degree. Our study of

---

[7] [Descartes 1637] p. 370: "Et mesme il est a remarquer, que par les plus simples on ne doit pas seulement entendre celles, qui peuvent le plus aysement estre descrites, ny celles qui rendent la construction, ou la demonstration du Problesme proposé plus facile, mais principalement celles, qui sont du plus simple genre, qui puisse servir a determiner la quantité qui est cherchée."

[8] Descartes did not give the equations of the curves traced by the Mesolabum. They are (cf. Figures 16.4 and 24.1):

$$x^{4n} = a^2(y^2 + x^2)^{2n-1}\,. \tag{25.1}$$

The degrees of the successive curves (starting with the one traced by *D*) are 4, 8, 12, etc., and thus their "genres" are 2, 4, 6, etc.

[9] [Descartes 1637] p. 371: "Mais pourceque la ligne courbe AD est du second genre, et qu'on peut trouver deux moyenes proportionelles par les sections coniques, qui sont du premier; et aussy pourcequ'on peut trouver quatre ou six moyenes proportionelles, par des lignes qui ne sont pas de genres si composés, que sont AF, et AH, ce seroit une faute en Geometrie que de les y employer. Et c'est une faute aussy d'autre costé de se travailler inutilement a vouloir construire quelque problesme par un genre de lignes plus simple, que sa nature ne permet."

his earlier thoughts on construction makes clear that the reader whom Descartes addressed here was his earlier self, who, at the time of the letter to Beeckman, explored the idea of universal instruments, the "new compasses," the mesolabum for mean proportionals and the multisector for angular sections. These instruments were conceived to supplement or even complete the arsenal for geometrical problem solving; the curves they traced were simple and the design of the instruments made the proofs evident and elementary. But the Descartes of the Mesolabum and the multisector was later confronted with the difficulty of generalizing the kinematic approach and he experienced the power of algebra and the structural wealth of Pappus' problem; as a result he reversed his earlier line of thought from the kinematical devices of curve tracing to the symbolic devices of equations. The reversal was indeed a choice for the strength and structural appeal of algebra, and thereby against the intuitive cogency of the kinematic interpretation of constructional procedures. That cogency was not lessened, no arguments against it had arisen. The passages quoted above contain, as it were, Descartes' *amende honorable* for this essential reversal in the development of his thoughts on geometry.[10]

25.3 Simplest possible curves and equations

Simplicity for fixed degree

With the degree as criterion of simplicity two further questions remained: How to choose the simplest curve among those of the same degree? And: what about curves (such as the circle among the conics), which, as means of construction, were less powerful than most other curves in the same class? Descartes did not explicitly discuss the first question. He chose the parabola for constructing the roots of third- and fourth-degree equations, and for those of fifth- and sixth-degree he used a special third-degree curve, the Cartesian parabola. These choices suggest that Descartes preferred curves that were simple in more than only the sense of having lowest possible degree. We have seen in connection with the 1625 version of Descartes' construction of the roots of third- and fourth-degree equations that the parabola was considered the simplest conic section (cf. Section 17.1). The Cartesian parabola also featured a simplicity within its class: Descartes considered it to be the simplest Pappus locus for five given lines (cf. Section 23.4). However, Descartes did not explicitly formulate further criteria of simplicity than the degree of the curve.

Exceptional curves

Descartes did discuss the fact that among curves of the same degree there could be particular ones that were essentially less powerful in constructions than the others. The obvious example was the circle, which as to its degree belonged to Descartes' first class of curves, but which had less constructive power

[10]The acceptance of the algebraic degree, rather than a more geometrical criterion, as the measure of simplicity was criticized by several later mathematicians, for instance, Newton and Jakob Bernoulli, cf. [Bos 1984] pp. 358–366.

than the other conic sections.[11] Indeed, for constructing solid problems a non-circular conic was required. Descartes investigated whether this phenomenon also occurred in the higher classes of curves. His formulations can only be interpreted as consistent if we assume that he thought of single-degree classes instead of classes of paired degrees. He wrote that within one class in principle all curves were equally complex and therefore could "serve for determining the same points and for constructing the same problems,"[12] except some essentially simpler curves whose power did not extend that far.[13] He mentioned the circle and, within the second class, the "ordinary conchoid which takes its origin from the circle."[14] Descartes apparently referred here to an observation he had made earlier in the *Geometry*,[15] namely, that the conchoid could be traced by a turning ruler and a moving circle.[16] He also knew that in classical geometry the conchoid served to perform the neusis construction (cf. Construction 2.5), and that all solid problems could be constructed by neusis. He may have concluded from this that the conchoid was nearer to the first class of curves, the conics, than to the class to which it belonged by virtue of its degree (which is four).

25.4 Reducibility

The simplest possible equation

In a general method for constructing the roots of an n-th degree equation, the product of the degrees of the constructing curves must be at least equal to n. This is a consequence of the fact, proved by Bezout (1779), that two curves of degrees m and n, respectively, intersect, in general, in $m \times n$ points. It is likely that Descartes was aware of this fact — the practice of elimination suggests it rather clearly — but he did not formulate it, let alone that he proved it. He was certainly aware of one consequence of the fact, namely, that in order to find constructing curves of minimal degree the equation to which the problem is reduced should have minimal degree as well.

Reducible equations

Yet in the practice of analysis it could easily happen that a geometrical problem was reduced to an equation with unnecessarily high degree, for instance, if trivial or otherwise known solutions entered as roots. In that case the equation was reducible, that is, it could be written as

$$F(x)G(x) = 0 , \qquad (25.2)$$

[11]In a later argument ([Descartes 1637] pp. 401–402, cf. Note 49 of Chapter 26) Descartes related the constructional superiority of non-circular conic sections over circles to the fact that a circle has only one curvature, whereas the curvature along a non-circular conic varies; I return to this argument in Section 26.6.

[12][Descartes 1637] p. 323: ". . . en sorte qu'elles peuvent servir a déterminer les mesmes poins, et construire les mesmes problesmes . . ."

[13]*Ibid.*: "qui n'ont pas tant d'estendue en leur puissance."

[14]*Ibid.*: "Conchoide vulgaire, qui a son origine du cercle."

[15][Descartes 1637] p. 322.

[16]Descartes did not explain his observation but if (cf. Figure 2.3) one considers a circle with center F and radius a as the curve moving along CD, and EG as the ruler moving around O and guiding the circle via point F, then this turning ruler and moving curve procedure produces precisely the conchoid.

where the roots of F were the additional ones and those of G provided the essential solutions of the problem.[17] The trivial or known additional roots would typically be given line segments in the figure or line segments constructible by straight lines and circles from the given ones (for instance, the hypotenuse of a right-angled triangle whose two other sides are given). This implied that the coefficients of F, and hence also those of G, could be constructed by straight lines and circles from those of the original equation. Descartes realized that his method should safeguard against inadvertently arriving at a reducible equation. He therefore provided algebraic techniques for checking whether the final equation was reducible and, if so, to find its irreducible components. These techniques constitute a large part of Descartes' contribution to the algebraic theory of equations to which Chapter 27 is devoted; I discuss them in Section 27.3.

25.5 Conclusion

Infiltration of algebra

The preceding sections have shown how in the case of the classification of curves, in the same way as with the demarcation question discussed in the previous chapter, Descartes finally adopted algebraic criteria, abandoning earlier attempts to classify according to methods of tracing. The process may be described as an infiltration of algebraic thinking into the interpretation of geometrical exactness. It is indeed a natural process: the intuitive understanding of geometrical exactness, especially with respect to construction and motion, is diffuse and not easily translated into a formal ordered mathematical structure; algebra, on the other hand, with its ordering of equations by their degrees, offers attractive formal structure. Forced, by his programmatic approach, to make choices, but failing to find convincing geometrical arguments on which to base them, Descartes could do little but to turn to algebra. Here, at least, the choices were clear, and apparently in the end this was sufficient reason for Descartes to accept the absence of clear arguments why geometrical exactness should be interpreted through algebraic criteria.

With these choices on acceptability and simplicity of curves Descartes had finally achieved a complete interpretation of exactness with regard to geometrical procedures of construction. The technical result was Descartes' final canon of geometrical construction to which I now turn.

[17] Another kind of reducibility arises if the equation can be written as $F(G(x)) = 0$ with degree $G > 1$. In that case the construction can be split in two lower-degree ones: $F(y) = 0$, yielding roots y_i, and $G(x) = y_i$. Descartes did not discuss this case separately. Cf. Note 24 of Chapter 27 and [Bos 1984] pp. 342–343.

Chapter 26

The canon of geometrical construction

26.1 The "construction of equations"

Beyond the fourth degree

In the first book of the *Geometry* Descartes explained how to construct the roots of quadratic equations. In order to move on to higher degrees he had to elaborate the interpretation of constructional exactness in the way explained in the previous chapters. In the present chapter I discuss the final result of this interpretation, Descartes' general canon for geometrical construction. Taking over terminology developed after Descartes (cf. Section 29.3) I use the term "construction of an equation" for the procedure of geometrically constructing the roots of an algebraic equation in one unknown.

By 1628 Descartes had simple and convincing constructions of the roots of equations of degrees 1–4. Especially, the general construction for equations of third and fourth degree by means of a parabola and a circle (cf. Chapter 17, Construction 17.1) was beautiful and constituted a marked improvement of the then extant methods. Yet a complete theory of geometrical construction demanded more, namely, a method for constructing any equation, whatever its degree. As we have seen, the writing of the *Rules* probably confronted Descartes with the need for such a general method of constructing equations, but at that time he had no answer.

Inspiration from Pappus' problem

In the *Geometry* Descartes provided an answer. It consisted of a general construction of the roots of fifth- and sixth-degree equations and the claim that this construction, together with the one for third- and fourth-degree equations, sufficiently indicated how the technique could be extended to higher-order equations. The construction for fifth- and sixth-degree equations used a circle and a Cartesian parabola, the latter generated by the procedure of a turning ruler and a moving parabola. I have argued earlier (cf. Chapter 19) that these ingredients, the Cartesian parabola, the procedure of its tracing, and the suggestive possibilities of iteration and reduction, were the fruits of Descartes' earlier studies of Pappus' problem. The Cartesian parabola, the locus in the simplest case of the five-line problem, was arguably the simplest curve beyond the conic sections; it was therefore natural to choose it as constructing curve for the next class of equations, those of degrees five and six. The fact that the curve was traced by the motion of a parabola provided additional reason to consider it as the successor of the parabola itself in the construction of the next group of equations. Moreover, the generation of the Cartesian parabola by the turning ruler and moving parabola procedure suggested generalizability: by iteration one might find the successive standard curves to be used in the construction of successive classes of equations of increasing degree.

Once the Cartesian parabola was chosen as constructing curve, the remaining task was calculation, which Descartes pursued until he had worked out the construction of fifth- and sixth- degree equations by the intersection of a Cartesian parabola and a circle. The exercise (which I discuss in more detail below) gave him the conviction that the process could be iterated for seventh- and eight-degree equations, etc. But he did not pursue the matter further — he merely stated that it would be easy to proceed in the same manner indefinitely.

26.2 The standard construction of third- and fourth-degree equations

Standard form of the equation

Descartes gave his standard constructions of equations of degree higher than two in the second half of Book III of the *Geometry* (sections III D–E in Table 20.1). The construction for third- and fourth-degree equations[1] was basically the same as the one he found c. 1625 and showed to Beeckman in 1628; I have discussed the latter in Section 17.1 (Construction 17.1). The version in the *Geometry* differed from the earlier one mainly in that it used another way of dealing with +/− case distinctions and that a proof was added.

Descartes assumed that the second term of the equation was removed in the usual fashion[2] and wrote the equation as[3]

$$z^4 = * \,.\, apzz \,.\, aaqz \,.\, a^3r \,. \tag{26.1}$$

The third-degree equation was subsumed in this form, namely, for $r = 0$. The "$*$" meant that a term was missing; the "." stood for either a $+$ or a $-$. Taking a as the unit, he rewrote the equation in non-homogeneous form:

$$z^4 = * \,.\, pzz \,.\, qz \,.\, r \,. \tag{26.2}$$

Descartes took p, q, and r to be positive line segments and kept track of the +/− distinctions by expressions as "if there is $+p$ in the equation," "if there is

[1][Descartes 1637] pp. 389–395; the margin title of the section is: "Facon generale pour construire tous les problesmes solides, reduits a une Equation de trois ou quatre dimensions."

[2]Cf. Chapter 27 Equation 27.8.

[3][Descartes 1637] p. 390.

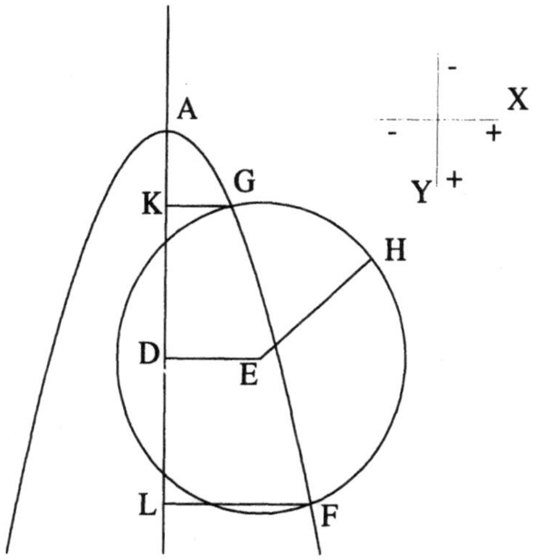

Figure 26.1: Descartes' standard construction of fourth-degree equations

$-r$," etc.[4] I write his equation as

$$x^4 = Px^2 + Qx + R\,, \tag{26.3}$$

with $|P| = p$, $|Q| = q$, $|R| = r$; Descartes' sign distinctions then correspond to marking P, Q, and R along directed axes according to whether they are positive or negative. I use this convention in my summary of the construction and in my Figure 26.1; I explain the details of his case distinctions in footnotes. For illustration I add, in Figure 26.2, the four drawings that Descartes needed to deal with the different cases.

Descartes' construction can now be summarized as follows: *The construction*

Construction 26.1 (Roots of a fourth-degree equation)[5]
Given the equation $x^4 = Px^2 + Qx + R$, it is required to construct its roots.

Construction:
1. Describe (see Figure 26.1) a parabola with vertical axis, *latus rectum* equal to 1, and vertex A as highest point (this implies that with respect to coordinate axes as drawn in the figure, its equation is $y = x^2$).

[4]E.g. [Descartes 1637] p. 391: "s'il y a $+p$."
[5][Descartes 1637] pp. 389–395.

2. Mark D on the vertical axis such that (oriented according to directions as indicated) $AD = \frac{1}{2}(P+1)$.
3. Draw $DE = \frac{1}{2}Q$ horizontally from D in the direction corresponding to its sign.
4. Construct a line segment EH equal to $\sqrt{\frac{1}{4}(1+P)^2 + \frac{1}{4}Q^2 + R}$ and draw a circle around E with its radius equal to that line segment.[6]
5. The circle intersects (or touches) the parabola in at most four points G, F, $\cdots$; draw perpendiculars GK, FL, $\cdots$ to the axis from each of these points.
6. The segments GK, FL, $\cdots$, with signs as indicated by their direction, are the roots of the equation.
[**Proof**: Descartes proved the correctness of the construction by setting $GK = x$ and calculating the value of the distance EG in two ways, one using that G was on the parabola, the other that G was on the circle; equating both expressions he arrived at the original equation.[7] Streamlined as to case distinctions in the same way as above, the proof comes down to the following: Put $GK = x$ and $AK = y$, then $y = x^2$ because G is on the parabola. G is also on the circle with center E whose coordinates are $y_E = \frac{1}{2}(P+1)$, $x_E = \frac{1}{2}Q$; the equation of the circle is $x^2 - Qx + y^2 - (P+1)y = R$. Inserting $y = x^2$ one finds from this the equation $x^4 = Px^2 + Qx + R$ as required.]

In his text corresponding to item **3** Descartes left it to the reader to chose a direction for $DE = \frac{1}{2}Q$; he afterwards adjusted directions by statements as "at the same side of the axis as E if there is $+q$ in the equation."[8] In his figures (cf. Figure 26.2) he consistently drew DE to the left, which meant that he had to interpret the horizontal direction to the right as positive if the sign of q in the equation was negative and *vice versa*.

Descartes noted that the circle might fail to intersect or touch the parabola in any point, which meant

> that there is no root at all in the equation either true or false, and that they are all imaginary.[9]

[6] Descartes performed the construction of the line segment EH explicitly (see the dotted lines in the drawings of Figure 26.2), using the fact that $EA = \sqrt{\frac{1}{4}(1+P)^2 + \frac{1}{4}Q^2}$. If $r = 0$ (the case of the cubic equation), the circle passes through A. If $r \neq 0$, the construction of the radius involves the unit and is perfomed by constructing right-angled triangles. The construction differs according to whether R is positive or negative.

[7] He did so only for one of his case distinctions (namely, $+p$, $-q$, and $+r$) and left the other cases to the reader.

[8] E.g. [Descartes 1637] p. 393: ". . . si la quantité q est marquée du signe +, les vrayes racines seront celles de ces perpendiculaires, qui se trouveront du mesme costé de la parabole, que E le centre du cercle . . ."

[9] [Descartes 1637] p. 393: ". . . cela tesmoigne qu'il n'y a aucune racine ny vraye ny fausse en l'Equation, et qu'elles sont toutes imaginaires."

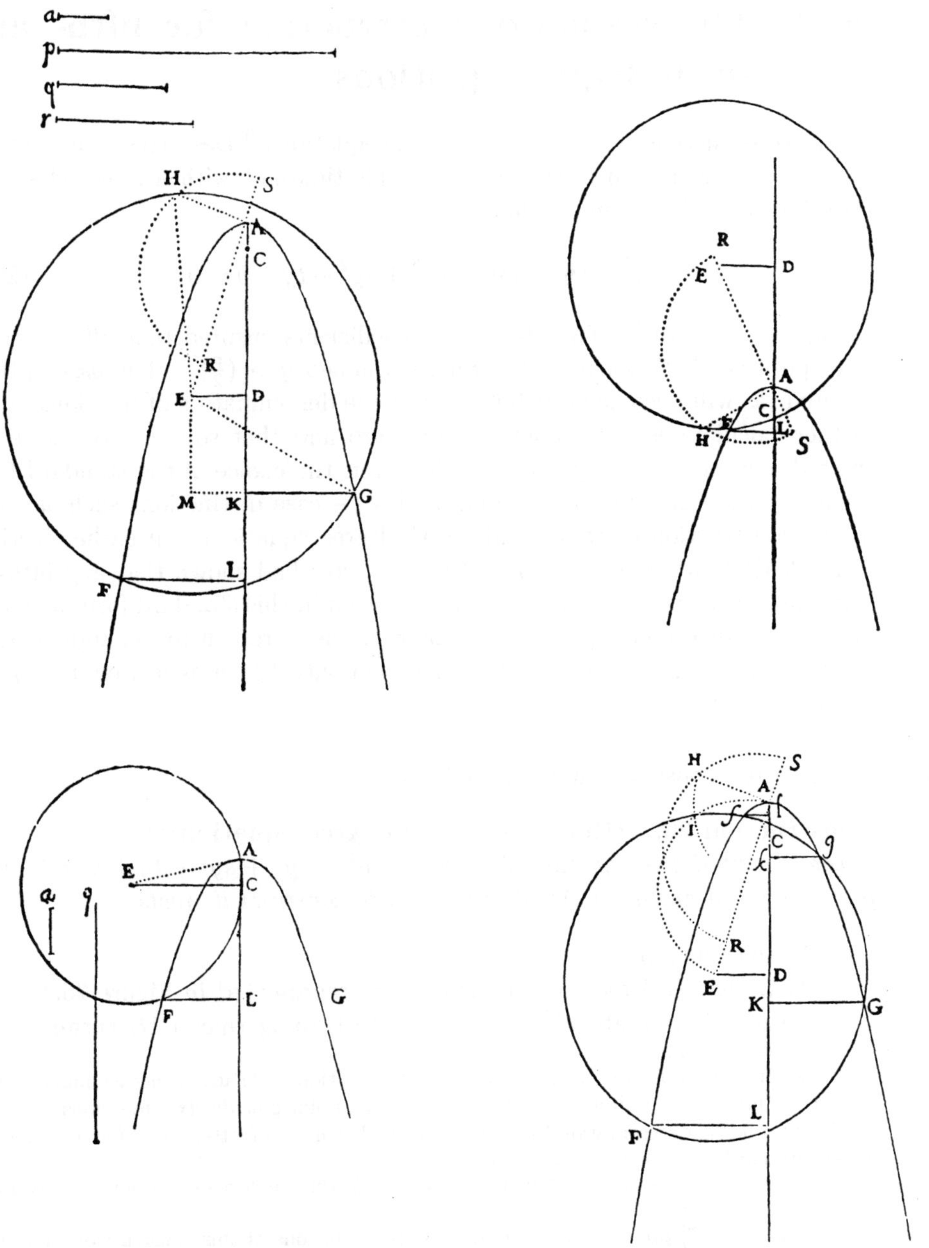

Figure 26.2: The standard construction of fourth-degree equations (*Geometry* pp. 390–392)

He did not comment on the possibility that the value under the root sign might be negative. In the case of the cubic equation the intersection at A corresponded to the solution $x = 0$, introduced by raising the equation to the fourth degree. Descartes went on to discuss special cases and examples, to which I will return in Section 26.6.

26.3 The standard construction for fifth- and sixth-degree equations

Standard form of the equation

For constructing fifth- and sixth-degree equations[10] Descartes again presupposed a standard form of the equation, this time not with a removed second term but with alternating coefficients:

$$y^6 - py^5 + qy^4 - ry^3 + sy^2 - ty + v = 0 \qquad (26.4)$$

$(p,\ q,\ \cdots,\ v > 0)$.[11] The alternating coefficients ensured that all real roots were positive.[12] Descartes furthermore assumed $q > (\frac{p}{2})^2$. The assumptions ensured (as will become clear below) that the denominators of fractions in the further calculations were different from zero and that square roots were extracted only of positive quantities. Moreover, the choice of the standard form as in Equation 26.4 avoided complicated $+/-$ case distinctions such as those in the construction of third- and fourth-degree equations. In earlier sections of Book III (part III-C of Table 20.1) Descartes had shown that any fifth- or sixth-degree equation could indeed be rewritten in this standard form by transformations corresponding to plane constructions. I return to his arguments in the next chapter, in particular in connection with the substitution $x = y - a$ (Equation 27.9).

The construction

Descartes' construction was as follows:

Construction 26.2 (Roots of a sixth-degree equation)[13]
Given a sixth-degree equation $y^6 - py^5 + qy^4 - ry^3 + sy^2 - ty + v = 0$, *with* $p,\ q,\ \cdots\ v > 0$ *and* $q > (\frac{p}{2})^2$, *it is required to construct its roots.*

Construction:
1. Draw (see Descartes' original figure represented in Figure 26.3) a Cartesian parabola[14] $QACN$ by the turning ruler (AE turning

[10][Descartes 1637] pp. 402–411; the margin of this section is: "Facon generale pour construire tous les problesmes reduits a une Equation qui n'a point plus de six dimensions."

[11]In the following presentation I keep to Descartes' own use of letters both for the algebraic quantities and for the points in the figure.

[12]For negative values of y the left-hand side is > 0; the result also agrees with Descartes' signrule, see Section 27.1.

[13][Descartes 1637] pp. 402–411. See also the presentations of this construction in Whiteside's note in [Newton 1967–1981] vol. 1 p. 495, Note 15, in [Rabuel 1730] pp. 566–577, and in [Galuzzi 1996].

[14]Descartes drew only one branch of the curve; the other is not involved in the construction.

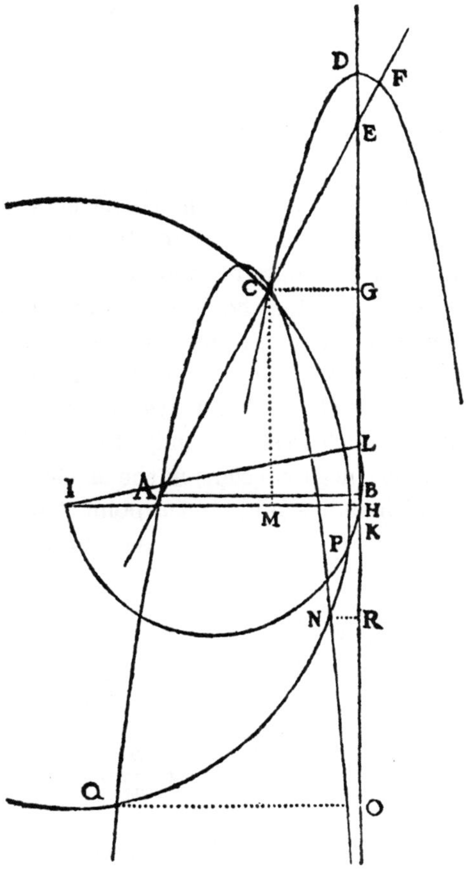

Figure 26.3: The standard construction of sixth-degree equations (*Geometry* p. 404)

around A) and moving parabola (CDF moving along the axis BED) procedure,[15] adjusting the parameters AB, DE, and the *latus rectum* n of the parabola (see Figure 26.3) as:

$$AB = \frac{p}{2}, \tag{26.5}$$

$$n = \sqrt{\frac{t}{\sqrt{v}} + q - \frac{p^2}{4}}, \tag{26.6}$$

$$DE = \frac{2\sqrt{v}}{pn}. \tag{26.7}$$

(Because of the assumptions about the coefficients, the quantities under the root signs are positive. Descartes took the axis of the parabola vertical, its vertex upward, and AB to the left of the vertical. These choices implied that he took the horizontal direction to the left and the vertical direction downward as positive directions. Accordingly I further describe his construction assuming an x, y-coordinate system along BO and BA, respectively. The directions of the line segments, which Descartes indicated explicitly, then correspond to the positive/negative signs in the algebraic expressions.)
2. Draw a circle with radius ρ around a point I with coordinates x_I, y_I ($x_I = BH$, $y_I = IH$), adjusting these values as follows:

$$y_I = \frac{t}{2n\sqrt{v}} - \frac{2\sqrt{v}}{pn}, \tag{26.8}$$

$$x_I = \frac{m}{n^2}, \quad \text{with } m = \frac{r}{2} + \sqrt{v} + \frac{pt}{4\sqrt{v}}, \tag{26.9}$$

$$\rho^2 = \frac{t}{2n\sqrt{v}} - \frac{s + p\sqrt{v}}{n^2} + \frac{m^2}{n^4}. \tag{26.10}$$

(Descartes explained that if the right-hand side of Equation 26.10 was negative, the roots of the given equations were "imaginary" — see below. Note that each of the lengths[16] in Equations 26.5–26.10 can be constructed by straight lines and circles.)
3. The y-coordinates CG, NR, QO, $\cdots$ of the points of intersection[17] C, N, Q, $\cdots$ of the circle and the Cartesian parabola are the required roots of the equation.
[**Proof**: Descartes proved the correctness of this construction by considering a point of intersection C (with coordinates x and y) and

[15]After giving the present construction Descartes noted that if this tracing seemed cumbersome ([Descartes 1637] pp. 407: "Que si la façon de tracer la ligne ACN par le mouvement d'une Parabole vous semble incommode . . ."), one could adopt a pointwise construction of the Cartesian parabola, which he then explained.

[16]All expressions apart from the one for ρ^2 were given explicitly by Descartes; for ρ^2 he described an equivalent construction.

[17]Note that Descartes did not mark the fourth intersection, near C, in his figure.

the pertaining line segment $CM = GH$. Now $CM = x_I - x$, and because C is both on the circle and on the Cartesian parabola, CM can be expressed in two ways in terms of y and constants. Equating the two expressions yields the original equation in y, so that the y-coordinate of C is indeed a root of the equation.]

Descartes stated that in this way all roots were found and that if the two curves intersected in less than six points, the remaining roots were "imaginary." If the curves did not intersect at all, all roots were "imaginary," and the same applied if the value for ρ^2 in Equation 26.10 was negative.[18]

Because all the roots were positive, only one branch of the Cartesian parabola was actually used in the construction, namely, the branch with the local extreme. As a consequence, a circle might intersect that branch of a Cartesian parabola in as many as six points. This feature, explicitly mentioned by Descartes,[19] was considered paradoxical by Roberval, who objected to it in 1638.[20]

How the construction was found

Descartes did not explain how he found the values of the parameters OM, n, etc. It is likely that he did so by writing out the equations of the Cartesian parabola and the circle with x_I, y_I, ρ, and OM, n, PV, respectively, as undetermined parameters,[21] eliminating x, and equating the coefficients of the resulting sixth-degree equation in y to $-p$, q, $-r$, etc. If we perform that procedure, we find indeed the values given in the construction. Moreover, it turns out that the easiest elimination of y proceeds by twice expressing $(x_I - x)^2$ in terms of y, first using the equation of the Cartesian parabola and then that of the circle. In Descartes' proof of the construction the value corresponding to $(x_I - x)^2$ plays a similar central role, which supports the assumption that he found the construction by the procedure with undetermined parameters.[22]

The choice of the Cartesian parabola

It should be stressed that Descartes could only find the present construction after having decided that a sixth-degree equation should be constructed by a Cartesian parabola and a circle, and that the values to be adjusted were the

[18] For Descartes' use of the term "imaginary" cf. Section 27.1.

[19] [Descartes 1637] p. 406; in Descartes' figure on p. 404 there were only four intersections.

[20] Cf. [Descartes 1964–1974] vol. 2 pp. 103–115 (Roberval contre Descartes, April 1638), in particular p. 114, and *ibid.* pp. 154–169 (Descartes to Mersenne 3-VI-1638) in particular pp. 156–157. Roberval thought that the circle would intersect the positive branch of the Cartesian parabola in at most four points, the other two being provided by the other branch. Descartes denied this and explained that for the example in the figure he had chosen a case in which two roots were imaginary because otherwise the intersections of the circle and the branch of the Cartesian parabola would be so oblique as to make the points of intersection indistinguishable.

[21] Such a version of the equation of the Cartesian parabola occurs in the *Geometry* in another context, namely, Descartes' use of the curve as an example for the application of his method of determining normals (Part II D in Table 20.1).

[22] In his notes in the 1659–1661 edition of the *Geometry* (cf. Note 15 of Chapter 17) Van Schooten showed how Descartes' construction of third- and fourth degree equations could be derived by a method of indeterminate coefficients. He added that the construction of fifth- and sixth-degree equations could be found in a similar way.

three parameters in the turning ruler and moving parabola procedure for tracing the Cartesian parabola, together with the three parameters that determine the position and the size of the circle. Evidently the decision was not based solely on algebraic considerations about the equations of these curves. If one starts without any premises and asks which two types of algebraic curves with sufficiently low degrees and a sufficient number of adjustable parameters are the best for constructing a sixth-degree equation, the choice is large and in no natural way is one guided to the Cartesian parabola. Indeed, even if one accepts that one of the curves should be a circle, the curves

$$y = x^3 + ax^2 + bx + c \tag{26.11}$$

provide a simpler alternative (as to the form of the equation) and the required elimination procedure in this case is certainly not more complicated than for the Cartesian parabola. These arguments support the conclusions already reached above, namely, that Descartes first decided on the Cartesian parabola as constructing curve for fifth- and sixth-degree equations, and apparently was prepared to accept considerable algebraic complication in working out that choice.

26.4 Constructing equations of higher degree

Generalizing the construction

Descartes was convinced that the construction procedures discussed above could be generalized to apply for equations of ever higher degrees. He stated so at the end of the book:

> Furthermore, having constructed all those [sc. problems] that are plane by letting a circle intersect a straight line, and all those that are solid by letting, again, a circle intersect a parabola, and finally all those that are one degree more complex by letting a circle intersect a curve one degree more complex than the parabola; one needs only to follow the same way to construct all those that are more complex to infinity. For with mathematical progressions it is so that once one has the first two or three terms, the others are not difficult to find.[23]

However, he gave no further particulars. Probably he envisaged that equations of degrees $2n-1$ and $2n$ should be constructed by the intersection of a circle and a curve C_n of degree n; C_1 was a straight line, C_2 a parabola, C_3 a Cartesian parabola, and generally (for $n > 2$) C_n was related to C_{n-1} in the same way as the Cartesian parabola was related to the parabola, that is, that C_n was

[23] [Descartes 1637] p. 413: "Puis outre cela qu'ayant construit tous ceux [sc. problems] qui sont plans, en coupant d'un cercle une ligne droite; et tous ceux qui sont solides, en coupant aussy d'un cercle une Parabole; et enfin tous ceux qui sont d'un degré plus composés, en coupant tout de mesme d'un cercle une ligne qui n'est que d'un degré plus composée que la Parabole; il ne faut que suivre la mesme voye pour construire tous ceux qui sont plus composés a l'infini. Car en matiere de progressions Mathematiques, lorsqu'on a les deux ou trois premiers termes, il n'est pas malaysé de trouver les autres."

generated by a turning ruler and moving curve procedure with C_{n-1} as the moving curve.[24]

It is difficult to judge how seriously Descartes meant the remark that the generalization to higher order was "not difficult to find." He certainly considered himself able to work out the extension to higher degrees as far as he wanted, but he was equally certain that he had other more important things to do.[25] On the other hand, he warned an enthusiastic reader of the *Geometry* who thought of working out the construction for seventh- and eighth-degree equations that perhaps there would be more difficulties in the project than one might have foreseen.[26]

It should be noted that Descartes' standard constructions and their generalization to higher degrees as detailed above are not smoothly compatible with his classification of curves by pairs of degrees (cf. Section 25.1). For successive classes of problems, grouped by pairs of degrees, the degree of the corresponding second constructing curve, is raised by one: straight line, parabola, Cartesian parabola, "etc." In this connection a classification of curves by degrees would be much more natural than Descartes' classification by pairs of degrees.

Status of the construction

Descartes' construction of fifth- and sixth-degree equations is, at least in modern eyes, remarkably complex. The determination of the parameters — not to speak of the tracing of the Cartesian parabolas themselves — requires so many intermediate constructions that it is difficult to conceive this procedure as the standard solution of a whole class of equations and the starting point for generalizations to higher degrees. It is therefore important to stress that Descartes himself considered the construction, and its supposed generalizability to higher degrees, as the crowning achievement within his theory of geometry.[27] Nor was Descartes alone in this appreciation of the construction; in fact, the canon of construction that he codified in the *Geometry* soon became, with only slight modifications, the paradigm of constructing in geometry. Indeed Descartes' canon carried so much conviction that in a relatively short time span it made mathematicians accept all algebraic curves and all problems leading to algebraic equations as solvable in principle and thereby of accidental interest only.

[24] This interpretation of Descartes' intention was expressed by a number of later seventeenth-century mathematicians, e.g., [Kinckhuysen 1660] pp. 63–65, [Hire 1679] p. 111, and [Bernoulli 1688] p. 349; cf. [Bos 1984] p. 345.

[25] See for instance his remarks in a letter to Mersenne of January 1638, [Descartes 1964–1974] vol. 1 pp. 492–493.

[26] Cf. Descartes to Haestrecht (?), October 1637, [Descartes 1964–1974] vol. 1 pp. 458–460, in particular p. 460: ". . . mais à cause qu'il s'y trouvera peut-estre plus de difficultez que vous n'en avez preveu . . ."

[27] Cf. [Descartes 1637] p. 413 and [Descartes 1964–1974] vol. 1, p. 492 (letter to Mersenne, January 1638).

26.5 The canon of construction

The canon

I am now able to formulate the canon of construction that Descartes presented in his *Geometry*. It was, as has become clear in the preceding sections, as follows:

> Construction in geometry should be performed by the intersection of curves. The curves had to be geometrically acceptable and simplest possible for the problem at hand. Geometrically acceptable curves were precisely the algebraic ones; their simplicity was to be determined by their degrees. With these premises the procedure for constructing problems was:
>
> **1.** Confronted with a problem, the geometer should first translate it into its algebraic equivalent, that is, an equation.
> **2.** If the equation involved one unknown only, the problem was a normal construction problem. In order to get the simplest construction, the geometer should reduce the equation to an irreducible one.
> **3.** Then he should rewrite it in a certain standard form appropriate to the standard construction to be used.
> **4.** In the case of equations of degrees six or less, the geometer could use standard constructions explicitly given by Descartes. These constructions then provided the geometrical solution of the original problem.
> **5.** In the case of higher-degree equations, he should work out a higher-order analog for Descartes' standard constructions. Descartes claimed that it should not be difficult to do so.
> **6.** If the equation arrived at in **1** contained two unknowns, the problem was a locus problem. The geometer could construct points on the locus by choosing an arbitrary value for one of the unknowns and dealing with the resulting equation (in which there was only one unknown left) according to items **2**–**5**, thus finding the corresponding value (or values) of the second unknown; the corresponding point (or points) on the locus could then be constructed.

Later changes of the canon

This canon pervaded much of the geometrical thinking in the hundred years after the publication of the *Geometry*. It remained unchanged in that period, apart from two aspects. The first was that mathematicians challenged the tenet that the simplicity of a curve corresponded to its degree. Several alternative criteria were suggested, but none of these were convincing enough to really replace the degree.[28] The second feature of the canon that was not taken over was the implicit suggestion that one of the constructing curves should always be a circle. In the system for choosing the degrees of the constructing curves that l'Hôpital published in 1707, equations of degree 8 were to be constructed by two curves of degree 3, rather than, as Descartes had suggested, a circle and a curve of degree 4.[29] Jakob Bernoulli argued that strict compliance with Descartes'

[28] Cf. [Bos 1984] pp. 358–366.
[29] [Hôpital 1707] pp. 346–347, cf. [Bos 1984] p. 349.

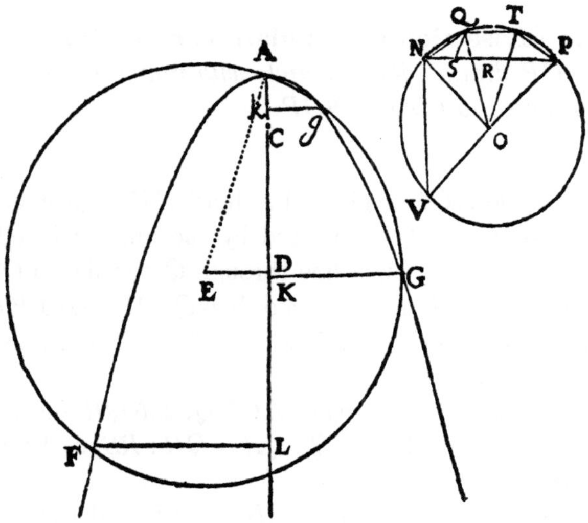

Figure 26.4: Descartes' trisection by the standard construction (*Geometry* p. 396)

rules led to constructing curves that had no natural relation to the problem at hand; he preferred curves naturally suggested by the problem even if these had higher degrees than necessary.[30]

26.6 Special Cases

Two mean proportionals and trisection

I now turn to various special topics which Descartes discussed in connection with his standard higher-order constructions. After explaining the standard construction of third- and fourth-degree equations, he first dealt with the two most obvious examples: he showed that both the determination of two mean proportionals and the trisection could be constructed by applying the general construction to the appropriate equations. Determining two mean proportionals between two line segments a and b led to the equation $x^3 = a^2b$; trisection of an angle to the equation $x^3 = 3x - a$ (in which a was the chord subtending the arc of the given angle in a circle with radius 1). For both cases Descartes explicitly elaborated the resulting construction.[31]

Descartes' trisection provides a clear example of how the standard construction of third- and fourth-degree equations (Construction 26.1) was applied to a special equation. It was as follows:

[30] [Bernoulli 1695] pp. 670–675; cf. [Bos 1984] pp. 358–359.
[31] [Descartes 1637] pp. 395–396 (two mean proportionals) and pp. 396–397 (trisection).

Construction 26.3 (Trisection by standard construction)[32]
Given the angle $\angle NOP$ drawn within a circle with radius 1 and center O (see Figure 26.4); it is required to trisect $\angle NOP$.

Analysis:
1. Because the angle NOP is given, its chord NP is given as well, call $NP = q$; assume $\angle NOP$ trisected by the lines OQ and OT; OQ intersects NP in R; draw a line through Q parallel to OT, its intersection with NP is S; draw the chords NQ, QT, and TP of the three equal parts of $\angle NOP$; call $NQ = z$, if z is constructed, the problem is solved.
2. It is easily seen that the triangles ΔONQ, ΔNQR and ΔQRS are similar, hence, $NO : NQ = NQ : QR = QR : RS$, or $1 : z = z : QR = QR : RS$, so $RS = z^3$.
3. It is also easily seen that $NP = 3NR - SR$ and that $NR = NQ$, hence, $q = 3z - z^3$; thereby the problem is reduced to the construction of the root or roots of the equation

$$z^3 = 3z - q\,. \qquad (26.12)$$

Construction (The numbers correspond to those in the standard Construction 26.1.) Multiplying the given equation 26.12 by z gives $z^4 = 3z^2 - qz$; comparing with the standard equation (26.3) $x^4 = Px^2 + Qx + R$ yields $P = 3$, $Q = -q$, and $R = 0$; applying the standard construction leads to the following steps:
1. Draw (see the fasimile of Descartes' figure in Figure 26.4) a parabola with vertical axis, *latus rectum* equal to 1, and vertex A as highest point.
2. Mark D on the axis below A such that $AD = \frac{1}{2}(P+1) = 2$.
3. Draw $DE = \frac{1}{2}Q = -\frac{1}{2}q$ horizontally to the left from D.
4. Draw EA and draw a circle through A with center in E. (Because $R = 0$ the radius $\sqrt{\frac{1}{4}(1+P)^2 + \frac{1}{4}Q^2}$ of the circle is equal to $\sqrt{(AD)^2 + (DE)^2}$, that is, to EA.)
5. The circle intersects the parabola in the points A, g, G, and F; draw the corresponding perpendiculars gk, GK, and FL to the axis. (The perpendicular at A is 0 and corresponds to the root introduced by increasing the degree of the equation to 4.)
6. The positive roots of the trisection equation are $z = kg$, $z = KG$; the negative one is $z = -FL$. (Thus the trisecting point Q on the arc NP is found by taking $NQ = kg$. Descartes notes that taking $NV = z = KG$ corresponds to trisecting the complement arc NVP, and that FL is equal to the sum of kg and KG.)
[**Proof**: The proof is implied in the proof of the general construction.]

[32][Descartes 1637] pp. 396–397.

Solid problems reduced to mean proportionals or to trisection

In the subsequent sections Descartes dealt with a result explained by Viète in his *Supplement to geometry* of 1593,[33] namely, that the solution of any third- or fourth-degree equation could be reduced to either the determination of two mean proportionals or to a trisection. He did not refer to Viète; rather he took the occasion to comment upon Cardano's formula[34] for the root of a cubic equation. He argued[35] as follows: Solid problems led to third- or fourth-degree equations; the latter could be reduced to second-degree ones by means of certain third-degree ones.[36] In third-degree equations the quadratic term could be removed. Thus ultimately any solid problem could be reduced to an equation of the form[37]

$$z^3 = Pz + Q\,. \tag{26.13}$$

Descartes then made a distinction corresponding to

$$1 : (Q/2)^2 > (P/3)^3 \quad \text{and} \quad 2 : (Q/2)^2 < (P/3)^3 \tag{26.14}$$

(he did not comment on the case of equality) and discussed the two cases separately. If the first condition held, the solution could be expressed by a rule "whose invention Cardano attributes to someone by the name of Scipio Ferreus,"[38] namely,[39]

$$z = \sqrt[3]{\frac{Q}{2} + \sqrt{(\frac{Q}{2})^2 - (\frac{P}{3})^3}} + \sqrt[3]{\frac{Q}{2} - \sqrt{(\frac{Q}{2})^2 - (\frac{P}{3})^3}}\,. \tag{26.15}$$

Indeed condition 1 of Equation 26.14 guaranteed that the quantities under the square root signs were positive, whereas in the case of condition 2 they were negative. (The latter case was known among algebraists as the "casus irreducibilis" in which Cardano's formula did not provide solutions because it involved uninterpretable square roots of negative quantities (cf. Section 16.2, Equations 16.3–16.4).) Thus, if condition 1 applied, the problem required the determination of cubic roots, that is, the determination of two mean proportionals between 1 and the (given) quantity under the cubic root sign. This, as Descartes had just shown, could be done by means of a conic section, namely, a parabola.[40]

---

33 [Viète 1593]. cf. Sections 10.2 and 10.3.

34 See Note 91 of Chapter 4.

35 [Descartes 1637] pp. 397–400; the margin title of this section is: "Que tous les problesmes solides se peuvent reduire a ces deux constructions."

36 Descartes probably had Ferrari's method in mind (cf. Note 18 of Chapter 10), it is not clear whether he was aware of Viète's method (cf. Chapter 10, Note 16). In the context of his techniques for checking reducibility, he had himself provided an alternative method that I discuss below in Section 27.3.

37 As above with $|P| = p$ and $|Q| = q$. Descartes here distinguished three cases according to the signs + or − on the right-hand side of the equations, leaving out the case $z^3 = -pz - q$ because he implicitly assumed that at least one solution was positive.

38 [Descartes 1637] p. 398: "la reigle dont Cardan attribue l'invention a un nommé Scipio Ferreus," that is, Scipione Ferro, see Note 91 of Chapter 4. At present the result is usually called Cardano's formula.

39 [Descartes 1637] pp. 398, 399; there are some printing mistakes in the three variants of the formula that Descartes gave.

40 [Descartes 1637] p. 398: ". . . sans avoir besoin des sections coniques pour autre chose, que pour tirer les racines cubiques de quelque quantité, données, c'est a dire, pour trouver deux moyennes proportionelles entre ces quantités et l'unité."

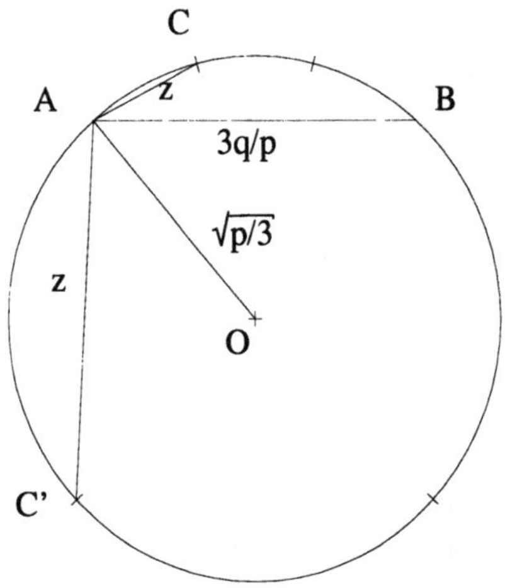

Figure 26.5: Construction of third-degree equation by reduction to trisection

In the case of condition 2 of Equation 26.14, Descartes showed (by giving the construction explicitly) that the solution of Equation 26.13 could be reduced to a trisection. Before discussing his further arguments about Cardano's solution (Equation 26.15) I give Descartes' reduction to trisection, that is, his construction of the roots of $z^3 = Pz + Q$ (with $(Q/2)^2 < (P/3)^3$) under the assumption that it is possible to trisect any angle.

It suffices to give one of the two cases he presented; I take the case[41] $z^3 = pz - q$, with the condition $(q/2)^2 < (p/3)^3$:

Construction 26.4 (Roots of third-degree equation, by reduction to trisection)[42]
Given the equation $z^3 = pz - q$, with p, q positive and $(\frac{q}{2})^2 < (\frac{p}{3})^3$, it is required to construct its roots.

Construction:
1. Draw (see Figure 26.5) a circle with center O and radius $\sqrt{\frac{p}{3}}$.
2. Draw a chord AB of length $\frac{3q}{p}$ in the circle (this is possible because the given inequality implies $\frac{3q}{p} < 2\sqrt{\frac{p}{3}}$).
3. Take (by trisection) two points C and C' on the two arcs defined by AB such that $\text{arc}AC = \frac{\text{arc}ACB}{3}$ and $\text{arc}AC' = \frac{\text{arc}AC'B}{3}$.
4. Draw the chords AC and AC'.

[41]The other case was $z^3 = +pz + q$ with $(q/2)^2 < (p/3)^3$.
[42][Descartes 1637] pp. 397–400.

5. AC and AC' are the two positive roots of the equation.[43]
[**Proof**: Descartes gave no proof. Deriving the trisection equation as in the **analysis** of Construction 26.3 but with radius ρ rather than 1 leads to $z^3 = 3\rho^2 z - a\rho^2$; adjusting ρ and a to the coefficients of the given equation $z^3 = pz - q$ yields the values $\rho = \sqrt{p/3}$ and $a = 3q/p$ used in the construction.]

Critique of Cardano's formula

As we have seen, Cardano's formula (Equation 26.15) was the basis of Descartes' demonstration that all solid problems were reducible to either the determination of two mean proportionals or the trisection of an angle. He also argued that his result was an essential improvement as compared with Cardano's algebraic result. For equations $z^3 = Pz+Q$ satisfying condition 1 in Equation 26.14, Descartes had given a geometric solution while the formula gave an algebraic one. For equations satisfying condition 2 Cardano's formula gave no solution (because it involved cubic roots of negative quantities); whereas Descartes did provide a solution, but a geometrical, not an algebraic one. The question, then, was in how far geometrical solutions (i.e., constructions or reductions to standard constructions) were better than algebraic ones (expressions of the roots involving radicals). Descartes first noted that the cubic roots involved in Cardano's formula, even if the quantity below the root sign was real, presupposed a geometrical interpretation: $\sqrt[3]{A}$ was the side of a cube whose content was known (namely A). This interpretation was

> . . . in no respect more understandable or simpler than expressing them [sc. the roots] by means of the relation which they bear to the chords of certain arcs or portions of circles, whose triplicate is given.[44]

Moreover, Descartes stated, one could easily create a new algebraic symbol denoting the chord of the third of an arc on a given chord; with such a symbol the expression of the roots of a third-degree equation in the "casus irreducibilis" would be considerably simpler than Cardano's.[45] He did not, however, develop this possibility.[46]

[43] Although Descartes did not explicitly mention it here, his treatment of the other case makes clear that he knew that the third root was negative with absolute value equal to $AC + AC'$, cf. Construction 26.3 item **6**.

[44] [Descartes 1637] p. 400: "Au reste il est a remarquer que cete façon d'exprimer la valeur des racines par le rapport qu'elles ont aux costés de certains cubes dont il n'y a que le contenu qu'on connoisse, n'est en rien plus intelligible, ny plus simple, que de les exprimer par le rapport qu'elles ont aux subtenduës de certains arcs, ou portions de cercles, dont le triple est donné."

[45] [Descartes 1637] pp. 400–401: "Mesme ces termes sont beaucoup moins embarassés que les autres, et ils se trouveront beaucoup plus cours si on veut user de quelque chiffre particulier pour exprimer ces subtenduës, ainsi qu'on fait du chiffre $\sqrt{C.}$ pour exprimer le costé des cubes."

[46] Using such a symbol would lead to expressions such as the following. Let $\heartsuit(a)$ denote the chord of the third of the arc with chord a in a circle with radius 1. The result of Construction 26.4 can then be expressed as follows: If p and q are positive and $(\frac{q}{2})^2 < (\frac{p}{3})^3$, then $z = \sqrt{\frac{p}{3}}\heartsuit(\frac{3\sqrt{3}q}{p\sqrt{p}})$ is a root of the equation $z^3 = pz - q$.

The role of the conic section

Descartes used the result about the reducibility of solid problems to mean proportionals and trisections in an explanation why solid problems in general could not be solved by straight lines and circles only, and why at least one nondegenerate conic section was necessary for their construction. The passage[47] is of interest because it was, as far as I know, the earliest attempt to prove or explain the *im*possibility of constructing certain problems (such as the trisection) with certain means (such as straight lines and circles).

Descartes noted that both the determination of two mean proportionals and the trisection required the determination of *two* points: the two endpoints of the mean proportionals or the two dividing points that trisected an arc. For that requirement the curvature ("courbure"[48]) of a circle was insufficient because the points of a circle had only one simple relation ("rapport") to a single point, in this case to the circle's center. Therefore the circle could only provide solutions of problems that required the determination of one point, such as bisecting an angle or determining one mean proportional. On the other hand, the curvature of a (nondegenerate) conic always depended on two different things, and therefore a conic sufficed for constructing problems that required the determination of two points.[49] Similarly, Descartes claimed,[50] problems depending on the determination of four mean proportionals or the section of an angle in five equal parts required a curve essentially more complicated than a conic.

Probably Descartes had focal properties in mind when he wrote about the two "things" ("choses") involved in the curvature of conics as opposed to the single relation (to the center) involved in the curvature of a circle. One may say that he saw the variability of the curvature as the essential feature that determined the power of curves in solving problems.

Such a conception of the variability of the curvature along a noncircular and non-degenerate conic section would also explain an assertion Descartes made at the beginning of the section on the standard construction of third- and fourth-degree equations, namely that any part, however small, of a conic would suffice for constructing solid problems.[51] He may have intuited that, given any part of a conic section along which the curvature varied, one might, by plane means, adjust the problem at hand in such a way that the available range of curvature provided the construction. However, Descartes left it at qualitative arguments; he did not work these out into a strict proof.[52]

---

[47][Descartes 1637] pp. 401–402; the margin title of the passage is: "Pourquoy les problesmes solides ne peuvent estre construits sans les sections coniques, ny ceux qui sont plus composés sans quelques autres lignes plus composées."

[48]Note that Descartes did not quantify the concept, this happened later in the seventeenth century.

[49]The argument is on pp. 401–402 of [Descartes 1637]; it ends: "Au lieu que la courbure des sections coniques, dependant tousiours de deux diverses choses, peut aussy servir a determiner deux poins differens."

[50][Descartes 1637] p. 402.

[51][Descartes 1637] pp. 389–390: ". . . on peut tousiours en [sc. a third- or fourth-degree equation] trouver la racine par l'une des trois sections coniques, laquelle que ce soit ou mesme par quelque partie de l'une d'elles, tant petite qu'elle puisse estre; en ne se servant au reste que de lignes droites, et de cercles."

[52]In the second edition of his *Mesolabum* ([Sluse 1668]) Sluse sketched a convincing proof

Special higher-order problems

After the explanation of the general construction of the roots of fifth- and sixth-degree equations by means of the Cartesian parabola and a circle, Descartes applied it to the problem of finding four mean proportionals between line segments a and b.[53] He added the remark that similarly the general rule provided easy constructions for the division of an angle in five equal parts and the construction of regular 11- and 13-gons.

Algebraic techniques

Thus Descartes' program was completed, he had indeed fulfilled the "incredibly ambitious" task formulated in his letter to Beeckman less than twenty years before. In my presentation of the canon of construction, however, I have left out the algebraic techniques related to reducibility and transformations. The next chapter deals with these issues.

that solid problems can be solved by any given conic; he was motivated to do so by a discussion in letters with Huygens on the assertions of Descartes; see [Bos 1985] pp. 153, 156.

[53] [Descartes 1637] pp. 411–412. The pertaining equation was $x^5 - a^4b = 0$, which Descartes changed into the required standard form (Equation 26.4) by multiplying with x and substituting $x = y - a$; the result was $y^6 - 6ay^5 + 15aay^4 - 20a^3y^3 + 15a^4yy - (6a^5 + a^4b)y + a^6 + a^5b = 0$.

Chapter 27

The theory of equations in the *Geometry*

27.1 Character of the theory and key ideas

Themes

Before completing his canon of construction in the last part of the *Geometry*'s third book, Descartes provided the necessary theory of equations in one unknown (sections III B–C in Table 20.1). In the present chapter I give a brief overview of his results.

Descartes' canon of geometrical construction required the constructions to be geometrical and simplest possible. These geometrical requirements induced algebraic ones: The problem had to be reduced to an equation that (1) was irreducible and (2) had a certain standard form. Thus, in order to avoid the twofold error of either constructing with too complicated means or trying to construct with too simple means,[1] the geometer needed a general theory of equations and a number of techniques to perform the necessary transformations and reductions.

Because of its special geometrical motivation, Descartes' theory of equations centered on three themes: roots of equations (their number and their signs), transformations of equations by linear substitutions, and reducibility of equations. The interest in the signs and the number of roots related to construction by the intersection of curves; the roots were constructed as ordinates of points of intersection and their signs were determined by the position of the intersections with respect to the relevant axis. Moreover, in Descartes' standard construction for sixth-degree equations, it was essential that all real roots were positive. The transformations of equations by linear substitutions were necessary for showing that any equation could be transformed to the appropriate standard form. Techniques for determining whether equations were reducible were necessary to

[1] Cf. Descartes' statement at the beginning of his sections on equations, [Descartes 1637] p. 371: "Or affin que ie puisse icy donner quelques reigles, pour eviter l'une et l'autre de ces deux fautes, il faut que ie die quelque chose en general de la nature des Equations . . ."

detect cases in which plane problems led to equations of degree higher than two (or solid problems to equations of degree higher than four, etc.). These equations had to be recognized as reducible and accordingly reduced to arrive at the proper simplest possible construction.

No proofs

Descartes' theory of equations contained no proofs. He merely stated a number of assertions about polynomial equations, their degrees, their coefficients, and their roots; he suggested that the attentive reader could find their proofs easily.[2] In several cases we may be fairly sure that he had no formal proof (as, e.g., in the case of the "sign rule," see below); it is impossible to decide whether he was aware that his assertions were not always simply obvious.

Fundamental theorem of algebra

It seems that most of Descartes' insights in the theory of equations[3] were based on the assumption that in principle any polynomial could be decomposed in linear factors:

$$ax^n + a_1 x_{n-1} + \cdots + a_n = a(x \pm x_1)(x \pm x_2) \cdots (x \pm x_n) \tag{27.1}$$

where the x_i were positive. In this decomposition the x_i were the roots of the equation, "true" roots if the sign was $-$, "false" roots if the sign was $+$.[4] He stated that an equation of degree n had at most n true or false roots; he showed how, if one of its roots, x_1, was known, an equation could be reduced to one of lower dimension by polynomial division by $(x \pm x_1)$; and he stated the rule of signs. All these statements seem to have been based on the assumption of decomposability into linear factors, and Descartes illustrated them with examples where such a decomposition indeed applied.

Only later in the text, in a short paragraph between the section on transformations and the one on reducibility, he mentioned the fact that the decomposition as in Equation 27.1 was not always possible. He wrote

[2]At the end of the section Descartes wrote, [Descartes 1637] p. 389: "Au reste i'ay omis icy les demonstrations de la plus part de ce que iay dit a cause qu'elles m'ont semblé si faciles, que pourvû que vous preniés la peine d'examiner methodiquement si iay failly, elles se presenteront a vous d'elles mesme: et il sera plus utile de les apprendre en cete façon, qu'en les lisant."

[3]The question of the origin of Descartes' algebraic ideas is an enigmatic one. Descartes himself did not acknowledge any debt to earlier writers; in particular, he denied being influenced by Viète's work. It has been noted that his treatment of equations in the *Geometry* shows similarities with the approaches developed by Roth, Faulhaber, Girard, and Harriot. These algebraists shared a particular interest in the decomposition of polynomials into factors; relevant statements by Roth, Girard, and Harriot are often mentioned as part of the prehistory of the fundamental theorem of algebra ([Tropfke 1980] pp. 489, 492 refers to [Roth 1608] p. B1^v, [Girard 1629] pp. E4^r–E4^v, and [Harriot 1631] pp. 3 ff.). In his recent biography of Faulhaber, Schneider has called attention to the "stupendous similarities" ("erstaunliche Übereinstimmungen" [Schneider 1993] pp. 104 and 107) between Descartes' theory of equations and parts of Faulhaber's *Miracula Arithmetica* ([Faulhaber 1622]). However, the extent of Descartes' familiarity with this approach, and especially the significance of a possible meeting with Faulhaber in 1619–1620 appears to be very difficult to assess. Schneider's book gives an up to date discussion of this theme (*op. cit.* pp. 171–198.); cf. also [Schneider 1991] and [Manders 1995].

[4][Descartes 1637] p. 372.

> For the rest neither the false nor the true roots are always real, sometimes they are only imaginary, that is to say that one may always imagine as many in any equation as I have said, but that sometimes there is no quantity corresponding to those one imagines.[5]

As an example he gave the equation[6]

$$x^3 - 6x^2 + 13x - 10 = 0\,, \tag{27.2}$$

with one real root (namely 2) and two imaginary ones. Together with the decomposition of Equation 27.1 these remarks about imaginary roots form Descartes' version of the fundamental theorem of algebra.[7]

Imaginary roots

Descartes also used the term "imaginary" in the letter to Beeckman of 1619 (cf. Section 16.2) and we have seen that its use there can tentatively be linked with Cardano's use of the term "imagine" in relation to square roots of negative numbers. Descartes was no doubt well aware that his "imaginary" roots involved square roots of negative quantities, and he may have chosen the term for that reason. The meaning of the term in the present context may be best rendered as: imagined and introduced for the purpose of making rules general.

It is more difficult to relate the term to Descartes' ideas about the function of the "imagination" in the achievement of knowledge, as expressed in the *Rules* (cf. Section 18.1). There the imagination appeared primarily as the mental faculty to form and contemplate two-dimensional images. But in the *Geometry* Descartes stated that no real quantities corresponded to the imaginary roots, so it seems that his use of the term here referred more to the symbolic representation of relationships by formulas than to the picturing of the real extension of bodies in the "imagination."

Sign rule

Descartes also explained his "sign rule" in connection with the decomposition of polynomials in linear factors. He formulated it as follows:

> From this [the fact that if x_0 is a root of $F(x) = 0$, the polynomial $F(x)$ is divisible by $(x - x_0)$] it can be concluded also how many true roots there can be in each equation, and how many false ones. Namely: there can be as many true ones as there are changes in the signs + and −, and as many false ones as two + signs or two − signs follow each other.[8]

[5][Descartes 1637] p. 380: "Au reste tant les vrayes racines que les fausses ne sont pas tousiours reelles; mais quelquefois seulement imaginaires; c'est a dire qu'on peut bien tousiours en imaginer autant que iay dit en chasque Equation, mais qu'il n'y a quelquefois aucune quantité, qui corresponde a celles qu'on imagine."

[6][Descartes 1637] p. 380.

[7]Cf. Note 3.

[8][Descartes 1637] p. 373: "On connoist aussy de cecy combien il peut y avoir de vrayes racines, et combien de fausses en chasque Equation. A sçavoir il y en peut avoir autant de vrayes, que les signes + et − s'y trouvent de fois estre changés; et autant de fausses qu'il s'y trouve de fois deux signes + ou deux signes − qui s'entresuivent." Cf. also [Bartolozzi & Franci 1993].

Descartes gave no proof. It may well be that he found the rule while studying the form of the equations required in his construction of fifth- and sixth-degree equations. This standard form (Equation 26.4), with alternating signs for the coefficients, ensured that no root could be negative; Descartes' sign rule can be seen as a generalization of this phenomenon.

27.2 The transformations

Standard forms

To achieve the standard forms for quadratic equations (Equations 22.2), no special transformations were required. Third- and fourth-degree equations had to be rewritten as (cf. Equation 26.2)

$$x^4 = \pm px^2 \pm qx \pm r \tag{27.3}$$

(x^3-term removed, p, q, and r positive or zero). Fifth- and sixth-degree ones had to be brought in the form (cf. Equation 26.4)

$$x^6 - px^5 + qx^4 - rx^3 + sx^2 - tx + v = 0 \tag{27.4}$$

(alternating coefficients, p, q, r, s, t, v positive and unequal to zero, $q > (\frac{p}{2})^2$).

Substitutions and their use

Descartes gave the techniques necessary for rewriting arbitrary equations in these standard forms. Most of these techniques involved a linear substitution of the form

$$x = \alpha y + a\ . \tag{27.5}$$

I now survey these substitutions, specifying in each case its precise form, Descartes' description of its effect, and its purpose within the canon of construction.

The substitution

$$x = -y \tag{27.6}$$

had the result that the false (positive) roots of the equations became true (negative) and the true ones false.[9]

The substitution

$$x = y \pm a \tag{27.7}$$

could be used to increase or decrease the roots without knowing them; when the true roots were increased, the false roots were decreased and vice versa.[10] Descartes illustrated the effects of substitutions 27.6 and 27.7 by several examples. They served as introductory explanation of the special substitutions that followed.

[9] [Descartes 1637] p. 373: "De plus il est aysé de faire en une mesme Equation, que toutes les racines qui estoient fausses devienent vrayes, et par mesme moyen que toutes celles qui estoient vrayes devienent fausses . . ."

[10] [Descartes 1637] p. 374 (margin title): "Comment on peut augmenter ou diminuer les racines d'une Equation, sans les connoistre;" *ibid.* p. 375 (margin title): "Qu'en augmentant les vrayes racines on diminue les fausses, et au contraire."

The substitution

$$x = y \pm \frac{a}{n} \tag{27.8}$$

served to remove the second term[11] in an equation $x^n \pm ax^{n-1} + \cdots = 0$; it was used for achieving the standard form of third- and fourth-degree equations and also in the technique for checking the (ir)reducibility of fourth-degree equations (see Section 27.3).

The substitution

$$x = y - a \tag{27.9}$$

had the result that "all false roots become true while the true roots do not become false," if the constant a was chosen larger than the "largest false root."[12] Descartes claimed that in the resulting equation no successive coefficients had the same sign and that, if a was chosen sufficiently large, the first coefficients of the resulting equation $y^n - py^{n-1} + qy^{n-2} - \cdots = 0$ satisfied $q > (\frac{p}{2})^2$. He evidently based this claim on an inversion of the sign rule (if there are no negative roots, there are no successive pairs of coefficients with equal sign). The argument is wrong (counterexample: $x^2 + 2x + 2 = 0$). The claims themselves are correct.[13] The procedure was necessary for arriving at the standard form for sixth-degree equations (Equation 27.4). Descartes claimed further that, although the false roots were unknown, it was easy to estimate their value and choose a larger value; he added an example.[14]

The same substitution

$$x = y - b \tag{27.10}$$

also achieved that, as Descartes expressed it, all places in the equations were filled,[15] which meant that none of the coefficients was zero; Descartes explained that, if $x = y - a$ yielded an equation in which one of the coefficients was equal to zero, the choice of a constant b just slightly larger than a would lead to an equation of the required form. Descartes mentioned in particular that the constant term of an equation could thus be made unequal to zero; the standard form (Equation 26.5) of the sixth-degree equation indeed presupposed the constant term to be unequal to zero. At this point he also explained that an $(n-1)$th-degree equation can be transformed in an nth-degree one by multiplying through with x. These two techniques enabled him to transform a fifth-degree equation in an appropriate sixth-degree one: multiplication with x yielded a sixth-degree

[11] [Descartes 1637] p. 376 (margin title): "Comment on peut oster le second terme d'une Equation."

[12] [Descartes 1637] p. 377 (margin title): "Comment on peut faire que toutes les fausses racines d'une Equation devienent vrayes, sans que les vrayes devienent fausses."

[13] They can be proved by writing out the new coefficients as functions of a, and considering their behavior for large a, cf. [Galuzzi 1996] pp. 323–324.

[14] Descartes may have had in mind an estimate like the following, which is easily derived: if the degree is n, the first coefficient is 1 and all coefficients are smaller than 10^k, then the absolute value of a root cannot be larger than 10^{k+n}. The example ([Descartes 1637] pp. 377–378), however, is difficult to interpret.

[15] [Descartes 1637] p. 378 (margin title): "Comment on fait que toutes les places d'une Equation soient remplies."

equation with constant term zero, a further substitution $x = y - b$ made this term unequal to zero.

The substitutions

$$x = cy \quad \text{and} \quad x = y/c \tag{27.11}$$

served to multiply or divide the roots without knowing them, to remove fractions from the coefficients of an equation, and in some cases to remove irrational coefficients.[16] Descartes explained the procedure by an example; he transformed the equation

$$x^3 - \sqrt{3}x^2 + \frac{26}{27}x - \frac{8}{27\sqrt{3}} = 0 \tag{27.12}$$

by the substitution $x = y/\sqrt{3}$ into

$$y^3 - 3y^2 + \frac{26}{9}y - \frac{8}{9} = 0\,, \tag{27.13}$$

and by a further substitution $y = z/3$ into

$$z^3 - 9z^2 + 26z - 24 = 0\,. \tag{27.14}$$

The substitutions in Equation 27.11 could also be used to "make the known quantity of one term of an equation equal to any other quantity,"[17] i.e., to give any required value to one of the coefficients (while keeping the first coefficient equal to 1). Descartes here gave the example

$$x^3 - b^2x + c^3 = 0\,, \tag{27.15}$$

in which the change of the coefficient b^2 into $3a^2$ is achieved by the substitution $y = x\sqrt{\frac{3a^2}{b^2}}$, yielding

$$y^3 - 3a^2y + \frac{3a^3b^3}{b^3}\sqrt{3} = 0\,. \tag{27.16}$$

The purpose of this technique is unclear. It reminds one, however, of Descartes' algebraic study of cubic equations in the *Private reflections* of 1619 (cf. Section 16.4) in which Descartes explored the effects of a substitution $x = \alpha y$.

Actually the substitutions $x = cy$ and $x = y/c$ cannot remove irrational factors from the coefficients in all cases. For instance, if the irrational factor $\sqrt{3}$ occurs in the coefficients of x and x^2 of Equation 27.12, it cannot be removed by a substitution as in Equation 27.11. Descartes seems to have been aware of this impossibility.[18] It appears from his further practice in the *Geometry* that Descartes assumed that the final equation could be achieved with first coefficient

---

[16] [Descartes 1637] p. 379: "De plus on peut, sans connoistre la valeur des vrayes racines d'une equation, les multiplier, ou diviser toutes, par telle quantité connuë qu'on veut. . . . Ce qui peut servir pour reduire a des nombres entiers et rationaux, les fractions, ou souvent aussy les nombres sours, qui se trouvent dans les termes des equations."

[17] [Descartes 1637] p. 380 (margin title): "Comment on rend la quantité connuë de l'un des termes d'une Equation esgale a telle autre qu'on veut."

[18] He speaks of removing irrationals "as far as possible;" cf. Note 27.

equal to 1. He did not explicitly mention this point, he may have considered it "easy to find" (cf. Note 27).[19]

27.3 Reducibility

The nature of equations resulting from geometrical problems

For understanding Descartes' theory of the reducibility of equations it is necessary to consider the general form of equations resulting from geometrical problems.

Descartes' canon instructed the geometer to reduce a problem to an equation in one unknown (cf. Section 26.5). To do so the given elements in the geometrical configuration about which the problem was posed were denoted by letters (from the beginning of the alphabet); in the resulting equation they appeared as indeterminates. There was one unknown only, hence, the coefficients were *known* in the geometrical sense (cf. Section 5.2), which meant that they could be constructed by straight lines and circles from the given elements of the initial figure. Thus, for instance, if the only given elements in a problem were the sides a and b of a rectangle, the coefficients of the resulting equation could well involve a square root like $\sqrt{a^2+b^2}$, because the diagonal of a given rectangle was constructible by plane means and therefore also given. However, the coefficients could not involve a cube root like $\sqrt[3]{a^2b}$ as a factor, because even if a and b were given, their two mean proportionals $\sqrt[3]{a^2b}$ and $\sqrt[3]{ab^2}$ were *not* given; they could not be constructed from a and b by straight lines and circles. Similarly, multiples αa of a given element a were also given if α was a rational number or an irrational number resulting from rational numbers by (possibly repeated) square-root extraction. If, in particular, the only given element of a configuration was a unit element e and the equation was written inhomogeneously by taking $e = 1$, then all coefficients of the resulting equations were rational numbers or irrational numbers involving square-root extraction only.

Consequently, an equation resulting from a geometrical problem by following the Cartesian canon of problem solving had the form

$$x^n + A_{n-1}x^{n-1} + \cdots + A_1x + A_0 = 0\,, \tag{27.17}$$

in which the A_i were algebraic expressions in terms of rational numbers and indeterminates (a, b, etc.); these expressions involved no other algebraic operations than the rational operations and square-root extraction.[20] I should add that Descartes himself did not express this conclusion explicitly; I do not know whether it was ever formulated later.[21]

---

[19] It can be done, for a polynomial equation $Ax^n + \ldots = 0$, by the substitution $x = y/A$, combined with a multiplication by A^{n-1}.

[20] In modern terms, the theory concerns the ring $\mathbf{L}[x]$ of polynomials with coefficients in $\mathbf{L}$, the field $\mathbf{L}$ being an extension by real square-root adjunctions of the field $\mathbf{Q}(a, b, \cdots)$ consisting of all rational functions of the indeterminates $a, b, \cdots$.

[21] The matter lost its interest when polynomial equations were no longer seen as primarily related to geometrical construction problems; this happened at the latest by the middle of the eighteenth century; cf. Section 29.3.

Reducibility and construction

The reducibility of equations was a crucial concept in Descartes' canon of problem solving, because, as I will illustrate in the next section by an example from the *Geometry*, there were usually several ways to choose the unknown in a problem, and the resulting equations differed according to this choice. In particular it could happen that for one choice the equation was of second degree (the problem therefore plane) whereas for another choice the equation turned out to be of degree higher than 2. The geometer who took the second choice might conclude that the problem was not plane, thereby committing what Pappus had called the "sin" of solving a plane problem by inappropriate means. Descartes had come to the conclusion that, if a plane problem led to an equation of higher degree than two, then this equation was reducible to a quadratic one.[22] Hence, a test for reducibility of higher-order equations was necessary to avoid inappropriate choices of means of construction on the basis of the degree of the equation. Descartes used here a concept of reducibility different from the modern one and therefore his ideas and techniques have to be discussed in some detail.

Reducibility

Descartes formulated his theory of reducibility in the form of a series of techniques with examples;[23] in my presentation I give the techniques in general terms (formulas) and I discuss his main example afterward in Section 27.4.

Descartes did not explicitly define reducibility of equations but from his methods we may conclude (cf. Section 25.4) that he considered an equation $F(x) = 0$ to be reducible[24] if the polynomial $F(x)$ could be written as a product $G(x)H(x)$ of two lower-degree polynomials G and H whose coefficients could be constructed by straight lines and circles from the coefficients of F.[25] Because the real roots of quadratic equations were constructible by straight lines and circles, a quadratic equation was reducible, in Descartes' construction-related sense of the term, if (and only if) its roots were real.

In the final part of the section on the theory of equations (III-C in Table 20.1) Descartes mainly discussed the reducibility of third- and fourth-degree equations. The margin titles of the relevant sections make clear their construction oriented nature:

> "The reduction of cubic equations when the problem is plane:"
> "Which problems are solid when the equation is cubic:"
> "The reduction of equations with four dimensions when the problem is plane, and which are those that are solid."[26]

[22] Descartes formulated this conclusion (without argument) with respect to third- and fourth-degree equations, but it appears that he was convinced it applied in general; cf. the quotation in Note 33.

[23] [Descartes 1637] pp. 380–389, cf. item III-C in Table 20.1.

[24] As mentioned in Note 17 of Chapter 25, another kind of reducibility arises in equations of the form $F(G(x)) = 0$. As will become clear below, cf. Equation 27.19, Descartes knew this phenomenon, but he did not separately discuss it.

[25] In modern terms and notation as in Note 20 this means that the polynomial F from $\mathbf{L}[x]$ is reducible over some extension $\mathbf{M}$ of $\mathbf{L}$ formed by adjunction of real square roots.

[26] [Descartes 1637] p. 380: "La reduction des Equations cubiques lorsque le problesme est

Third-degree equations

For checking the reducibility of third-degree equations arising from geometrical problems Descartes prescribed first applying the substitutions he had explained earlier and various other means "easy to find," to reduce the equation "as far as will be possible" to an equation whose coefficients involved no irrational or fractional numbers.[27] He then claimed that if the resulting equation was of third degree and if it was reducible, one of its roots was a factor of the constant term. Hence, if none of these factors was a root, the equation was irreducible.[28]

It is not clear how Descartes arrived at this condition of reducibility; in fact, it is difficult to determine what the condition meant.[29] At this point[30] Descartes explained the technique of dividing two polynomials.

Fourth-degree equations

For fourth-degree equations Descartes suggested the same technique — checking the factors of the constant term — to find constructible roots and reduce the equation to a third- or lower-degree one. He also discussed the case that the fourth-degree polynomial decomposed in two quadratic factors.[31] To check if reduction can be achieved in that way he suggested first to remove the second

---

plan:" *ibid.* p. 383: "Quels problesmes sont solides, lorsque l'Equation est cubique:" "La reduction des Equations qui ont quatre dimensions, lorsque le problesme est plan. Et quels sont ceux qui sont solides."

[27] [Descartes 1637] pp. 380–381: "Or quand pour trouver la construction de quelque problesme, on vient a une Equation, en laquelle la quantité inconnuë a trois dimensions; premierement si les quantités connuës, qui y sont, contienent quelques nombres rompus, il les faut reduire a d'autres entiers, par la multiplication tantost expliquée; Et s'ils en contienent de sours, il faut aussy les reduire a d'autres rationaux, autant qu'il sera possible, tant par cete mesme multiplication, que par divers autres moyens, qui sont assés faciles a trouver."

[28] [Descartes 1637] p. 381: "Puis examinant par ordre toutes les quantités, qui peuvent diviser sans fraction le dernier terme, il faut voir, si quelqu'une d'elles, iointe avec la quantité inconnué par le signe + ou −, peut composer un binome, qui divise toute la somme; et si cela est le problesme est plan, c'est a dire il peut estre construit avec la reigle et le compas . . ." and p. 383: "Mais lorsqu'on ne trouve aucun binóme, qui puisse ainsi diviser toute la somme de l'equation proposee , il est certain que le problesme qui en depend est solide."

[29] If F is a polynomial with integer coefficients and first coefficient equal to 1, and if the equation $F(x) = 0$ has a rational root x_0, then this root is itself an integer and it divides the constant term of F; thus in that case $F(x) = G(x)(x - x_0)$ for some polynomial G with integer coefficients. This theorem, which can easily be proved by elementary means, implies Descartes' condition if reducibility is interpreted as factorization with integer coefficients. Indeed if a cubic polynomial with integer coefficients $x^3 + Ax^2 + Bx + C$ is reducible in this sense, one of its factors is a linear one, $x - P$, with P a factor of C. However, the concept of reducibility involved in the theorem above is different from Descartes' construction-related conception of reducibility. As explained earlier, a polynomial equation $F(x) = 0$ is reducible in Descartes' sense if it can be factored as $G(x)H(x) = 0$ in which the coefficients of the polynomials G and H are constructible by ruler and compass from those of F. Thus (cf. Notes 20 and 25) the coefficients of the polynomial F are in an extension $\mathbf{L}$ of $\mathbf{Q}(a, b, \ldots)$ and those of its factors G and H are in a further extension $\mathbf{M}$ of $\mathbf{L}$, both extensions being formed by adjunction of square roots. In this situation the concept of integer is ambiguous and it is unclear what it means to check all the factors of the constant term. Indeed the rings arising from the ring of integers in $\mathbf{Q}(a, b, \ldots)$ by adjunction of square roots lack unique prime factorization. I have not been able to formulate a theorem in modern terms that is convincingly analogous to what Descartes meant here; as a result I do not know whether and in what sense it is true.

[30] [Descartes 1637] pp. 381–383.

[31] [Descartes 1637] pp. 383–387.

term:[32]

$$x^4 + Px^2 + Qx + R = 0\,. \tag{27.18}$$

He then considered the following cubic equation in y^2:

$$y^6 + 2Py^4 + (P^2 - 4R)y^2 - Q^2 = 0\,. \tag{27.19}$$

If the latter equation was reducible, then, by his earlier argument discussed above, some factor S of the constant term Q^2 was a root of Equation 27.19 as a cubic equation in y^2 and hence $y_1 = \sqrt{S}$ was a root of the equation. In that case, Descartes asserted, the left-hand side of Equation 27.18 could be written as product of two quadratic factors:

$$\begin{aligned} &x^4 + Px^2 + Qx + R = \qquad\qquad (27.20)\\ &= (x^2 - y_1x + \frac{1}{2}y_1^2 + \frac{1}{2}P + \frac{Q}{2y_1})(x^2 + y_1x + \frac{1}{2}y_1^2 + \frac{1}{2}P - \frac{Q}{2y_1})\,. \end{aligned}$$

whereby (because y_1 was constructible) the problem was shown to be plane. Descartes added that if Equation 27.19 was not reducible, one could be sure that the original problem was not plane but solid.[33]

Descartes did not explain the origin of Equation 27.19, but from his examples it appears that he found it by a method of undetermined coefficients. Indeed if one writes

$$x^4 + Px^2 + Qx + R = (x^2 - yx + u)(x^2 + yx + v)\,, \tag{27.21}$$

and eliminates u and v from the equations that result by comparing coefficients, one arrives at Equation 27.19 for y^2 and at the decomposition in Equation 27.20.

The procedure showed the reducibility of fourth-degree equations to third-degree ones to which Descartes had referred earlier as an argument for grouping curves in classes with degrees 3–4, 5–6, etc. (cf. Section 25.1). The reduction is usually considered as an important contribution by Descartes to the algebraic solution of equations,[34] on a par with similar reductions by Ferrari[35] and Viète.[36] Yet it should be noted that the procedures did not facilitate the algebraic solution of equations, either in this case or when Descartes discussed Ferrari's reduction. Here the reduction served the technique of determining whether an apparently solid problem was in fact plane; when he referred to Ferrari's reduction, it was to prove that all solid problems could be reduced to the

---

[32] As elsewhere in the *Geometry*, Descartes took the coefficients to be positive and allowed both + and − signs in the equations, explaining at some cost of words the behavior of the signs in the different cases. In representing his equations together with his case distinctions I use the convention introduced in Section 26.2 (Equation 26.3).

[33] [Descartes 1637] pp. 384–385: "Aprés que l'equation est ainsi reduite a trois dimensions, in faut chercher la valeur d'*yy* par la methode desia expliquée; et si celle ne peut estre trouvée, on n'a point besoin de passer outre; car il suit de là infalliblement, que le problesme est solide." Cf. Note 22.

[34] Cf. [Tropfke 1980] p. 458.

[35] See Note 18 of Chapter 10.

[36] Cf. [Viète 1615] pp. 149–150 (translation: [Viète 1983] pp. 286–289); cf. Note 15 of Chapter 10.

two geometrical problems of the trisection and the determination of two mean proportionals. Descartes was in fact little interested in the algebraic solution of equations; his primary interest was a geometrical one.

Higher-order equations

As to the reducibility of higher-order equations Descartes merely stated[37] that one should check in a similar manner (that is, presumably, by undetermined coefficients and/or by checking the factors of the constant term) whether the polynomial can be written as the product of two polynomials, working through all possible cases for the degrees of the factors.

27.4 Descartes' example: a problem from Pappus

A reducible equation

In presenting the techniques for transforming and decomposing equations Descartes gave various examples, most of them with numerical coefficients. He presented one example with indeterminate coefficients, namely, the equation

$$x^4 - 2ax^3 + (2a^2 - c^2)x^2 - 2a^3x + a^4 = 0\,, \qquad (27.22)$$

which first appeared in connection with the use of the substitution for removing the second term of an equation (Equation 27.8).[38] In this case $x = z + \frac{1}{2}a$ yielded

$$z^4 + (\frac{1}{2}a^2 - c^2)z^2 - (a^3 + ac^2)z + \frac{5}{16}a^4 - \frac{1}{4}a^2c^2 = 0\,. \qquad (27.23)$$

Then Descartes used this equation to illustrate the test of reducibility into quadratic factors (Equations 27.18–27.21) The pertaining cubic in y^2 turned out to be[39]

$$y^6 + (a^2 - 2c^2)y^4 + (c^4 - a^4)y^2 - a^6 - 2a^4c^2 - a^2c^4 = 0\,. \qquad (27.24)$$

The reader had met this equation two pages before where Descartes used it to illustrate the method for finding linear factors of a cubic polynomial by checking the factors of the constant term. In this case the factorization of the constant term was

$$a^6 + 2a^4c^2 + a^2c^4 = a^2(a^2 + c^2)^2\,, \qquad (27.25)$$

and Descartes showed that $a^2 + c^2$ was a root of Equation 27.24 as a cubic in y^2. Setting $y_1^2 = a^2 + c^2$ he arrived, following the general rule (cf. Equation 27.21), at the decomposition of Equation 27.23 into two quadratic equations (note that the coefficients involve square roots but remain constructible by straight lines

[37] [Descartes 1637] p. 389.
[38] [Descartes 1637] p. 377.
[39] [Descartes 1637] p. 384.

and circles from the given line segments a and c)

$$z^2 - z\sqrt{a^2+c^2} + 3\frac{a^2}{4} - \frac{1}{2}a\sqrt{a^2+c^2} = 0\,, \tag{27.26}$$

$$z^2 + z\sqrt{a^2+c^2} + 3\frac{a^2}{4} + \frac{1}{2}a\sqrt{a^2+c^2} = 0\,. \tag{27.27}$$

Descartes noted that the two roots of the Equation 27.26 were[40]

$$z = \frac{1}{2}\sqrt{a^2+c^2} \pm \sqrt{-\frac{1}{2}a^2 + \frac{1}{4}c^2 + \frac{1}{2}a\sqrt{a^2+c^2}}\,, \tag{27.28}$$

and hence two roots of Equation 27.22 "for the finding of which we performed all these operations"[41] are known:

$$x = z + \frac{1}{2}a\,. \tag{27.29}$$

Geometrical origin of the equation Only after all these operations Descartes explained why he had chosen Equation 27.22 as an example: it occurred in the analysis of a problem from Pappus' *Collection.* Descartes explained the problem and gave Pappus' construction:

Construction 27.1 (Plane neusis problem — Pappus)[42]
Given a square $OABC$ with side a, and a length c (see Figure 27.1); the side CB of the square is prolonged; it is required to construct a straight line through O intersecting AB and CB prolonged such that the interval between the intersections is equal to c.

Construction:
1. Mark $OD = c$ along OC; draw AD $[AD = \sqrt{a^2+c^2}]$.
2. Prolong OA until E, with $AE = AD$.
3. Draw a semicircle on diameter OE; it intersects CB prolonged in F.
4. OF is the required line; it intersects AB in G and $GF = c$.
[**Proof**: Descartes gave no direct proof; Pappus first derived the equality

$$AE^2 = OA^2 + GF^2\,. \tag{27.30}$$

He did so by noting that $\triangle BFG$ is similar to $\triangle AOG$, which itself is congruent to $\triangle HFE$ (this folows easily from the equalities of angles indicated in the figure); hence $FE = OG$. Because

---

[40]Descartes did not discuss the roots of the second equation; they are $z = -\frac{1}{2}\sqrt{a^2+c^2} \pm \sqrt{-\frac{1}{2}a^2 + \frac{1}{4}c^2 - \frac{1}{2}a\sqrt{a^2+c^2}}$; for $c < 2\sqrt{2}a$ these roots are, in Descartes' terms, imaginary.

[41][Descartes 1637] p. 387: ". . . pour la connaissance de laquelle nous avons fait toutes ces operations . . ."

[42][Pappus Collection] VII, Props 71–72, pp. 605–608; in [Pappus 1986] pp. 202–205; construction as in [Descartes 1637] p. 387.

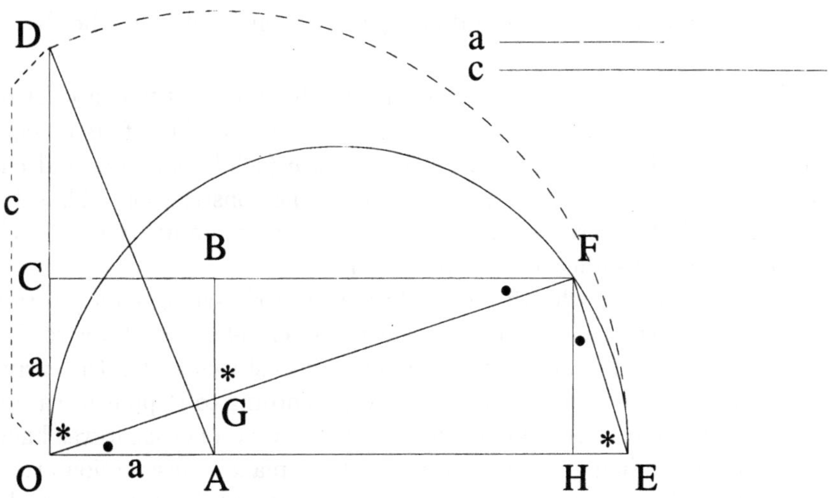

Figure 27.1: Plane neusis problem — Pappus

$\angle GAE = \angle EFG = 90^0$, the points G, A, E, F are on a circle, whence $OG \cdot OF = OA \cdot OE$ ($\star$). Because $\triangle OFE$ is a right-angled triangle and $FE = OG$, we have $(OA + AE)^2 = OF^2 + FE^2 = OF^2 + OG^2 = (OG + GF)^2 + OG^2 = 2OG^2 + 2OG \cdot GF + GF^2 = 2OG \cdot OF + GF^2$. ($\star\star$). Equations ($\star$) and ($\star\star$) combined yield $(OA + AE)^2 - GF^2 = 2(OA \cdot OE)$ from which it follows that $AE^2 = OA^2 + GF^2$, i.e., Equation 27.30.
Now by construction $AE^2 = AD^2 = OA^2 + OD^2$, from which, in combination with Equation 27.30 it follows that $GF = OD$, as required.]

Descartes then explained that this particular construction could have been found by following his canon, namely, by choosing AE (cf. Figure 27.1) as unknown and deriving the pertaining equation.[43] But, as he noted, that choice of the unknown was not an obvious one, it was more natural to choose BG or AG. He

[43]With $x = AE$ and $GF = c$, Equation 27.30 in Pappus' proof yields the equation $x^2 = a^2 + c^2$, which is of second degree, showing that the problem is indeed plane; the resulting construction is the same as Pappus'. Descartes suggested that this equation for $x = AE$, though simple, was difficult to find; indeed one would have to argue along the lines of Pappus' proof, which, for an analysis, is rather complicated. Yet Pappus probably found the construction by such an analysis.

then showed that the latter choice, $x = AG$, led precisely to Equation 27.22,[44] which, with help of all the techniques of transformation and reduction, was reduced to Equations 27.28 and 27.29, from which a plane construction could easily be derived. Descartes did not give this construction explicitly.

Significance of the example

The example was well chosen for illustrating the need to check for reducibility; otherwise, one would not notice the reducibility of the fourth-degree equation 27.22 and one would construct x by the (solid) parabola and circle construction, which was inferior to Pappus' plane construction. Thus Descartes' techniques indeed prevented the geometer from committing the "sin" of constructing with insufficiently simple means.

Another reason why the example was well chosen was its position in the literature. Although Descartes made no other references than to Pappus, it seems likely that he was aware of the problem's status within the early modern tradition of geometrical problem solving. According to Pappus it was one of the problems treated by Apollonius in his lost treatise on neusis; it was known to be plane; Ghetaldi had given an alternative (also plane) construction arrived at by classical analysis;[45] he had also claimed that the problem could not be solved by algebra.[46] Descartes might well have argued that if others would try to solve it by his new methods, they might arrive at the fourth-degree equation, forget the reducibility argument, and conclude that the new method was not powerful enough. It was a challenge which he could hardly ignore.

27.5 Conclusion

Algebra in the service of construction

In connection with Descartes' "new compasses" from 1619 I have noted that he was then comparatively a stranger to algebra (cf. Section 16.4). No longer so in 1637! He had successfully tackled the conceptual problem of reinterpreting the algebraic operations so as to apply for general magnitudes, and he had developed a versatile new notation incorporating indeterminates as well as unknowns. In both respects he had chosen approaches different from the prevailing ones of Viète. On top of that, as the present chapter shows, he had achieved a clear and effective general conception of equations and he had developed a series of techniques for transforming equations and for investigating their reducibility.

Yet it appears that Descartes was not interested in algebra for its own sake. The fascination with equations, as evident in the work of algebraists like Van Roomen and Viète (especially their studies on the equations related to angular

---

[44]Call (see Figure 27.1) $AG = x$ and $OG = v$ and note that $(a - x) : c = x : v$ (similarity of $\triangle BGF$ and $\triangle AGO$) and that $v^2 = a^2 + x^2$ (because $\triangle OAG$ is right angled); eliminating v from these two equations and rewriting yields the equation $x^4 - 2ax^3 + (2a^2 - c^2)x^2 - 2a^3x + a^4 = 0$, i.e., Equation 27.22.

[45]In fact Ghetaldi treated the variant in which the square $OABC$ is a rhombus; I have discussed his analysis and his construction in Section 5.4 (Analysis 5.5 and Construction 5.6).

[46]Cf. Section 5.4 The problem and its history is discussed in detail in [Brigaglia & Nastasi 1986]; see in particular pp. 125–127 for a discussion of Descartes' solution.

sections), was absent in Descartes' investigations. Each of the special algebraic techniques he explained in the *Geometry* had its purpose within the geometrical rationale of the book and was not developed further than necessary for that purpose. We may therefore characterize Descartes' algebra as subservient to geometry, more precisely to the canon of construction that Descartes elaborated in order to solve "all the problems of geometry."

Chapter 28

Conclusion of Part II

28.1 Forces and obstacles

The Geometry *as endpoint*

In the previous chapters I have followed the development of Descartes' ideas about geometry from 1619 till 1637. That year saw the publication of the *Geometry* in which Descartes formulated his convictions about geometrical exactness, presented his canon for geometrical problem solving, and explained the techniques of algebraic analysis he had developed for translating problems into equations and for constructing the roots of these equations. After 1637 Descartes occasionally returned to geometrical matters but he did not essentially develop the results reached in the *Geometry* — it appears that he considered his project of geometrical investigation completed.[1] In the letter to Beeckman of 1619 he had written that he intended to achieve a "completely new science by which all questions in general may be solved";[2] this goal he now had reached for geometry, the science which from the beginning inspired his vision of the scientific method.

So this is the point to conclude the present study — there will be an epilogue on developments emanating from the *Geometry* in Chapter 29. I do not recapitulate the main findings of the previous chapters; for that purpose I refer to Chapters 14, 15, 20 and 26. My primary interest has been to describe the development of Descartes' geometrical ideas and the genesis of the *Geometry*. In the present concluding chapter I focus on the principal dynamics of these developments, the main forces and impediments which stimulated and obstructed them. In that connection I also highlight a number of issues which, I feel, need some final emphasis.

[1] Cf. e.g. Descartes to Plempius, 3-X-1637, [Descartes 1964–1974] pp. 409-412, i.p. p. 411: "Non ignoro Geometriam meam paucissimos lectores habituram; nam cùm ea scribere neglexerim quae ab aliis sciri suspicabar, et paucissimis verbis multa (imò omnia quae unquam in illâ scientiâ poterunt inveniri) vel complecti vel saltem attingere sim conatus, lectores non modo peritos eorum omnia quae hactenus in Geometriâ et Algebrâ cognita fuere, sed etiam valdè laboriosos, ingeniosos et attentos desiderat."

[2] Cf. Section 16.1, Note 6.

Principal dynamics The creation and adoption of (finite[3]) algebraic analysis as a tool for geometry constituted the principal dynamics of the developments within the early modern tradition of geometrical problem solving discussed in Part I. The interpretation of constructional exactness was an important issue for the practitioners of geometrical problem solving before Descartes, but it did not constitute a stimulating force within the tradition.

In the case of Descartes the dynamical balance was different; the potential of algebraic analysis for geometry and the challenge of the questions of method and of geometrical exactness were stimuli of comparatively equal strength for his geometrical achievements.

28.2 Descartes' transformation of the art of geometrical problem solving

Breakthroughs From the publication of Pappus' *Collection* till c. 1635 the main impetus for the developments in the field of geometrical problem solving was the new use of algebraic methods. However, by 1635 these developments had lost momentum, and I characterised (cf. Section 14.5) the field at that time as waiting for essential breakthroughs with respect to the following three issues: (1) clarifying the general objective of problem solving by unifying, ordering, and if necessary extending the procedures for construction beyond the use of circles and straight lines, (2) understanding the relations between problems, equations and constructions, in order to direct the procedures of algebraic analysis, and (3) establishing a clear and complete interpretation of constructional exactness.

The characterisation, of course, is based on hindsight: these were the issues for which Descartes provided effective answers. By his answers the stagnation was lifted, the field acquired new incentives and new directions. Descartes indeed transformed the art of geometrical problem solving and this transformation may best be summarised according to the three issues mentioned above.

Construction In the *Geometry* Descartes proposed the following canon of construction (cf. Section 26.5): Constructions should be performed by the simplest possible, geometrically acceptable curves. To be acceptable the curves had to be algebraic; simplicity was measured by the degree. Descartes' standard constructions showed how, in principle, proper constructions could be achieved for any problem, that is, for any equation in one unknown.

The canon indeed offered a clear and complete general objective for the field of geometrical problem solving. It explained and ordered the algebraic techniques necessary for arriving at constructions of problems and it strongly suggested that by basically straightforward iterative algebraic procedures all geometrical problems could be solved. Not only did he present solutions for all problems leading to equations of degree ≤ 6, he also theoretically charted the

[3]Cf. Chapter 1 Note 17.

complete frontier area of higher-order problem solving, be it without actually exploring the terrain.

The clarity provided by Descartes' canon of problem solving primarily concerned the solution of ordinary problems, leading to equations in one unknown. The canon was less definite with respect to the construction of curves. Descartes showed that all geometrically acceptable curves could in principle be constructed pointwise, namely by the repeated solution of ordinary problems. He also argued that pointwise construction was equivalent to tracing by well coordinated motion. However, his arguments for this equivalence were not completely persuasive, and several passages from the *Geometry* implied that tracing methods provided more appropriate knowledge of curves than pointwise construction. Yet Descartes did not offer a general technique for finding tracing methods for curves with given equation. As a result the *Geometry* left a certain ambivalence about the merit of equations for representing curves.

Algebraic analysis

Descartes' theory of equations, discussed in Chapter 27, shows his awareness of the complexity of the relations between problems, equations and constructions. Since Viète, the translation of geometrical requirements into algebraic language was well understood. Descartes realized that this translation was in fact insufficient for directing the algebraic procedures involved in geometrical problem solving; it had to be supplemented by techniques to check whether the resulting equations were reducible and to reduce them if necessary. He provided such techniques, as well as the methods to transform equations into forms suited for the standard constructions which he also provided. These techniques were based on a deep, if tacit, understanding of equations in general. Yet it was not from a purely algebraic interest that Descartes acquired this understanding; all his new algebraic techniques served the purpose of geometrical construction.

Geometrical exactness

Descartes' mature interpretation of geometrical exactness, as given in the *Geometry*, was based on the premise that construction should be performed by the intersection of simplest possible algebraic curves. He sustained this premise by complex and perceptive arguments concerning the requirements for clear and distinct knowledge about motion and curves. The various interpretations of constructional exactness previously proposed within the early modern tradition of geometrical problem solving had been restricted, little persuasive and on the whole ineffectual. In contrast, Descartes' interpretation covered in principle all problems, it was argued with seriousness and depth, and it sustained an effective canon of construction. Thus it put the arguments about exactness on a higher qualitative level.

28.3 The path to the *Geometry*

Dynamics and chronology

Descartes' geometrical interpretation of the algebraic operations and his new algebraic techniques provided the technical basis of his transformation of the art of geometrical problem solving. But the transformation involved more than algebraic innovations; it crucially depended on Descartes' long standing interest in method, and on his philosophical approach to geometrical exactness. As mentioned above, the potential of algebraic analysis for geometry and the challenge of the questions of method and geometrical exactness were stimuli of comparatively equal strength for Descartes' geometrical achievements. In the present section I review the development of Descartes' geometrical thinking with particular emphasis on the aspects of method and exactness.

I briefly recall the chronology of Descartes' geometrical findings (cf. Section 20.1). By 1620 he had developed a clear programmatic vision for a 'new science' in which arithmetical and geometrical problem solving provided the paradigm for classifying and solving scientific problems in general. In geometry he explored the implications of using instruments which generalized the workings of rulers and compasses. At that time he was apparently little aware of the potential of algebra for geometrical analysis. However, by c. 1625 he had become aware of the effectiveness of algebraic analysis, as is evident from his general construction of the roots of third- and fourth-degree equations. In the *Rules*, written in the 1620's and abandoned after 1628, Descartes attempted to elaborate the epistemological base of his program, exploring in particular the certainty of the mental processes corresponding to the algebraic operations. His study of Pappus' problem in the winter of 1631-32 gave him new ideas about higher-order construction and the relation between curves, their equations and the methods to trace them. In the subsequent years these ideas developed into Descartes' mature interpretation of geometrical exactness and the corresponding canon of geometrical problem solving which he presented in the *Geometry* in 1637.

Expectations

It is of interest to consider Descartes' path towards his mature conception of geometry in terms of his earlier expectations. Measured against his own programmatic statements his achievements were great, but incomplete. In two instances Descartes gave up essential earlier ambitions. First, when abandoning the *Rules* he relinquished the hope that the elementary steps in certain, scientific reasoning could be identified with the basic algebraic operations applied to continuous or discrete magnitudes. Secondly, in the period between 1632 and 1637 he gave up the hope of developing explicit curve tracing methods which would convincingly and directly show that all algebraic curves could be traced by coordinated continuous motions and *vice versa*. In both cases he had to abandon expectations and to adjust his program or strategy accordingly.

The Rules

These adjustments should not be considered as failures to find the right answers; rather it appears that in the directions Descartes had in mind, there were no answers to be found.

In the case of the *Rules* his program required the elaboration of clear and distinct procedures, consisting of movements of geometrical objects (line segments or rectangles) in the plane, equivalent to algebraic operations. For addition, subtraction, multiplication and division these could be found, although they involved the difficult question of the dimension jumps in the case of multiplication and division. But Descartes' studies of the relations between geometrical problems and algebraic equations had made him aware that solving problems required more than merely adding, subtracting, multiplying and dividing. What was necessary as well was equation solving, that is, square root extraction, general root extraction and solving equations generally. Square root extraction was equivalent to determining a mean proportional, which involved, as Descartes termed it in the *Rules*, dividing by an unknown divisor. The performance of this operation in the plane involved more than the movement of line segments and rectangles which served for performing the elementary arithmetical operations; it required a geometrical construction by the intersection of circles and straight lines. Descartes may have considered the motions involved in this construction sufficiently clear and distinct to be acceptable, but he must have realized that square roots were not enough to solve problems generally; higher-order root extraction was also required — not to mention the solution of general higher-order equations — and the geometrical constructions corresponding to these operations involved the intersection of curves at least as complicated as the conics. Such procedures could hardly be considered as immediately obvious to the intellect contemplating their execution in the imagination.

No doubt the episode made him realize how complex was the geometrical interpretation of the algebraic operations and of the solution of equations. It also showed him the importance of motion processes in geometrical constructions. Thus the results of the *Rules* provided him with a basis from which to build up his later interpretation of the algebraic procedures. But a loss accompanied this gain: he had to give up the high ambitions of the *Rules* and he had to loosen the links between geometry and general scientific reasoning; his pursuit of geometry became an independent program.[4]

Could he have maintained his earlier program? The question comes down to a technical mathematical one: Are there motion procedures for achieving roots of polynomial equations which may be imagined in such a clear and distinct way as to be immediately convincing and acceptable? Later research has produced some procedures for mechanically achieving roots of equations,[5] but these clearly do not satisfy Descartes' criteria. The question, then, has to be answered in the negative; in other words, Descartes was compelled to abandon his expectations by the force of mathematical circumstances.

Tracing algebraic curves

The experience with the *Rules* forced Descartes to give up the hope of interpreting all scientific problem solving as equivalent to geometrical or arithmetical problem solving. If the conjecture which I elaborated in Chapter 19 is correct, a similarly discouraging experience later compelled him to rely more on algebra in his geometry than fitted his earlier ambitions.

Descartes' first studies of Pappus' problem in 1631-1632 showed that all loci which satisfied the requirement of a Pappus problem were algebraic curves. I

[4]Cf. Section 18.5.

[5]Cf. the instruments surveyed in [Frame 1943–1945].

have argued that Descartes probably believed that these loci could be traced by the iteration of the 'turning ruler and moving curve' procedure. This tracing method he deemed acceptable in geometry and he considered it as essentially different from the methods by which curves such as the quadratrix and the spiral were generated and which he considered unacceptable in true geometry. He found a strict correspondence between the number of given lines in a Pappus problem and the degree of its solution, and he probably believed that the degree also directly related to the number of iterations necessary to trace the locus. Moreover, he convinced himself that any algebraic curve was a locus to some Pappus problem, and could therefore be traced by motions which he deemed acceptable.

Thus the episode of Pappus' problem suggested a ready and convincing answer to the demarcation question of geometry: geometrically acceptable curves were precisely the algebraic curves; they were acceptable because they could be traced by acceptable motions; these motions were simpler in as much as the degree of the curve was lower; construction in geometry should be performed by the intersection of acceptable, i.e. algebraic, curves of lowest possible degree.

However, in working out this vision of geometry Descartes found it impossible to argue directly that any algebraic curve could be traced by the motions which he considered geometrically acceptable. Thus one cornerstone of his edifice, namely the identification of geometrical acceptability with algebraicity and simplicity with the algebraic degree, remained without direct proof; the argument for it in the *Geometry* (cf. Section 24.1), though impressive, was ultimately unconvincing. In fact Descartes' demarcation of geometry was persuasive by its simplicity rather than by the arguments he provided. In that sense Descartes came to rely more on algebra than fitted his earlier convictions; algebra provided a formulation of the demarcation whose simplicity concealed the absence of direct geometrical argument.

Here, as earlier in the case of the *Rules*, we may ask whether Descartes could have kept to his earlier program and built the demarcation directly on the nature of the motions which trace algebraic curves. Again the question comes down to a technical mathematical one: are there procedures for tracing algebraic curves which involve linkages of rulers like in Descartes' Mesolabe or in the 'turning ruler and moving curve' procedure and which are structurally analogous to the equations of the curves? The structural analogy should involve in particular a correspondence between the degree of the equation and the complexity of the procedure; preferably the tracing procedure should be an iterative one, and the number of iterations should correspond to the degree.

As far as I know, the mathematical literature from Descartes till the present offers only one general procedure for tracing (segments of) algebraic curves, the one devised by Kempe and published in 1876.[6] We may conclude from the intricacy and the late date of Kempe's method that a general tracing method for algebraic curves was not within Descartes' reach, let alone one which satisfied his further criteria.

Failures?

The arguments above imply that the adjustments which Descartes had to make in his programs cannot reasonably be considered as failures. Yet we may ask whether Descartes himself saw the adjustments as failures, and whether his readers noticed the absence of the results which Descartes had hoped for.

Although the modification of his program after giving up the *Rules* was a significant one, it is difficult to imagine that Descartes considered it as a failure. He loosened the links between general scientific reasoning and geometry and he pursued the latter as an independent program; but these changes did not hamper him in the further elaboration of his philosophical program, nor did they affect the basic tenets of his philosophical program such as the criteria for certain understanding.

The *Rules* were not effectively published before 1701 (a Dutch translation appeared in 1684). Descartes' earlier private texts on his program for a general science remained unknown for much longer. In his published writings Descartes stressed the importance of mathematics as a source of inspiration for his ideas and he published his *Geometry* as an 'essay' of the new method. Thus his contemporaries could not be aware that in an earlier phase the link between algebraic procedures and the general method for the direction of the mind had been much closer and that Descartes had been forced to modify his hopes and expectations in this respect.

As to the absence of general methods for tracing algebraic curves it should be noted that readers of the *Geometry*, notably in the circle around Van Schooten, showed a lively interest in the curve tracing methods; Van Schooten's own treatise on tracing conics[7] provides a characteristic example of this interest. Yet it appears that the question of a general tracing method for all algebraic curves did not attract attention. Nor was Descartes criticised for failing to provide such a general method in support of his demarcation of geometry. When, later, the demarcation was challenged, notably by Leibniz, it was not because a general tracing method was lacking. Rather the dissatisfaction with the Cartesian interpretation of geometrical exactness was caused by its exclusion of non-algebraic curves whose importance became more and more recognized in post 1637 mathematics .

---

[6] [Kempe 1876]. Kempe substitutes $x = a\cos\theta + b\cos\varphi$, $y = a\cos(\theta - \pi/2) + b\cos(\varphi - \pi/2)$ in the equation $F(x, y) = 0$ of the given algebraic curve, which yields

$$C + \Sigma A_{r,s}\cos(r\varphi + s\theta + \alpha) = 0\,; \quad \star$$

a and b are constants, $\alpha = \pi/2$ or 0, p and r are integers, $A_{p,r}$ and C are constants depending on the coefficients of F. He then constructs a concatenation of separate linkages. Each corresponds to a summand in ($\star$); its type and dimension are determined by the values of r, s, $A_{r,s}$ and α. Kempe then proves that if a point at the beginning of the linkage system is moved along a straight line, a point at the end of the system describes the required curve.

[7] [Schooten 1646].

Dissatisfaction? Descartes ultimately based his demarcation of geometry and his classifications of problems and curves on the simplicity of algebra rather than on direct arguments about curve tracing. He elaborated his canon of construction in detail only for lower-degree equations and merely claimed that it would be easy to proceed. Was his satisfied with the result? The closing sentences of the *Geometry* certainly did not bespeak dissatisfaction, nor are there later statements which suggest disappointment about the outcome of his geometrical program. Yet I find the question compelling because the reliance on algebra and the complexity of the standard constructions appear to be incompatible with the requirements of clarity and distinctiveness which are so fundamental in Descartes' thinking. It is not at all clear why the simplicity of algebra should carry the same intuitive clarity and distinctiveness as the simplicity of motions which Descartes invoked in his examples of geometrically acceptable curve tracing. And Descartes' standard constructions were algebraically simple only with respect to the degree of the constructing curves. Apart from that they involved very complicated algebraic manipulations whose geometrical effectuation, though not difficult in principle, was far from simple. Indeed the construction for 5th- and 6th-degree equations seems hardly convincing as canonical solution for a class of geometrical problems. Descartes may have believed that, as he stated at the end of the *Geometry*, it was possible to find general constructions for higher-order equations in the same way, but it seems highly unlikely that he expected these constructions to be simple, elegant or otherwise satisfactory as solutions.

I raise the question of Descartes' own appraisal of his final results on demarcation and construction because his situation with respect to these issues was a remarkable one. Not merely was he confronted with questions to which he could not find the answers; indeed there were no answers. The obviously rational and meaningful questions: 'What is exactness in geometry?' and: 'When is a geometrical problem adequately solved?' turned out to be unanswerable. These questions were the geometrical versions of the general question 'When do we know?'. Hence within the very science he had taken as paradigm for scientific understanding, Descartes was confronted with an instance in which it appeared that the decisive philosophical question 'When do we know?' had no answer.

We do not know whether Descartes was aware of this implication of his theory of construction and the demarcation of geometry. Yet it is important to point out how near he was to a realization that decisive philosophical questions may be without answers, and that in particular the question of when the thinking mind has reached the truth may be unanswerable. Such a realization would be at variance with the most fundamental tenet of Descartes' rational philosophy, namely that there is a method for finding the truth in the sciences. Clearly, if Descartes consciously realized this consequence of his geometrical explorations it must have been a highly disturbing experience.

I may add that, personally, I find it likely that to some extent Descartes realized these implications and was disturbed by them. Thus I sense a strain of tragedy blended with the evident greatness and success in the story of Descartes' geometrical investigations. His philosophical acumen made him understand better than any of his mathematical contemporaries the problems involved in applying algebraic analysis to the practice of geometrical problem solving. It forced him to explore in depth the questions (of simplicity and exactness) involved in the creation of a general canon of geometrical construction. Doing so he entered an area where straightforward generalization of geometrical arguments (concerning curve tracing in particular) proved impossible and where algebra allowed no adequate translation of geometrical criteria and provided no effective alternatives either. He came through by accepting algebraic criteria more generally than would have fitted his earlier convictions and by leaving to others the laborious (if not impossible) task of elaborating general higher-order constructions. This was success at a price, and I find it probable that Descartes was aware of the price.

The fact that his success had a price does not diminish the greatness of Descartes' geometry as intellectual achievement — on the contrary, the combination of success and adjustment to the loss of earlier expectations makes it the more admirable and impressive.

28.4 Descartes and the interpretation of exactness

Attitudes and strategies

In his *Geometry* Descartes redefined geometrical exactness. In Chapter 1 I have categorized a number of attitudes and strategies which mathematicians may adopt in dealing with the interpretation of exactness. I now discuss the question how Descartes' redefinition of geometrical exactness fits in this classification. The categories I introduced were (cf. Section 1.6):

1 Appeal to authority and tradition
2 Idealization of practical methods
3 Philosophical analysis of the geometrical intuition
4 Appreciation of the resulting mathematics
5 Refusal, rejection of any rules
6 Non-interest

Before turning to Descartes I summarize what I wrote earlier about the attitudes and strategies towards exactness adopted by Pappus, Clavius, Viète, Kepler, Molther and Fermat.

Pappus, Clavius, Viète

As far as we can reconstruct it from his *Collection*, Pappus' attitude to the interpretation of exactness of geometrical constructions was ambivalent (cf. Section 3.4). On the one hand he found it important not only to explain the classification of problems (plane, solid and line-like) but also to warn that it was a "considerable sin" among geometers to present solutions inappropriate to the class of the problem. He supported this precept with a clear appeal to tradition in geometry and to the anonymous authority of the "ancients". For the early modern mathematical audience, the authority of Pappus himself added to the strength of this argument. On the other hand, elsewhere in the *Collection* he presented, apparently without restraint, several solutions of problems which violated the very precept he gave in connection with the classification of problems.

To his early modern readers, Pappus' passages on the "sin" made more impression than the fact that he himself 'sinned' too; I have found no references to this inconsistency in the texts from the tradition of geometrical problem solving, whereas there were many references to the "sin".

We find the first explicit early modern discussion of exactness in geometrical constructions in Clavius' treatise of 1589 on the quadratrix (cf. Chapter 9). He asserted that his pointwise construction of the quadratrix was acceptable in pure, exact geometry because in practice the corresponding constructions, by ruler and compass, were very precise. The assumption implicit in this argument was that pure geometry was an idealization of practical geometry and that therefore exactness of geometrical constructions should be analogous to precision in practical geometry. Clavius further strengthened his argument with an appeal to authority (Apollonius, Archimedes and Menaechmus) and tradition.

Viète epitomized an approach to the interpretation of exactness totally different from Clavius' (cf. Chapter 10). He realized that his 'new algebra' provided a means to classify the totality of solid problems and chart their dependence upon two basic constructions: the trisection and the determination of two mean proportionals. These were solid problems whose standard constructions were then generally considered to be 'mechanical' and not geometrical. Thus, in order to incorporate his new results within the confines of legitimate geometry, the status of these basic constructions had to be reassessed; an argument was needed to assert their exactness. Viète provided such an argument: He gave the neusis construction (by means of which both the trisection of an angle and the determination of two mean proportionals could be performed) the status of a postulate. Thus he remedied, in one stroke of his pen, a defect of geometry and legitimized his newly discovered results on solid problems. He did not support this legitimation by arguments; in a way he simply asserted the authority of the geometer, in particular himself, to lend postulate status to certain powerful results. Viète's attitude and strategy were simple: the quality of the resulting mathematics justified the geometer's choices.

Kepler, Molther, Fermat

Viète's aim was to enlarge the domain of legitimate geometry. Kepler provides us with an example of an interpretation of exactness with the opposite purpose, namely to defend the strict adherence to the Euclidean postulates which legitimized no other constructions than those by circles and straight lines. His choice for this restrictive interpretation of constructional exactness was induced by his philosophy of harmony (cf. Chapter 11); constructibility in the strict Euclidean sense provided a demarcation between harmonic and non-harmonic ratios without which his philosophy would lose its meaning. His attitude to the interpretation of exactness was based on extra-mathematical, in this case philosophical, considerations. He defended his choice in the matter by appeal to tradition and authority, in particular the authority of Euclid and Proclus. To this he added an acute analysis of the available methods of higher-order construction showing that none of them was convincingly certain.

In his book of 1619 on the Delian problem Molther treated the interpretation of constructional exactness with much less depth than Kepler. He addressed the question with the candid simplicity of an amateur; the exactness of neusis constructions had not yet been convincingly established and fame could be gained by doing so, so he did it. His strategy was the idealization of practice, which he elaborated a bit further than Clavius. While Clavius simply noticed the practical precision of his constructions, Molther attempted to idealize the practical operations with instruments into a kind of mental operationalism in which motion was a legitimate ingredient of geometry and the geometer's inner sense could regulate the motions of abstract rulers and stop them exactly at the moment in which a required position had been obtained.

Fermat (cf. Chapter 13) had little if any interest in questions of geometrical exactness. When he decided to choose the intersection of conics — rather than the neusis as Viète had done — for constructing solid problems, he did so as a matter of course, sufficiently legitimized by the classical authority, for instance of Pappus.

Descartes

Descartes redefined geometrical exactness in his *Geometry*. In the previous chapters his attitude to the matter of exactness, the canon of geometrical construction which he elaborated and his strategy for convincing his audience about his choices have been discussed in considerable detail. Descartes created a new position for himself in the spectrum: the 'philosophical analysis of the geometrical intuition' as basis of the interpretation of geometrical exactness. In the post-classical development of mathematics no philosophical analysis of mathematical understanding had been developed with such profundity and at the same time with such strong direct consequences for the practice of mathematics.

But it was not only the nature of his choice which made Descartes' position a special one. He was much more intensely concerned about the interpretation of geometrical exactness than the other mathematicians of his time, with the possible exception of Kepler, whose philosophical starting point, however, implied a restrictive interpretation and thereby precluded a further analysis of the geometrical intuition. As a result Descartes' arguments were of an essentially higher intellectual quality than those of the other mathematicians I have discussed. Also, his interpretation of exactness, together with the canon of geometrical construction based on it, effectively eclipsed the earlier approaches to geometrical construction and set a standard which informed mathematical thinking about construction for about a century to come.

28.5 Success and failure in the interpretation of exactness

An ineffectual strategy

In order to arrive at a further assessment of Descartes' interpretation of exactness it is useful to ask whether the classification and the examples we have been discussing, admit conclusions about the relative efficacy of the various strategies.[8]

It appears that the least successful of the strategies was 'idealization of practical methods', exemplified by Clavius' pointwise construction of the quadratrix and Molther's instruments for legitimating the neusis construction. As interpretations of the exactness of geometrical procedures they were unconvincing. Clavius himself eventually lost faith in his arguments and I know of no mathematicians who accepted Molther's legitimation. Moreover, Clavius and Molther did not open up new mathematical territories by their arguments; the constructions by means of the quadratrix were already known from Pappus and the constructions legitimized by Molther's arguments were known from Pappus and Viète. As I noted in Section 12.5 the essential weakness of the idealization of practice as an interpretation of exactness is that it attempts to derive a strict demarcation — pure geometry or not — from an essentially continuous quality namely the precision of practical procedures.

Successful strategies

The categories which I termed 'Appreciation of the resulting mathematics' and 'non interest' (combined, if necessary, with 'appeal to authority and tradition') may be qualified as successful strategies. Among my examples these categories are represented by Viète and Fermat. Their interpretations of exactness accompanied new and significant mathematical achievements. Characteristically for this approach, Viète and Fermat used few, if any, explicit arguments in defending their interpretations of exactness (the neusis postulate in Viète's case, the choice of conics as constructing curves in Fermat's); their results were sufficient justification. Although the postulate status of the neusis was little discussed, Viète's approach to solid problems was very influential in the seventeenth-century Vietean tradition of geometrical problem solving. Fermat's analytical method for finding the constructing conics of solid problems, although overshadowed later by Descartes' approach, constituted a significant starting point for the analytical study of curves.

It appears that the success of the two strategies 'appreciation of the resulting mathematics' and 'non interest', is related to a fundamental implied message which they have in common: don't spend too much effort on matters of principle such as exactness, but move on and let your effort be judged on its mathematical results. If the results are indeed interesting enough, then it is an effective strategy to take the hurdle of redefining exactness by a simple postulational approach.

[8] In [Bos 1993] I have discussed this question in somewhat more detail, taking into account some additional examples concerning the interpretation of the exactness of curve constructions c. 1700.

Descartes' strategy

Descartes' approach to the interpretation of exactness was the 'philosophical analysis of the geometrical intuition'. The question whether or not this approach should be counted as a successful strategy is more involved that in the two cases discussed above. There is no doubt that the *Geometry* was highly influential on the technical mathematical level. But for Descartes, unlike Viète and Fermat, the interpretation of exactness was not merely a hurdle to be taken quickly before presenting an interesting mathematical structure. For him the redefinition of exactness and the demarcation of geometry were matters of the first importance. Was he successful in convincing his audience of his views? In a sense he was; his demarcation of geometry — the rejection of all non-algebraic curves — was generally acknowledged, if not accepted, and it was even anchored in the mathematical terminology: 'geometrical curves' was the term by which algebraic curves were designated during the period c. 1650 – c. 1750. However, in the 1650's one of the cornerstones of Descartes' demarcation, namely the belief that ratios between curved and straight line segments could never be found, was invalidated by the first rectifications of algebraic curves. Moreover, many mathematicians started exploring the realm of non-algebraic curves which Descartes had banished from geometry. Those mathematicians who kept to Descartes' demarcation of geometry mostly accepted it as dogma and showed little interest in the way how Descartes had based it on the certainty of the intuition of combined motions. The second essential element of Descartes' redefinition of exactness, namely the tenet that the simplicity of a curve was determined by its degree, was not generally accepted.[9]

Thus Descartes' philosophical analysis of the geometrical intuition attracted very little active interest. His audience eagerly accepted the system of geometry he presented (the analytical techniques and classifications concerning geometrical problem solving and the study of algebraic curves) but generally disregarded the philosophical ideas that had been so fundamental in creating the system. Indeed one may wonder whether the success of the *Geometry* wouldn't have been almost the same if Descartes had adopted the strategy 'appreciation of the resulting mathematics' in writing the book, just spelling out the canon of construction and the demarcation of geometry as postulates and concentrating on the wealth of technical results he could offer. I think the preceding chapters have sufficiently proved that Descartes could not have done so; his mathematics was a philosopher's mathematics and in that mathematics one cannot postulate without argument. But the reception of the book revealed a disparity between the aims of the writer and the effects of the book. Descartes' aim was to give geometry its definitive form. That aim, though exalted, was essentially conservative; Descartes' view of geometry as the art of solving geometrical problems was based on the contemporary tradition of geometrical problem solving, which by the 1630's was no longer a vigourous field and was soon afterwards superseded by other mathematical interests. Yet the influence of the *Geometry* was far from conservative; on the contrary, it was the most innovative treatise in mathematics of the first half of the seventeenth century.

[9]Cf. Notes 24, 28 and 29 of Chapter 26.

The *Geometry*, then, exerted its innovative influence despite the conservative methodological interest which guided its author and which is so strongly reflected in its structure. Thus Descartes' approach to the interpretation of exactness was not an effective strategy if we look at the reception of his philosophical analysis of the geometrical intuition. Was it an ineffectual strategy? If so, then certainly not of the somewhat shallow kind we met in the examples of Clavius and Molther. Descartes' strategy suffered defeat in the sense that he failed to convince the mathematical community of his philosophy of the mathematical understanding; but it was a glorious defeat because of the intellectual quality of the arguments and because of the wealth of new mathematics which Descartes worked out in pursuing his objective of giving geometry its definitive form.

And here again, as at the end of my sketch above of Descartes' path to the *Geometry*, I perceive a tragic element in Descartes' mathematical quest. As I speculated there, he may have come near to an intimation that the basic question 'When do we know?' could be unanswerable (cf. Section 28.4). In addition, he encountered little interest and less understanding for his program of securing the certainty of mathematics by a philosophical analysis of the geometrical understanding. Again one may wonder: could he have promoted his ideas more successfully? I think not. It seems — but this impression should be checked by further study of historical examples — that the absence of success is inherent to the strategy 'philosophical analysis of the geometrical intuition', particularly if the resulting interpretation of exactness is a restrictive one (excluding the transcendental curves in Descartes' case). Methodological restrictions of mathematics, however solidly rooted in tradition, authority or philosophical argument, tend to fade, and they fail when interesting mathematics appears to await investigation beyond the frontiers of orthodoxy. As far as Descartes' redefinition of geometrical exactness extended the frontiers (by admitting *all* algebraic curves) it was easily taken over; his restrictive precept (algebraic curves *only*) did not stop later seventeenth-century mathematicians to explore the realm of transcendental curves and in due time the philosophically inclined (Leibniz in this case) provided the arguments for incorporating these curves within legitimate geometry.

28.6 What does exactness mean?

An exact science Let me now return to the question raised at the beginning of this book (Section 1.1): What does exactness mean in mathematics? I have investigated the meaning independently of the term 'exactness', focusing on the conglomerate of qualities of mathematical procedures common to such terms as 'correct', 'acceptable', 'legitimate', 'rigorous', 'precise', 'certain', 'exact', and 'properly mathematical'.[10] In that sense 'exactness' has enjoyed no fixed interpretation; several times in history the requirements for mathematical procedures to be

[10]Cf. Section 1.2 Note 6.

exact have been readjusted in answer to developments within or around mathematics. I have reported on the early modern discussions on exactness with respect to geometrical problem solving. The interpretations of exactness which were proposed proved unsatisfactory in the long run — none of them acquired a lasting place among the rules of mathematics. Yet they were important at the time; they gave direction and structure to the research in geometry and algebraic analysis. Moreover, in the case of Descartes, the questions around geometrical exactness constituted one of the main forces that shaped the development of his geometrical ideas. Indeed the program and the structure of the *Geometry* was to a large extent determined by the endeavor to redefine geometrical exactness.

The importance of the interpretation of exactness as a factor in the development of early modern mathematics does not imply that there was clarity and consensus about the matter. Exactness was a fluid concept; there was often a tension between orthodoxy and practice and for the historian the attitudes and strategies of the protagonists are at least as informative and revealing as their actual ideas and arguments. The interpretation of exactness was an ongoing concern, with all the associated fluctuations and unclarities. But mathematicians were actively concerned about exactness and in that sense the mathematics of the period I've been describing may rightly be considered an exact science.

Chapter 29

Epilogue

29.1 Pre-Cartesian geometrical problem solving

Construction — the end of a period

In Chapter 1 I structured the story of geometrical construction in early modern mathematics by distinguishing two overlapping periods, c. 1590 – c. 1650 and c. 1635 – c. 1750. I dealt with the first period in Part I. The overlap of the two periods was dominated by Descartes, whose ideas on construction I discussed in Part II. (The second period is not treated in the present study.[1]) Thus I have now reached the end of the first period, the era (cf. Section 1.5) of the early modern tradition of geometrical problem solving. In this epilogue I draw the lines of influence emanating from that tradition, concentrating on the concept of construction and the interpretation of exactness.

The epilogue is meant primarily as a sketch of the aftermath of the ideas and techniques discussed in the previous chapters; it may also serve as an introduction to the second period in the history of the concept of construction and as a survey of themes for further research.

With and without algebra

By 1635 (cf. Chapter 14) the first generation of mathematicians active in the early modern tradition of geometrical problem solving had passed away. In their time the major innovation in the field was Viète's use of his new algebra. Some mathematicians, Clavius, for instance, paid no attention to this innovation; Kepler even rejected the use of algebra in geometry. But it seems that by 1635 the practice of geometrical problem solving without algebra (exemplified by Clavius' division of a triangle, Construction 4.18), and the pertaining classical method of analysis by the concept of "given" (cf. Sections 5.2 and 5.5), had vanished from the scene of active mathematical investigation. It is, however, of interest to explore how much of the techniques of this non-algebraic approach to geometry nevertheless remained known in the period after 1635, for instance,

[1] I have studied aspects of that period in [Bos 1974], [Bos 1984], [Bos 1985], [Bos 1987], [Bos 1988], [Bos 1989], [Bos 1993], [Bos 1996].

in connection with the recovery and restitution of ancient Greek mathematical treatises.

Viète's use of algebra in geometry was promoted and developed by a small but active group of mathematicians (in particular, Anderson and Ghetaldi).[2] Thus started a Vietean tradition that was soon enriched by Fermat's techniques for studying curves by their equations. Although later overshadowed by Cartesian geometry, it remained alive throughout the seventeenth century; a typical representative was Sluse.[3] The main characteristics of the tradition were the adherence to Vietean notation and to homogeneity of equations, and a tendency to interpret and classify equations (in one unknown) in terms of corresponding geometrical problems involving proportionalities (cf. Section 4.4). Thus the concept of "constitutive problems" for classes of equations can be traced to as late as 1702.[4] A special study of this Vietean influence on later seventeenth-century geometrical conceptions may provide an important addition to our understanding of the mathematics of that period.

29.2 Early reactions to the *Geometry*

Overall influence

It is hardly possible to overestimate the impact of Descartes' *Geometry* on later mathematics. The core of its influence consisted in the spread of Descartes' insights and techniques about the relation between curves and their equations or, more generally, about the interplay between figures and formulas. Without these insights and techniques the later seventeenth-century developments in mathematics (the infinitesimal calculus, the exploration of non-algebraic curves) are unthinkable.

My analysis of the *Geometry* in the preceding chapters has shown, however, that Descartes' main motivation in writing the book was not to expose the equivalence of curve and equation. Rather, it was to provide an exact, complete method for solving "all the problems of geometry." This program informed the structure and a large part of the contents of the book. Thus, as I stated in Section 28.5, the main influence of the book did not concur with its program. Indeed the *Geometry* exerted its main influence *despite* its primary motivation.[5]

Because my interest in the present study is primarily in those aspects of the *Geometry* that were germane to Descartes' program, I concentrate here on the reception and the aftermath of the ideas and techniques related to construction and the interpretation of exactness.

Acclaim and criticism

During Descartes' lifetime the *Geometry* evoked a relatively small number of reactions. Shortly after its appearance a small group of Dutch mathematical amateurs enthusiastically studied the book; their activity resulted in a short

[2] Van Ceulen's exploration of the use of (quadratic) algebra in solving geometrical problems (cf. Section 8.6) apparently attracted no interest.

[3] Cf. [Bos 1985].

[4] Cf. Note 17 of Chapter 4.

[5] Cf. [Bos 1990] p. 368 (p. 52 in ed. [Bos 1993c]).

introductory treatise that circulated in manuscript.[6] It was written by one or more members of the group, but Descartes saw it and apparently approved. In the fall of 1643 Descartes had the pleasure of witnessing how the Princess Elizabeth had learned enough from his *Geometry* to successfully apply the new method to Apollonius' problem.[7] In recognition of the quality of her solution he wrote down for her a detailed account of his own ideas about how to choose the known and the unknown elements of a problem so as to arrive at the simplest equations.[8] Frans Van Schooten, who had been involved in preparing the *Geometry* for the press, kept a lively interest in it; he wrote a large number of explanatory notes, and translated the text into Latin. This work formed the basis of his 1649 Latin edition of the *Geometry*, which, apart from Van Schooten's annotations, also included a sizable collection of short notes to the work by Florimond Debeaune.[9] The 1649 edition,[10] published one year before Descartes' death, marked the beginning of the effective spread of his geometrical ideas and techniques.

Debeaune wrote his notes to the *Geometry* shortly after it appeared. In February 1639 Descartes received a copy of the notes; in a return letter he expressed his appreciation for the openness with which Debeaune approached the novelties of the *Geometry*.[11]

Descartes praised Debeaune's open attitude because he felt that in most other reactions from France such an attitude was painfully lacking. The *Geometry* indeed provoked acrimonious disputes with French mathematicians. There was an extended epistolary debate (via Mersenne) with Fermat, first on Descartes' derivation of the law of refraction in the *Dioptrics*[12] and later on Fermat's method of extreme values and its relation to Descartes' tangent method as explained in the *Geometry*.[13] The tone in this debate was bitter; it was set primarily by Descartes who was extremely sensitive about his intellectual property and about his reputation as a mathematician and natural philosopher. Roberval and Beaugrand, who had their own grudges against Descartes, also entered the dispute. They directed their attacks particularly against the *Geometry* and, based on elementary misunderstandings and misinterpretations, accused Descartes of plagiarism (from Viète and Harriot) and algebraic incompetence.[14]

No interest in Descartes' program

A striking feature of these first reactions to the *Geometry*, both positive and negative, is the lack of interest in Descartes' overall program for geometry. Descartes wanted to institute, once and for all, the proper canon of geometrical construction and to provide a complete analytical method by which all problems

[6] [Anonymus Calcul].

[7] I discussed the problem in Part I, see Construction 5.8.

[8] Descartes to Elizabeth, November 1643, [Descartes 1964–1974] vol. 4, pp. 45–50.

[9] [Debeaune 1649].

[10] [Descartes 1649].

[11] Descartes to Debeaune, 20 II 1639, [Descartes 1964–1974] vol. 2 pp. 510–523, esp. 510.

[12] The first "essai" accompanying the *Discourse*, cf. [Descartes 1637b].

[13] Cf. three letters of Descartes to Mersenne from January 1638 (?), [Descartes 1964–1974] vol. 1 pp. 481–486, 486–496, and 499–504.

[14] Cf. [Descartes 1964–1974] vol. 2, pp. 82, 103–115, 457–461, and 508–509.

of geometry could be solved. Most of the reactions during his lifetime concerned elementary algebraic techniques, notation, homogeneity (Debeaune started his *Notes* with a lengthy comment on Descartes' use of a unit line segment[15]), and the determination of normals and tangents.

Three episodes in the discussions between Descartes and his contemporaries about the *Geometry*, however, came near to the methodological core of the *Geometry* and should therefore be briefly discussed here. They were: a comment by Roberval on the construction of equations, a treatise by Fermat on how to choose the proper constructing curves, and Descartes' own confrontation with problems concerning transcendental curves.

Roberval Soon after the *Geometry* had appeared, Roberval criticized the crowning technical achievement in Descartes' program, the construction of the roots of sixth-degree equations. The construction (Construction 26.2) used only one branch of the Cartesian parabola. Roberval claimed that no circle could intersect this branch in more than four points, whereas a sixth-degree equation could have six real roots. He explained the situation by assuming that the remaining two roots would appear through the circle's intersections of the second branch of the Cartesian parabola. Descartes refuted the comment easily by an explicit example.[16] Yet the episode is of interest because it shows how new the domain of higher-order curves was; the phenomenon of a circle intersecting a convex curve in more than four points was unusual and apparently in conflict with primary intuitions about curves.[17]

Fermat Fermat commented on Descartes' canon of problem solving in a treatise, the *Tripartite dissertation*, which he probably wrote in the early 1640s. Descartes never saw the treatise; it remained in manuscript until 1679.[18] Fermat had come to the conclusion that according to Descartes' canon of construction the roots of equations of degree $2n$ and $2n-1$ should be constructed by curves of degrees $2n-3$ and $2n-4$. It is not clear how he arrived at this conclusion; Descartes' own ambiguity about his classification of curves by "genres"(cf. Section 25.1) may have played a role. The numbers in Fermat's interpretation fit for fifth- and sixth-degree equations ($n = 3$), which Descartes indeed constructed by curves of degree 3 (the Cartesian parabola) and 2 (the circle). But Fermat's numbers do not apply for Descartes' constructions of lower-degree equations. Moreover, as I have argued above (Section 26.4), Descartes probably envisaged the construction of equations of degrees $2n-1$ and $2n$ by one curve of degree n

[15] Cf. Note 14 of Chapter 21.

[16] Cf. "Roberval contre Descartes', April 1638, [Descartes 1964–1974] vol. 2 pp. 103–115, and Descartes to Carcavi 17 VIII 1649, *ibid.* vol. 5 pp. 391–401, esp. pp. 397–399.

[17] As late as 1713 Rolle used the term "paradox" for the phenomenon that two graphs, both concave and increasing, may intersect each other in arbitrarily many points, cf. [Bos 1984] p. 369, note 83.

[18] [Fermat DissTrip], on its date cf. [Mahoney 1994] p. 130, note 94; excerpts of it were published in 1657 (cf. *ibid.* p. 141, note 110), the full text appeared in the *Varia Opera* of 1679.

(traced by some iteration of the "turning ruler and moving curve" procedure) and one of degree 2, namely a circle. Fermat's interpretation suggested that Descartes used constructing curves of too high degree; he stressed this negative judgment by invoking Pappus' precept:

> Certainly it is an offence against the more pure Geometry if one assumes too complicated curves of higher degrees for the solution of some problem, not taking the simpler and more proper ones; for it has often been declared already, both by Pappus and by more recent mathematicians, that it is a considerable error in geometry to solve a problem by means that are not proper to it.[19]

Fermat intended to show that he could do better than Descartes. He presented a general procedure that, for any given equation of degree $2n$ or $2n-1$ in one unknown, supplied the equations of two curves of degree n, by whose intersections the roots could be found. In terms of the simplicity of curves, as measured by the degree, this procedure was clearly better than what Fermat thought Descartes meant, but worse than what Descartes probably did mean.

Fermat concluded his treatise by giving examples of equations with degree n whose roots could be constructed by means of curves of degrees of the order of $\sqrt{n}$. He expressed the importance of this result by saying that, if such equations could be found for arbitrarily large degree n, there existed problems "whose degree has to the degrees of the curves used in [their] construction a ratio larger than any given ratio"[20] — a result meant to dwarf Descartes' achievement, for which, in Fermat's opinion, this ratio, would become 1 for large n. It was in this connection that Fermat mentioned (and needed) his famous conjecture that numbers of the form $2^{2^k}+1$ are prime.[21]

It is unlikely that, had Descartes seen Fermat's treatise, a constructive discussion between the two would have ensued. Descartes would probably have been put off because Fermat adopted the least favorable interpretation of the scheme for the degrees in the canon of construction. But more important, Fermat did not show any interest in the programmatic aspects of the book. He accepted without further question that curves were simplest and appropriate for

[19] [Fermat DissTrip] p. 121: "Puriorem certe Geometriam offendit qui ad solutionem cujusvis problematis curvas compositas nimis et graduum elatiorum assumit, omissis propriis et simplicioribus, quum jam saepe et a Pappo et a recentioribus determinatum sit non leve in Geometria peccatum esse quando problema ex improprio solvitur genere."

[20] [Fermat DissTrip] p. 131: "problema construemus cujus gradus ad gradum curvarum ipsius solutionem inservientum rationem habet data quavis majorem."

[21] Fermat took $n = 2^{2^k}+1$ for $k = 1, 2, 3$, etc., and considered the equations $x^n = a^{n-1}b$ (these are the equations for the first of $n-1$ mean proportionals between a and b). For shortness I write $l = 2^{2^{k-1}}$, so $n = l^2+1$. Fermat then considered two curves with the equations $x^{l+1} = y^l b$ and $x^{l-1}y = a^l$, respectively. Eliminating y from the two equations leads to $x^{l^2+1} = a^{l^2}b$, that is, $x^n = a^{n-1}b$. Hence the x-coordinates of the points of intersection of the two curves satisfy the original equation. — In the context of the construction of roots of equations it was essential for Fermat to find large n's that were prime, because for non-prime n the equation $x^n = a^{n-1}b$ is reducible. As is well known, Euler later refuted Fermat's conjecture by showing that $2^{2^5}+1$ is not prime.

a class of problems if they were algebraic and had lowest possible degree. On that basis he directed his critical attention solely to the techniques in Descartes' book and used the occasion to show his algebraic skill and his knowledge of arcane number theoretical results. He disregarded Descartes' arguments for his canon of higher-order geometrical problem solving, in particular the philosophical and kinematical arguments that Descartes had developed in order to justify this interpretation. Fermat gave no thoughts to the question how curves were geometrically generated. For him the intersection of algebraic curves was simply a matter of eliminating one unknown from a pair of equations; finding constructing curves for a given equation was the inverse operation: finding two equations from which by elimination the given equation would result.

The pattern of response to the *Geometry* exemplified in Fermat's treatise, concentration on technical algebraic results and disregard for methodological aspects, was to be repeated by many seventeenth-century readers of the book.

The cycloid

One of the key elements of Descartes' geometrical doctrine was the demarcation between truly geometrical curves, on the one hand, and non-geometrical ones ("mechanical" in Descartes' terminology), on the other hand. He asserted that the generation of mechanical curves by motion was so inexact that certain knowledge about them could not be achieved. Hence these curves ought not to be used in solving geometrical problems. When he formulated these principles, he knew few examples of such curves; apart from the classical quadratrix and spiral, he probably had gained some insights about the logarithmic curve and he may have known about the cycloid.

Soon after the *Geometry* appeared Descartes found himself studying two "mechanical" curves in considerable detail. The first was the cycloid. In April 1638 Mersenne wrote to Descartes about Roberval's quadrature of the cycloid and related problems such as the determination of tangents to the curve. Descartes checked the quadrature, provided a proof, and solved the problems.[22] It turned out, he wrote to Mersenne, that neither in the proof nor in the solutions of the problems could he use the methods from the *Geometry*. Instead he developed an ingenious but ad hoc approach to the problem. He also realized why his method was not applicable: the curve was not "geometrical."

> It should also be noted that the curves described by rolling circles are entirely mechanical lines and belong to those which I have rejected from my *Geometry*; that is why it is no wonder that their tangents cannot be found by the rules I put in it.[23]

Debeaune's problem

Shortly after his encounter with the cycloid, the discussions in Parisian math-

[22]Mersenne to Descartes 28 IV 1638, [Descartes 1964–1974] vol. 2 pp. 116–122; Descartes to Mersenne 25 V, 27 VII and 23 VIII 1638, *ibid.* pp. 134–153, 253–280, and 307–343.

[23]Descartes to Mersenne 23 VIII 1638, [Descartes 1964–1974] vol. 2 p. 313: "Il faut aussy remarquer que les courbes descrites par des rouletes sont des lignes entierement mechaniques, et du nombre de celles que i'ay reietés de ma Geometrie; c'est pourquoy ce n'est pas merveille que leur tangentes ne se trouvent point par les regles que i'y ay mises."

ematical circles confronted Descartes with another "mechanical" curve. Debeaune had circulated a number of problems, one of which required the determination of a curve from a given property of its tangents. In modern terms the problem implies the solution of a first-order differential equation; it was one of the first of its type to arise in mathematics. Later seventeenth-century mathematicians referred to the type as "inverse tangent problems." In the case of Debeaune's problem the required curve was a logarithmic one. Descartes gave two solutions, both of which he considered insufficient. One was a approximative procedure that yielded points lying arbitrarily close to the curve; the other specified a kinematic process by which the curve could be traced. The first was insufficient because it only supplied approximations. The motion specified in the other solution was a combination of two rectilinear motions, one with constant velocity and the other with a velocity inversely proportional to $1 - s$, s being the distance traversed by the first motion.[24] Descartes commented:

> But I believe that these two movements are so incommensurable that they cannot be regulated by each other in an exact way; and therefore that this line belongs to those which I have rejected from my Geometry as being only Mechanical, which is the reason why I am no longer surprised that I could not find it by the other artifice which I had used, because that only applies to geometrical lines.[25]

Limitations

Thus, again, Descartes encountered a problem that transcended the power of the methods of the *Geometry*; and, again, he had to conclude that the reason why these methods were inapplicable lay in the non-"geometrical," that is, the non-algebraic nature of the curve.

Yet, uninteresting these problems were not; hardly a year after banning curves like the cycloid from geometry Descartes was sufficiently captivated by them to spend considerable energy in their study. It is remarkable how accurately these two encounters with transcendental curves marked the limitations of Descartes' new approach to geometry: it did not cover quadrature problems, its method for determining tangents only applied to algebraic curves and it was of little help for the general problem of determining a curve from its tangents. The fact that these problems crossed Descartes' path so soon after the appearance of the *Geometry* was almost symbolic; it foreshadowed the turn mathematicians were soon to make toward problems that fell outside the domain and the power of Descartes' new methods. In the second half of the seventeenth century the tradition of geometrical problem solving, which had so strongly informed Descartes' geometrical doctrine, moved to a peripheral position in mathematics, whereas quadratures, tangents, inverse tangent problems, and transcendental

[24]Descartes to Debeaune 20 II 1639, [Descartes 1964–1974] vol. 2 pp. 510–523, esp. pp. 514–518. On Debeaune's problem see also [Whiteside 1960–1962] pp. 368–370 and [Scriba 1961].

[25][Descartes 1964–1974] vol. 2 pp. 517: "Mais ie croy que ces deux mouvemens sont tellement incommensurables, qu'ils ne peuvent estre reglez exactement l'un par l'autre; et ainsi que cette ligne est du nombre de celles que i'ay rejettées de ma Geometrie, comme n'estant que Mechanique; ce qui est cause que ie ne m'estonne plus de ce que ie ne l'avois pû trouver de l'autre biais que i'avois pris, car il ne s'etend qu'aux lignes Geometriques."

curves dominated the center of activity; indeed they inspired the creation of the differential and integral calculus.

29.3 The construction of equations

An early modern research program

The early reactions to the *Geometry* show that Descartes' methodological and philosophical arguments on geometry attracted little interest. Yet the resulting canon of problem solving, and the related algebraic techniques, were accepted by many mathematicians, especially through the two Latin editions of the *Geometry*. Stripped of its philosophical and methodological components (exactness, acceptability of curves and constructions, interrelation of geometry and algebra), and expressed in modern notation, Descartes' doctrine of problem solving reduces to the following general problem:

Problem 29.1 (Construction of Equations)[26]
Given: an equation in one unknown

$$H(x) = a_n x^n + a_{n-1} x^{n-1} + \cdots + a_1 x + a_0 = 0 \,, \tag{29.1}$$

it is required to find two algebraic curves Φ and Γ of minimal degrees, such that all (real) roots of the equation $H(x) = 0$ occur as X-coordinates of the points of intersection of Φ and Γ.
As these coordinates are determined by eliminating y from the two equations of Φ and Γ, the requirement becomes:
To find two polynomial equations in two unknowns with minimal degrees

$$F(x,y) = 0 \quad \text{and} \quad G(x,y) = 0 \tag{29.2}$$

such that the roots of $H(x) = 0$ occur among the roots of

$$R_{F,G}(x) = 0 \,, \tag{29.3}$$

in which $R_{F,G}$ is the resultant *of F and G, that is, the polynomial resulting from eliminating y from the two equations in 29.2.*

The problem deviates from Descartes' examples in the sense that later geometers usually did not assume a circle for one of the constructing curves, but rather two curves of approximately the same degree.

Problem 29.1 was called the *construction* of the equation $H(x) = 0$ and the general theory about how to find the required curves or equations was called the *Construction of Equations.*[27] From c. 1650 to c. 1750 the construction of equations constituted a small but recognizable mathematical research program; a considerable number of books devoted substantial sections to the subject and prominent mathematicians, such as Fermat, de la Hire, Newton, Jakob Bernoulli, l'Hôpital, Euler, and Cramer, contributed to it. During the second half of the seventeenth century the construction of equations absorbed the Vietean tradition in geometrical problem solving. After 1750 the subject gradually slipped into oblivion and died.

[26] Cf. [Bos 1984] p. 343.
[27] Cf. Section 26.1

Rise and decline

I have given elsewhere[28] a survey of this mathematical subfield, its development and its later decline. With Descartes, the construction of equations was the natural outcome of a comprehensive program for geometrical problem solving by means of algebraic analysis. In particular, it provided the translation of geometrical criteria of adequacy of constructions into algebra, based on Descartes' own interpretation of geometrical exactness. In algebraic terms the "construction" of an equation was adequate if the constructing curves were algebraic and of minimal degree. Precisely here lay the reason for the subject's eventual decline and death: the algebraic formulation of the geometrical criteria was both too liberal (for one equation $H(x) = 0$ there were many pairs of curves Φ and Γ that satisfy the requirements) and ultimately unconvincing (especially the tenet that the simplicity of curves should be measured by their degrees was often criticized — but no workable alternative criterion of geometrical simplicity was found). The construction of equations made sense as long as mathematicians realized its original geometrical motivation. But few mathematicians adopted Descartes' philosophical rationale for the procedure and later attempts to provide alternative geometrical motivations of the procedure failed. But as a purely algebraic problem the construction of equations made little sense. Thus, together with the geometrical criteria, the subject lost its meaning and died.

For further details I refer to my earlier study. Here let it suffice to note that the "construction of equations" was the main mathematical field that, be it in a primarily technical sense, developed the program and the exactness-related concerns of the *Geometry*. It was a solid but rather unimaginative mathematical venture and, as such, an ambivalent heir to Descartes' imposing philosophical doctrine of geometry. Its fate shows that Descartes' interpretation of geometrical exactness was ultimately unconvincing and lacked sufficient innovative power to stay alive. Thus the development of the construction of equations in the century after the publication of the *Geometry* actualized the tragedy of which, as I suggested in the previous chapter (cf. Section 28.3), Descartes may have had some premonition when he felt himself forced to accept algebraic rather than geometrical criteria in his interpretation of exactness.

29.4 The interpretation of exactness

The investigation of curves after 1650

The "construction of equations" remained a peripheral enterprise in mathematics. But Descartes' doctrine of construction also affected the core of mathematical activity after 1650. The most vigorous and productive mathematical developments in that period occurred in connection with curves. Curves were of central importance within mathematics; their investigation by finite and infinitesimal analysis was the context for the creation of the differential and integral calculus by Newton and Leibniz. Curves were also the medium through which advanced mathematical methods could be applied in astronomy, mechanics, optics, and natural science generally.

[28] [Bos 1984].

In Chapter 1 I have characterized the principal dynamics within the study of curves as "the emancipation of (finite and infinitesimal) analysis from its geometrical context," or the "de-geometrization of analysis." Analysis in that period meant the use of algebraic formalisms, including notations for limit-related operations such as integration and differentiation.

This analysis proved to be a very powerful tool for solving problems. The most consequential of these were quadratures and inverse tangent problems (evolving gradually to their analytical analogs: integration and the solution of differential equations). These problems often had non-algebraic curves as their solutions and thus mathematicians had to decide how to *represent* these curves, that is,[29] how to identify the curve in question in such a way that other mathematicians would consider it adequately determined and the problem thereby properly solved. It was here that Descartes' ideas on construction and exactness exerted a marked influence on the study of curves. The influence was two-sided; Descartes' achievements facilitated the understanding and use of algebra as analytical tool, but his authority impeded the acceptance of non-algebraic relations and curves.

Around 1700 mathematicians had internalized Descartes' doctrine to such a degree that the construction of algebraic curves was generally considered to be elementary and uninteresting. After that time such constructions were hardly ever spelled out in detail — a mere statement of the form "construct the curve with this equation" sufficed. Explicit references to Descartes were seldom given, yet the presupposed procedure obviously was Descartes' generic pointwise construction of algebraic curves (cf. Sections 23.2 and 24.3), based on the assumption that the roots of any equation (in one unknown) could be constructed. Thus fairly soon after Descartes' *Geometry* mathematicians were so far habituated to algebraic curves that the equation of such a curve no longer presented a problem (how to construct the curve with that equation); rather it represented an object (the curve with that equation).

Exactness and non-algebraic curves

The habituation to non-algebraic curves took more time. This was partly because the representation of such curves was far from trivial; there were (at least until c. 1700) very few notational means available to express their equations. In the absence of analytical means of representation, a non-algebraic curve could only be imagined and talked or written about in terms of a geometrical procedure to construct or trace it. In the case of non-algebraic curves these procedures involved combinations of motions, or pointwise constructions, which Descartes had expressly banned from genuine geometry because, in the case of non-algebraic curves, they did not provide proper knowledge of the objects.

A number of mathematicians felt that a reinterpretation of geometrical exactness was needed, overcoming the obstacle of the restrictive Cartesian orthodoxy. Thus in the second half of the seventeenth century Descartes' ideas about genuine geometrical knowledge induced a new debate on the interpretation of exactness in connection with the proper representation of non-algebraic curves.

[29] Cf. Section 1.3.

Similarity of the debates on exactness

I have argued elsewhere[30] that a study of the arguments in and the origins of the debate on the representation of curves is essential for understanding the process by which analysis gradually became independent of its strong roots in geometrical imagery.

I hope to show in later relevant studies that the debate on the interpretation of exactness of construction and representation of curves in the period c. 1635 – c. 1750 followed a course similar to that of the debate on constructing geometrical problems in the period c. 1590 – c. 1650. In my present study I have used a distinction of archetypical attitudes and strategies of mathematicians confronted with the interpretation of mathematical exactness (cf. Section 1.6). The same attitudes were evident in the later debate on the representation of curves; some mathematicians argued by idealizing practical methods, some appealed to authority, etc. In both cases also the debates were inconclusive. What remained of Descartes' forcefully argued canon of construction was merely the tacit conviction that algebraic relations in geometry did not present problems with respect to exact knowledge. Similarly the later arguments about the proper representation of non-algebraic curves by geometrical constructions lost their urgency, while, in the first half of the eighteenth century, mathematicians gradually accepted analytical expressions as sufficient representation of curves. Yet, despite their inconclusiveness, the two debates decisively influenced the direction of mathematical research in the two periods.

29.5 Conclusion — metamorphoses

Geometrical problem solving

The sections above show that the story of construction in the early modern period is one of perpetual change. Indeed its main message seems to me to be that nothing remains quite the same in mathematics. Therefore the present epilogue is appropriately concluded by a brief overview of the principal metamorphoses that have occurred with respect to the concept of construction in the early modern period.

The early modern tradition of geometrical problem solving underwent a remarkable metamorphosis. First, primarily through Viète's "new algebra," it changed from a mainly geometrical endeavor to a largely algebraic one. Then Descartes' reinterpretation of the concept of construction practically reduced geometrical problem solving to a single algebraic problem: the "construction of equations." As a result, the endeavor moved from a central to a peripheral position in mathematics. In this position it stayed alive for about a century and then disappeared for lack of meaning. All these changes — algebraization, reduction to one problem, shift to a peripheral position, and demise — were linked to what I have called the principal dynamics in the tradition up to Descartes, namely, the creation and adoption of (finite) analysis as a tool for geometry. Indeed the final demise of the "construction of equations" can be understood as a belated reaction to the introduction of finite analysis as a tool in geometry; ultimately the original geometrical aims of the field — genuinely geometrical

[30] [Bos 1974], [Bos 1987], [Bos 1988], and [Bos 1996].

solution of all plane and higher-order problems — could not be translated into algebraic terms, and thereby the endeavor lost its meaning.

Investigation of curves The investigation of curves, which was the context of the developments concerning the concept of construction in the second period, c. 1635 – c. 1750, also underwent a strong metamorphosis, but different from geometrical problem solving. Until around 1700, curves were studied by means of finite and infinitesimal analysis, but they were considered known or understood only if a geometrical construction procedure was explicitly formulated. Gradually mathematicians became less concerned about these constructions; they learned to understand the curves from their equations. As a result, after 1700 the central object in this endeavor was no longer the curve but the analytical expression (formula) that represented the curve. It was taken for granted that from this expression a construction of the curve could be derived; to do so was no longer seen as necessary or enlightening. Habituation to analytical expressions made the focus shift from explicitly constructed curves to relations between variables explicitly or implicitly defined by analytical equations. I have called this process the "de-geometrization" of analysis, or the emancipation of analysis from its geometrical context. This was the principal dynamics in the field and, as in the case of geometrical problem solving, it completely transformed the aims and the criteria of adequacy in the field. Yet, contrary to the case of geometrical problem solving, these changes did not result in a shift to a peripheral position and subsequent oblivion. Indeed, the analytical study of relations between variables became the very core of eighteenth-century mathematics.

Analytic geometry The present discussion of metamorphoses of mathematical fields invites a comment on analytic geometry. Descartes' *Geometry* gave to mathematics the insight that a curve and its equation are in a sense equivalent; this insight is usually considered to be the essence of analytic geometry. How, then, did analytic geometry fare among the different shifts of analytical and geometrical endeavors discussed above?

The relation between a curve and its equation was crucial in the developments in mathematics after Descartes. However, there was no recognizable subfield of mathematics that coincided with what at present is called analytic geometry, and therefore the question of its development does not arise. This is not surprising because "analytic geometry," like "calculus," is a category of mathematical topics defined not by active research but by didactical practice, notably the twentieth-century practice of propaedeutic mathematics teaching at universities.

Construction The changes, discussed above, in the two contexts in which the concept of geometrical construction functioned show that the concept itself also underwent a profound metamorphosis. Starting from the appearance in print of Pappus' classification of problems in 1588, the concept was made more precise, it was extended beyond the confines of solid problems, and it was also extended to cover the construction of curves. Descartes codified the concept for problems leading to polynomial equations (in one unknown) and for algebraic curves. Later mathematicians explored means to construct non-algebraic curves and thereby — contra Descartes — to legitimize them.

An even more fundamental change occurred in the role of the concept of construction. At first, construction was the exclusive means to solve problems (in geometry) and to know objects (like curves). It was, thereby, the principal instrument to attain mathematical understanding and certainty. This crucial role it lost; by 1750 analytical expressions provided as much certainty and understanding to mathematicians as the earlier constructions. Thus the decline of constructions mirrored the rise of analytical expressions and equations: By 1650 an equation was a problem whose solution was a construction; by 1750 problems as well as their solutions were couched in terms of equations or analytical expressions. As a result the criteria of adequacy for solutions had to be redefined, and they were.

Interpretation of exactness

Together with the concept of construction, the interpretation of exactness underwent a metamorphosis in the early modern period. For a time Pappus' precept was the standard interpretation of geometric exactness and debates occurred about the legitimacy of non-plane constructions. Descartes settled and practically silenced these debates by his interpretation of constructional exactness. This interpretation soon became an orthodoxy, adopted or opposed without much interest in Descartes' philosophical arguments for it. Attempts to reinterpret constructional exactness so as to include the non-algebraic curves did not lead to a new canon of construction. Rather they accompanied a process of habituation by which mathematicians came to accept analytical expressions as sufficiently exact to represent the objects.

Dynamics

It is of interest to consider the forces behind the metamorphoses in early modern mathematics discussed above. The principal dynamics consisted in the introduction of analytical methods. But one should note two other forces at work whose influence in the development of mathematics were considerable. The one was habituation. By habituation, the equations of analysis lost their character of problem or task (namely: construct the roots of, or the curve defined by the equation) and acquired the status of a solution or answer (the roots of this equation, the curve defined by this equation). The transition occurred not because mathematicians learned essentially new things about the equations, but because they became so used to analytical expressions that they accepted them as answers.

The other force at work is what I would call the erosion of restrictive methodologies. Kepler's restriction of geometry to plane constructions, however conscientiously derived from authority and convictions about the harmony of the creation, did not restrain later mathematicians in exploring non-plane constructions. Also Descartes' interpretation of exactness, diligently based on philosophical convictions, was a restrictive one; non-algebraic relations were beyond the confines of legitimate geometry. Yet it did not take mathematicians long to overcome the Cartesian inhibition and explore the realm of non-algebraic relations. These transgressions of restrictive interpretations of exactness illustrate the success of the strategy which I have called "Appreciation of the resulting mathematics' (cf. Section 1.6). In other words, if the mathematics beyond is alluring, the border will be crossed, whatever restrictive methodology guards it.

List of problems, analyses, constructions, and instruments

This list refers only to the formal presentations of the items in Proposition-form; for further references to these and other problems, analyses, etc., see the subject index.

The items arranged as to methods of construction:

A. Constructions of problems

By circles and straight lines: 4.1, 4.2, 4.3, 4.4, 4.8, 4.16, 4.18, 4.21, 5.2, 5.4, 5.6, 8.1, 21.1, 21.2, 21.3, 22.1, 22.4, 27.1.

By shifting rulers: 2.2, 2.3, 12.1.

By approximation: 4.9.

By the intersection of conic sections: 3.1 (parabola-hyperbola), 3.8 (circle-hyperbola), 17.1 (circle-parabola), 26.1 (idem), 26.3 (idem).

By neusis: 2.6, 3.9, 10.1, 10.2, 12.2.

By special curves: 2.5 (conchoid), 3.4 (quadratrix), 3.6 (cissoid), 4.11 (proportionatrix), 16.2 (curve traced by trisector), 16.4 (curve traced by mesolabum), 4.6.

By reduction to trisection: 4.19, 26.4.

By reduction to constructing two mean proportionals: 4.15.

By reduction to a Vietean standard solid construction: 4.22, 5.10, 5.8.

By intersection of higher-order curves: (circle-cartesian parabola) 26.2.

B. Constructions of curves

Pointwise: 3.5 (cissoid), 4.10 (proportionatrix), 9.1 (Quadratrix).

By combination of motions: 3.2 (spiral), 3.3 (quadratrix).

By the intersection of moving curves: (turning ruler and moving parabola): 19.4

(five lines locus), 23.5 (idem), 19.4 (idem).

By reduction to Apollonius' construction by the intersection of a cone and a plane: 23.3 (ellipse), 23.3 (idem).

Bibliography

[Anderson 1612] Anderson, Alexander, *Supplementum Apollonii redivivi, sive analysis problematis hactenus desiderati ad Apollonij Pergaei doctrinam ⟨peri neuseoon⟩ à Marino Ghetaldo Patritio Ragusino hucusque, non ita pridem restitutam . . . Huic subnexa est variorum problematum practice, eodem auctore.*, Paris, 1612.

[Anderson 1615] Anderson, Alexander, *⟨Aitologia⟩ pro zetetico Apolloniani problematis a se iam pridem edito in Supplemento Apollonij Redivivi . . .*, Paris, 1615.

[Anderson 1617] Anderson, Alexander, *Animadversionis in Franciscum Vietam a Clemento Cyriaco nuper editae brevis diakrisis*, Paris, 1617.

[Anderson 1619] Anderson, Alexander, *Exercitationum mathematicarum decas prima . . .*, Paris, 1619.

[Anonymus Calcul] Anonymus, *Calcul de Mons. Des Cartes : Introduction à sa Géométrie* (text composed 1638, probably under supervision of Descartes, published in [Descartes 1964–1974] vol. 10 pp. 659–680).

[Apollonius Conics] Apollonius, *Conics* (references by Book- and Proposition-number; standard editions: [Apollonius 1891–1893] (books I–IV), [Apollonius 1990] (books V–VII); translations: [Apollonius 1963] (French), [Apollonius 1961] (English, modernized mathematical notation)).

[Apollonius 1537] Apollonius, *Apollonii Pergaei philosophi mathematicique excellentissimi Opera . . .* (tr. ed. Giovanni-Battista Memmo), Venice, 1537.

[Apollonius 1566] Apollonius, *Conicorum libri quattuor, una cum Pappi Alexandrini lemmatibus et commentariis Eutocii Ascalonitae* (tr. annot. ed. F. Commandino), 2 vols., Bologna, 1566.

[Apollonius 1891–1893] Apollonius, *Apollonii Pergaei quae Graece exstant cum commentariis antiquis* (tr. ed. J.L. Heiberg; Greek text of Conics I–IV and Eutocius' commentary, with Latin translation), 2 vols., Leipzig, Teubner, 1891–1893.

[Apollonius 1961] Apollonius, *Treatise on conic sections* (ed. tr. intr. T.L. Heath, mathematical notation modernized, reprint of ed. Cambridge 1896), Cambridge, Heffer, 1961.

[Apollonius 1963] Apollonius, *Les coniques d'Apollonius de Perge* (tr. ed. P. Vereecke; repr. of ed. 1923), Paris, Blanchard, 1963.

[Apollonius 1990] Apollonius, *Conics : Books V to VII : the Arabic translation of the lost Greek original in the version of the Banu Musa* (Arabic text and English traslation, ed. tr. comm. G.J. Toomer), 2 vols., Sources in the history of mathematics and physical sciences 9, New York, Springer, 1990.

[Archibald 1915] Archibald, Raymond Claire, ed., *Euclid's book on divisions of figures (⟨peri diaireseoon biblion⟩) with a restoration based on Woepcke's text and on the Practica geometriae of Leonardo Pisano*, Cambridge, Cambridge University Press, 1915.

[Archimedes Lemmas] Archimedes, *Book of lemmas* (standard edition: [Archimedes 1880–1881] vol. 2 pp. 427–446; page references to translation in [Archimedes nd], pp. 301–318; other translation: [Archimedes 1960] pp. 521–542).

[Archimedes nd] Archimedes, *The works of Archimedes* (ed. intr. tr. Thomas L. Heath; mathematical notation modernized; reprint of ed. 1897, with suppl: "the method" of 1912), New York, Dover.

[Archimedes SphrCyl] Archimedes, *On the sphere and cylinder* (standard edition: [Archimedes 1880–1881] vol. 1 pp. 1–255; page references to translation in [Archimedes nd], pp. 1–90; other translation: [Archimedes 1960] pp. 1–124).

[Archimedes Spirals] Archimedes, *On spirals* (standard edition: [Archimedes 1880–1881] vol. 2 pp. 1–139; page references to translation in [Archimedes nd], pp. 151–188; other translation: [Archimedes 1960] pp. 237–299).

[Archimedes 1544] Archimedes, *Opera . . . omnia . . . nuncque primum et Graece et Latine in lucem edita . . . Eutocii Ascalonitae in eosdem Archimedis libros commentaria item Graece et Latine . . .* (ed. Th. Geschauff), Basel, 1544.

[Archimedes 1615] Archimedes, *⟨Archimedous panta soozomena⟩ : Archimedis Opera quae extant. Novis demonstrationibus illustrata* (ed. D. Rivault), Paris, 1615.

[Archimedes 1880–1881] Archimedes, *Archimedis opera omnia cum commentariis Eutocii* (tr. annot. ed. J.L. Heiberg), 3 vols., Leipzig, 1880–1881.

[Archimedes 1960] Archimedes, *Les oeuvres complètes d'Archimède suivies des commentaires d'Eutocius d'Ascalon* (tr. ed. P. Vereecke; re-ed. of 1921 version), 2 vols., Liege, 1960.

[Armogathe 1990] Armogathe, Jean-Robert, "La publication du Discours et des Essais," in: [Belgioioso 1990], pp. 17–25.

[Barbour 1961] Barbour, J.M., *Tuning and temperament, a historical survey*, East Lansing (Mich.), 1961.

[Baron 1969] Baron, M.E., *The origin of the infinitesimal calculus*, Oxford, Pergamon, 1969.

[Bartolozzi & Franci 1993] Bartolozzi, Margherita & Franci, Raffaella, "La regola dei segni dall'enunciato di R. Descartes (1637) alle dimostrazione di C.F. Gauss (1828)," *Archive for history of exact sciences*, 45, 1993, pp. 335–374.

[Becker 1933] Becker, Oskar, "Eudoxos-Studien II : Warum haben die Griechen die Existenz der vierten Proportionale angenommen?," *Quellen und Studien*, 2B, 1933, pp. 369–387.

[Beeckman 1939–1953] Beeckman, Isaac, *Journal tenu par Isaac Beeckman de 1604 à 1634* (ed. C. de Waard), 4 vols., The Hague, 1939–1953.

[Behboud 1994] Behboud, Ali, "Greek geometrical analysis," *Centaurus*, 37, 1994, pp. 52–86.

[Belgioioso 1990] Belgioioso, Giulia, et al., eds., *Descartes: il metodo e i saggi; Atti del convegno per il 350o anniversario della pubblicazione del Discours de la Méthode e degli Essais*, 2 vols., Florence, Armando Paoletti, 1990.

[Benedetti 1599] Benedetti, Joh. Bapt., *Speculationum liber*, Venice, 1599.

[Bernoulli 1688] Bernoulli, Jakob, "Animadversio in geometriam Cartesianam et constructio quorundam problematum hypersolidorum," *Acta Eruditorum*, Leipzig, June, 1688, pp. 323–330 (page references to edition in [Bernoulli 1967] vol. 1 pp. 343–351).

[Bernoulli 1695] Bernoulli, Jakob, "Notae et animadversiones tumultuariae in geometriam Cartesii" (first published in the *Geometria* edition Frankfurt 1695), in [Bernoulli 1967] pp. 665–717.

[Bernoulli 1967] Bernoulli, Jakob, *Opera* (facs. of ed. 1744), Brussels, 1967.

[Besson 1582] Besson, J., *Theatrum Instrumentorum et Machinarum*, Lyon, 1582.

[Bockstaele 1976] Bockstaele, Paul, "The correspondence of Adriaan van Roomen," *Lias*, 3, 1976, pp. 85–129, 249–299.

[Bombelli 1572] Bombelli, Rafael, *L'algebra, parte maggiore dell'arithmetica divisa in tre libri*, Bologna, 1572 (page references to edition in [Bombelli 1966]).

[Bombelli 1579] Bombelli, Rafael, *L'algebra*, Bologna, 1579.

[Bombelli 1929] Bombelli, Rafael, *l'Algebra; libri IV e V compredenti "la parte geometrica" inedita tratta dal manoscritto B. 1569 della Biblioteca dell' Archiginnasio di Bologna* (ed. Ettore Bortolotti), Bologna, 1929.

[Bombelli 1966] Bombelli, Rafael, *L'Algebra* (introd. U. Forti, pref. E. Bortolotti), Milano, Feltrinelli, 1966 (edition of the text of [Bombelli 1572] together with books 4 and 5 of the *Algebra* which were first published in 1929).

[Bos 1974] Bos, Henk J.M., "The lemniscate of Bernoulli," in: *For Dirk Struik* (R. S. Cohen, et al., eds., Dordrecht, Reidel, 1974), pp. 3–14 (repr. in [Bos 1993c] pp. 101–111).

[Bos 1974b] Bos, Henk J.M., "Differentials, higher-order differentials and the derivative in the Leibnizian calculus," *Archive for history of exact sciences*, 14 (1974), pp. 1–90.

[Bos 1981] Bos, Henk J.M., "On the representation of curves in Descartes' Géométrie," *Archive for history of exact sciences*, 24, 1981, pp. 295–338.

[Bos 1984] Bos, Henk J.M., "Arguments on motivation in the rise and decline of a mathematical theory; the "construction of equations," 1637 – ca. 1750," *Archive for history of exact sciences*, 30, 1984, pp. 331–380.

[Bos 1985] Bos, Henk J.M., "The significance of Sluse's Mesolabum within seventeenth-century geometry and algebra," in: *Actes du Colloque International "René-François de Sluse" Amay-Liège-Visé, 20–22 Mars 1985 (Bull. Soc. Roy. Sci. Liège, 55, fasc. 1, 1986; Introd. R. Halleux)* (1985), pp. 145–163.

[Bos 1987] Bos, Henk J.M., "The concept of construction and the representation of curves in seventeenth-century mathematics," in: *Proceedings of the International Congress of Mathematicians, August 3–11, 1986 Berkeley, California, USA* (A. Gleason, ed., Providence, American Mathematical Society, 1987), pp. 1629–1641 (repr. in [Bos 1993c] pp. 23–36).

[Bos 1988] Bos, Henk J.M., "Tractional motion and the legitimation of transcendental curves," *Centaurus*, 31, 1988, pp. 9–62.

[Bos 1989] Bos, Henk J.M., "Recognition and wonder; Huygens, tractional motion and some thoughts on history of mathematics," *Tractrix, yearbook for the history of science, medicine, technology and mathematics*, 1, 1989, pp. 3–20 (repr. in [Bos 1993c] pp. 1–21).

[Bos 1990] Bos, Henk J.M., "The structure of Descartes' Géométrie," in: [Belgioioso 1990], pp. 349–369 (repr. in [Bos 1993c] pp. 37–57; French translation "La structure de la Géométrie de Descartes" in *Revue d'histoire des sciences* 51 (1998), pp. 291–317).

[Bos 1992] Bos, Henk J.M., "Descartes, Pappus' Problem and the Cartesian Parabola: a conjecture," in: *An investigation of difficult things — Essays on Newton and the History of the Exact Sciences* (P. M. Harman & Alan E. Shapiro, eds., Cambridge, Cambridge University Press, 1992), pp. 71–96.

[Bos 1993] Bos, Henk J.M., "On the interpretation of exactness," in: *Philosophy of mathematics ; Proceedings of the 15th International Wittgenstein-Symposium, 16th to 23d August 1992, Kirchberg am Wechsel (Austria), vol I* (Johannes Czermak, ed., Vienna, Hölder-Pichler-Tempsky, 1993), pp. 23–44.

[Bos 1993b] Bos, Henk J.M., "Johann Molther's "Problema Deliacum," 1619," in: *Vestigia Mathematica : Studies in medieval and early modern mathematics in honour of H.L.L. Busard* (M. Folkerts & J. P. Hogendijk, eds., Amsterdam, Rodopi, 1993), pp. 29–46.

[Bos 1993c] Bos, Henk J.M., *Lectures in the history of mathematics*, Providence, R.I., American Mathematical Society, 1993.

[Bos 1996] Bos, Henk J.M., "Johann Bernoulli on exponential curves, ca. 1695 : innovation and habituation in the transition from explicit constructions to implicit functions," *Nieuw Archief voor Wiskunde*, 14, 1996, pp. 1–19.

[Bos 1996b] Bos, Henk J.M., "Tradition and modernity in early modern mathematics : Viète, Descartes and Fermat," in: *L'Europe mathématique, histoires, mythes, identités : Mathematical Europe, history, myth, identity* (Catherine Goldstein et al., eds., Paris, Maison des sciences de l'homme - Bibliothèque, 1996), pp. 183–204.

[Bos & Reich 1990] Bos, Henk J.M. & Reich, Karin, "Der doppelte Auftakt zur frühneuzeitlichen Algebra: Viète und Descartes," in: *Geschichte der Algebra: eine Einführung* (Erhard Scholz, ed., Mannheim etc., Wissenschaftsverlag, 1990), ch. 7, pp. 183–234.

[Boyer 1959] Boyer, Carl B., "Descartes and the geometrization of algebra," *Americal mathematical monthly*, 66, 1959, pp. 390–393.

[Breger 1986] Breger, Herbert, "Leibniz' Einführung des Transzendenten," *Studia Leibnitiana* (Sonderheft 14, 300 Jahre "Nova Methodus" von G.W. Leibniz (1684–1984), Wiesbaden, Franz Steiner) 1986, pp. 119–132.

[Brigaglia & Nastasi 1986] Brigaglia, Aldo & Nastasi, Pietro, "Le riconstruzioni apolloniane in Viète e in Ghetaldi," *Bollettino di storia delle scienze matematiche*, 6, 1986, pp. 83–134.

[Buteo 1559] Buteo, Johannes, *De quadratura circuli libri duo, ubi multorum quadraturae confutantur et ab omnium impugnatione defenditur Archimedes*, Lyon, 1559.

[Bynum 1984] Bynum, W.F., et al., eds., *Dictionary of the history of science*, London, Macmillan Press, 1981.

[Cardano 1545] Cardano, Girolamo, *Ars magna*, Nürnberg, 1545 (also in [Cardano 1966] vol. 4 pp. 221–302; page references to translation in [Cardano 1968]).

[Cardano 1966] Cardano, Girolamo, *Opera Omnia* (facs. of ed. 1663, intr. August Buck), 10 vols., Stuttgart, Friedrich Fromann, 1966.

[Cardano 1968] Cardano, Girolamo, *The great art or the rules of algebra* (ed. tr. Richard Witmer), Cambridge, MA, 1968.

[Ceulen 1615] Ceulen, L. van, *De arithmetische en geometrische fondamenten . . . In veele verscheydene constighe questien, soo geometrice door linien als arithmetice door irrationale ghetallen, oock door den regel coss ende de tafelen sinuum ghesolveert*, Leiden, 1615.

[Ceulen 1615b] Ceulen, L. van, *Fundamenta arithmetica et geometrica cum eorum usu in varii problematis geometricis, partim solo linearum ductu, partim per numeros irrationales et tabulas sinuum et algebram solutis* (tr. W. Snellius), Leiden, 1615.

[Ceulen 1619] Ceulen, L. van, *De circulo et adscriptis liber* (tr. annot. W. Snellius), Leiden, 1619.

[Clavius 1604] Clavius, Chr., *Geometria Practica*, Rome, 1604 (other edition: [Clavius 1611–1612] vol. 2 pp. 3–230).

[Clavius 1608] Clavius, Chr., *Algebra*, Rome, 1608.

[Clavius 1611–1612] Clavius, Chr., *Opera Mathematica*, 4 vols., Mainz, 1611–1612.

[Cohen 1984] Cohen, H. Floris, *Quantifying music; the science of music at the first stage of the scientific revolution, 1580–1650*, Dordrecht, Reidel, 1984.

[Costabel 1969] Costabel, Pierre, "La mathématique de Descartes antérieure à la Géométrie," *Actes du symposium international "la géométrie et l'algèbre au début du XVIIe siècle" a l'occasion du quantrième centenaire de la naissance de Marin Getaldic, Dubrovnik, 29. IX - 3. X 1968* (Z. Dadic, et al., eds., Zagreb, 1969), pp. 127–137.

[Costabel 1983] Costabel, Pierre, "L'initiation mathématique de Descartes," *Archives de philosophie*, 46, 1983, pp. 637–646.

[Crapulli 1969] Crapulli, Giovanni, *Mathesis universalis; genesi di una idea nel XVI secolo*, Rome, Edizioni dell'Ateneo, 1969.

[Cusanus 1980] Cusanus, Nicolaus, *Die mathematische Schriften* (tr. Josepha Hofmann, intr. annot. Joseph E. Hofmann, first published 1952), Hamburg, Felix Meier, 1980.

[Cyriacus 1616] Cyriacus, Clemens, *Problemata duo nobilissima quorum nec analysis geometricam videntur tenuisse Ioannes Regiomontanus et Petrus Nonius, nec demonstrationem satis accuratam repraesentasse Franciscus Vieta et Marinus Ghetaldus . . .*, Paris, 1616.

[Debeaune 1649] Debeaune, Florimond, "Notae breves" (tr. F. van Schooten), in [Descartes 1649] pp. 119–161 (also in [Descartes 1659–1661] vol. 1, pp. 107–142).

[Dee & Commandino 1570] Dee, John & Commandino, Federico, eds., *De superficierum divisionibus liber Machometo Bagdedino ascriptus*, Pesaro, 1570.

[Descartes CogPriv] Descartes, René, *Cogitationes Privatae* (page references to edition in [Descartes 1964–1974] vol. 10 pp. 213–256).

[Descartes Rules] Descartes, René, *Regulae ad directionem ingenii* (page references to edition in [Descartes 1964–1974] vol. 10 pp. 359–469; other edition: [Descartes 1966]; translations: [Descartes 1985–1991] vol. 1 pp. 7–78 (English), [Descartes 1977] (French), [Descartes 1973] (German)).

[Descartes 1637] Descartes, René, "La géométrie," in: *Discours de la méthode pour bien conduire sa raison et chercher la verité dans les sciences; plus la dioptrique les meteores et la geometrie qui sont des essais de cete methode* (Leiden, 1637), pp. 297–413 (page references to facsimile edition in [Descartes 1954]; other edition: [Descartes 1964–1974], vol. 6, pp. 367–485; translations: [Descartes 1954], [Descartes 1965]).

[Descartes 1637b] Descartes, René, *Discours de la methode pour bien conduire sa raison et chercher la verité dans les sciences; plus la dioptrique les meteores et la geometrie qui sont des essais de cete methode*, Leiden, 1637 (page references to edition in [Descartes 1964–1974] vol. 6 pp. 1–80; translation (Discours): [Descartes 1985–1991], vol. 1, pp. 109–151; other translation: [Descartes 1965]).

[Descartes 1649] Descartes, René, *Geometria* (tr. F. van Schooten, annot. F. Debeaune), Leiden, 1649.

[Descartes 1659–1661] Descartes, René, *Geometria* (sec. enl. ed. by F. van Schooten of [Descartes 1649], 2 vols.; includes separate treatises by van Schooten, Hudde, Heuraet, Debeaune, and De Witt), Amsterdam, 1659–1661.

[Descartes 1695] Descartes, René, *Geometria* (reprint of ed. 1659–1661 by F. van Schooten, with notes by Jakob Bernoulli), Frankfurt/M, 1695.

[Descartes 1954] Descartes, René, *The geometry of René Descartes* (tr. ed. D.E. Smith and M.L. Latham), New York, Dover, 1954.

[Descartes 1964–1974] Descartes, René, *Oeuvres de Descartes* ("nouvelle présentation" of the ed. 1897–1913, ed. Charles Adam & Paul Tannery), 12 vols., Paris, Vrin, 1964–1974.

[Descartes 1965] Descartes, René, *Discourse on method, optics, geometry and meteorology* (tr. intr. Paul J. Olscamp), The library of the liberal arts 211, Indianapolis - New York, Bobbs-Merill, 1965.

[Descartes 1966] Descartes, René, *Regulae ad directionem ingenii* (ed. Giovanni Crapulli, with 17th-century Dutch version), The Hague, Nijhoff, 1966.

[Descartes 1973] Descartes, René, *Regulae ad directionem ingenii : Regeln zur Ausrichting der Erkenntniskraft* (tr. ed. Heinrich Springmeyer, Lüder Gäbe, Hans Günther Zekl), Hamburg, Philosophische Bibliothek, 1973.

[Descartes 1977] Descartes, René, *Règles utiles et claires pour la direction de l'esprit et la recherche de la vérité; traduction selon le lexique Cartésien, et annotation conceptuelle* (tr. Jean-Luc Marion), Archives internationales d'histoire des idées, nr 88, The Hague, Martinus Nijhoff, 1977.

[Descartes 1985–1991] Descartes, René, *The philosophical writings of Descartes* (ed. tr. J. Cottingham, R. Soothoff, D. Murdoch, and Anthony Kenny), 3 vols., Cambridge, Cambridge University Press, 1985–1991.

[Descartes 1987] Descartes, René, *Exercices pour les éléments des solides : Essai en complément d'Euclide : Progymnasmata de solidorum elementis* (ed. intr. trad. annot. Pierre Costabel), Paris, Presses universitaires de France, 1987.

[Diophantus Arithmetica] Diophantus, *Arithmetica*.

[Diophantus 1575] Diophantus, *Rerum arithmeticarum libri sex, quorum duo adjecta habent scholia Maximi Planudis, item liber de numeris polygonis seu multangulis* (ed. G. Xylander (W. Holzmann)), Basel, 1575.

[Egmond 1985] Egmond, Warren van, "A catalog of Viète's printed and manuscript works," in: *Mathemata: Festschrift für Helmuth Gericke* (Menso Folkerts & Uta Lindgren, eds., Stuttgart, Franz Steiner, 1985), pp. 359–396.

[Euclid Data] Euclid, *Data* (references by Proposition-number; standard edition: [Euclid 1883–1916] vol. 6 pp. 1–187; translations: [Euclid 1962] (German), [Euclid 1980] and [Euclid 1993] (English)).

[Euclid Elements] Euclid, *Elements* (references by Book- and Proposition-number; standard edition: [Euclid 1883–1916] vols.1–4; translation: [Euclid 1956]).

[Euclid 1505] Euclid, *Euclidis Megarensis . . . Elementorum libri XIII . . . cum expositione Theonis . . .* (ed. B. Zamberti, contains also the Data), Venice, 1505.

[Euclid 1533] Euclid, ⟨*Eukleidou stoicheioon . . .*⟩ *. . . Adiecta praefatiuncula in qua de disciplinis mathematicis nonnihil* (ed. S. Grynaeus; contains also Proclus' commentary), Basel, 1533.

[Euclid 1574] Euclid, *Elementorum libri XV, accessit XVI de solidorum regularium comparatione* (ed. Chr. Clavius), Rome, 1574.

[Euclid 1589] Euclid, *Euclidis elementorum libri XV accessit XVI de solidorum regularium cuiuslibet comparatione* (ed. Chr. Clavius), 2 vols., Rome, 1589.

[Euclid 1603] Euclid, *Elementorum libri XV. Accessit liber XVI de solidorum regularium cuiuslibet intra quodlibet comparatione . . .* (ed. Chr. Clavius), Rome, 1603.

[Euclid 1625] Euclid, ⟨*Eukleidou dedomena . . .*⟩ *Euclidis Data . . . Adiectus est . . . Marini philosophi commentarius Graecè et Latinè* (ed. Cl. Hardy), Paris, 1625.

[Euclid 1645] Euclid, *Elementorum geometricorum libros tredecim Isidorum et Hypsiclem et recentiores de corporibus regularibus et Procli propositiones geometricas; immisionemque duarum rectarum linearum continue proportionalium inter duas rectas* (ed. Claude Richard), Antwerp, 1645.

[Euclid 1883–1916] Euclid, *Opera omnia* (ed. J.L. Heiberg and H. Menge), 8 vols., Leipzig, 1883–1916.

[Euclid 1956] Euclid, *The thirteen books of Euclid's Elements* (tr. intr. comm. Thomas L. Heath; reprint of ed. 1926), 3 vols., New York, Dover, 1956.

[Euclid 1962] Euclid, *Die Data von Euklid* (tr. ed. Cl. Thaer), Berlin etc., Springer, 1962.

[Euclid 1980] Euclid, *The medieval Latin translation of the Data of Euclid* (ed. tr. Shuntaro Ito), Tokyo, University of Tokyo Press, 1980.

[Euclid 1993] Euclid, *The Data of Euclid : translated from the text of Menge* (tr. George L. McDowell & Merle A. Sokolik, intr. Richard Delahide Ferrier), Baltimore, Union Square Press, 1993.

[Eutocius CommSphrCyl] Eutocius, *Commentaries on Archimedes' "On the sphere and cylinder"* (standard edition: [Archimedes 1880–1881] vol. 3 pp. 1–225; page references to translation in [Archimedes 1960], vol. 2, pp. 553–696).

[Faulhaber 1622] Faulhaber, Johannes, *Miracula arithmetica*, Augsburg, 1622.

[Fermat Appendix] Fermat, Pierre de, *Appendix ad isagogen topicam, continens solutionem problematum solidorum per locos* (standard edition: [Fermat 1891–1922] vol. 1 pp. 103–110; other edition: [Fermat 1969] pp. 9–11).

[Fermat DissTrip] Fermat, Pierre de, *De solutione problematum geometricorum per curvas simplicissimas et unicuique problematum generi proprie convenientes dissertatio tripartita* (standard edition: [Fermat 1891–1922] vol. 1 pp. 118–133; other edition: [Fermat 1969] pp. 110–115).

[Fermat Isagoge] Fermat, Pierre de, *Ad locos planos et solidos isagoge* (standard edition: [Fermat 1891–1922] vol. 1 pp. 91–103; other edition: [Fermat 1969] pp. 1–8; translation: [Fermat 1923]).

[Fermat 1679] Fermat, Pierre de, *Varia opera mathematica . . . accesserunt selectae quaedam ejusdem epistolae . . .*, Toulouse, 1679.

[Fermat 1891–1922] Fermat, Pierre de, *Oeuvres* (ed. Paul Tannery & Charles Henry, suppl. vol. ed. C. de Waard), 4+1 vols., Paris, Gauthiers-Villars et Fils, 1891–1922.

[Fermat 1923] Fermat, Pierre de, *Einführung in die ebenen und körperlichen Örter* (tr. H. Wieleitner), Ostwald Klassiker 208, Leipzig, 1923.

[Fermat 1969] Fermat, Pierre de, *Varia opera mathematica . . . accesserunt selectae quaedam ejusdem epistolae . . .* (Photomech. reprint of edition 1679), Brussels, 1969.

[Fibonacci 1857–1862] Fibonacci, Leonardo Pisano, *Scritti di Leonardo Pisano* (ed B. Boncompagni), 2 Vols., Rome, 1857–1862.

[Field 1988] Field, J. V., *Kepler's geometrical cosmology*, Chicago, University of Chicago Press, 1988.

[Fine 1544] Fine, Oronce, *Quadratura circuli, tandem inventa et clarissimè demonstrata*, Paris, 1544.

[Forbes 1977] Forbes, E. G., "Descartes and the birth of analytic geometry," *Historia Mathematica*, 4, 1977, pp. 141–151.

[Frame 1943–1945] Frame, J.S., "Machines for solving algebraic equations," *Mathematical tables and aids to computation* 1, 1943–1945, pp. 337–353.

[Freguglia 1981] Freguglia, Paolo, "La Géométrie di Descartes e la nascita della Geometria analitica," *Miscellanea filosofica* (Firenze, Le Monnier, 1981), pp. 50–70.

[Freguglia 1988] Freguglia, Paolo, *Ars analytica : mathematica e methodus nella seconda metà del cinquecento*, Busto Arsizio, Bramante, 1988.

[Freguglia 1989] Freguglia, Paolo, "Algebra e geometria in Viète," *Bollettino di storia delle scienze matematiche*, 9, 1989, pp. 49–90.

[Gäbe 1972] Gäbe, Lüder, *Descartes' Selbstkritik; Untersuchungen zur Philosophie des jungen Descartes*, Hamburg, Meiner, 1972.

[Galuzzi 1985] Galuzzi, Massimo, "Recenti interpretazioni della Géométrie di Descartes," *Scienza e Filosofia : saggi in honore di Ludovico Geymonat* (C. Mangioni, ed., Milan, Garzanti, 1985), pp. 643–663.

[Galuzzi 1996] Galuzzi, Massimo, "La soluzione dell'equazione de sesto grado nella Géométrie di Descartes," in *Per una storia critica della scienza* (Beretta, M. et al. eds., Bologna, Cisalpino, 1996) (Quaderni di Acme 26), pp. 315–330.

[Garibaldi 1991] Garibaldi, Antonio C., "Sulle ovali di Descartes," *Giornate di storia della matematica* (Massimo Galuzzi, ed., Commenda di Rende, 1991), pp. 191–215.

[Garibaldi 1992] Garibaldi, Antonio C., "Vecchia e nuova analisi di problemi geometrici da Viète a Ghetaldi," in: *La matematizzazione dell'universo : momenti della cultura matematica tra '500 e '600* (Lino Conti, ed., Assisi, Edizioni Porziuncola, 1992), pp. 165–204.

[Ghetaldi 1607] Ghetaldi, Marin, *Variorum problematum collectio*, Venice, 1607 (other edition: [Ghetaldi 1968], pp. 101–172).

[Ghetaldi 1607b] Ghetaldi, Marin, *Apollonius redivivus seu restituta Apollonii Pergaei inclinationum geometria*, Venice, 1607 (other edition: [Ghetaldi 1968], pp. 191–221).

[Ghetaldi 1613] Ghetaldi, Marin, *Apollonius redivivus seu restituta Apollonii Pergaei inclinationum geometriae liber secundus*, Venice, 1613 (other edition: [Ghetaldi 1968], pp. 223–351).

[Ghetaldi 1630] Ghetaldi, Marin, *De resolutione et compositione mathematica libri quinque opus posthumus*, Rome, 1630 (other edition: [Ghetaldi 1968], pp. 353–707).

[Ghetaldi 1968] Ghetaldi, Marin, *Collected works* (facsimile, ed. Z. Dadic), Zagreb, 1968.

[Girard 1629] Girard, Albert, *Invention nouvelle en algèbre*, Amsterdam, 1629 (facs. ed. [Girard 1884]).

[Girard 1884] Girard, Albert, *Invention nouvelle en algèbre* (facsimile of ed. 1629, ed. D. Bierens de Haan), Leiden, 1884.

[Giusti 1990] Giusti, Enrico, "Numeri, grandezze e Géométrie," [Belgioioso 1990], pp. 419–439.

[Giusti 1992] Giusti, Enrico, "Algebra and geometry in Bombelli and Viète," *Bollettino di storia delle scienze matematiche* 12, 1992, pp. 303–328.

[Grisard 1969] Grisard, Jean, "l'Animadversio in Franciscum Vietam de C. Cyriaque de Mangin," in: *Actes du symposium international "la géométrie et l'algèbre au début du XVIIe siècle" a l'occasion du quantrième centenaire de la naissance de Marin Getaldic, Dubrovnik, 29. IX - 3. X 1968* (Z. Dadic, et al., eds., Zagreb, 1969), pp. 185–188.

[Grosholz 1991] Grosholz, Emily, "Descartes' Geometry and the classical tradition," *Revolution and continuity : essays in the history and philosophy of early modern science* (Peter Barker, Roger Ariew, eds., Washington DC, Catholic University of America press, 1991), pp. 183–216.

[Gulley 1958] Gulley, N., "Greek geometrical anaysis," *Phronesis*, 3, 1958.

[Hara 1985] Hara, Kokiti, "Comment Descartes a-t-il découvert ses ovales?," *Historia scientiarum*, 19, 1985, pp. 51–82.

[Harriot 1631] Harriot, Th., *Artis analyticae praxis*, London, 1631.

[Heath 1921] Heath, Thomas L., *A history of Greek mathematics*, 2 vols., Oxford, Clarendon, 1921.

[Heath 1949] Heath, Thomas L., *Mathematics in Aristotle*, Oxford, Clarendon, 1949.

[Hintikka & Remes 1974] Hintikka, J. & Remes, U., *The method of analysis, its geometrical origin and its general significance*, Dordrecht, Reidel, 1974.

[Hire 1679] Hire, Philippe de la, *Nouveaux elemens des sections coniques, les lieux geometriques, la construction ou effection des equations*, Paris, 1679.

[Hofmann 1960] Hofmann, Joseph E., "Ueber die Quadrisectio Trianguli," *Mathematische Zeitschrift*, 74, 1960, pp. 105–118.

[Hogendijk 1984] Hogendijk, Jan P., "Greek and Arabic Constructions of the Regular Heptagon," *Archive for History of Exact Sciences*, 30, 1984, pp. 197–330.

[Hogendijk 1991] Hogendijk, Jan P., "Desargues' Brouillon Project and the Conics of Apollonius," *Centaurus*, 34, 1991, pp. 1–43.

[Hôpital 1707] Hôpital, G.F. de l', *Traité analytique des sections coniques et de leur usage pour la resolution des equations dans les problêmes tant déterminez qu'indéterminez, ouvrage posthume*, Paris, 1707.

[Huygens 1654] Huygens, Christiaan, *De circuli magnitudine inventa, accedunt eiusdem problematum quorundam illustrium constructiones*, Leiden, 1654 (other edition: [Huygens 1888–1950] vol. 12 pp. 91–237).

[Huygens 1888–1950] Huygens, Christiaan, *Oeuvres complètes de Christiaan Huygens; publiées par la société hollandaise des sciences*, 22 vols., The Hague, Martinus Nijhoff, 1888–1950.

[Israel 1990] Israel, Giorgio, "Dalle Regulae alle Géométrie," [Belgioioso 1990], pp. 441–474.

[Israel 1997] Israel, Giorgio, "The analytical method in Descartes Géométrie" (revised english version of [Israel 1990]), *Analysis and synthesis in mathematics ; history and philosophy* (Michael Otte, Marco Panza., eds., Dordrecht etc., Kluwer, 1997), pp. 3–34.

[Itard 1956] Itard, Jean, *La géométrie de Descartes*, Conférences du Palais de la Découverte, sér. D, 39, Paris, 1956.

[Jullien 1996] Jullien, Vincent, *Descartes La Géométrie de 1637*, Paris, Presses Universitaires de France, 1996.

[Kempe 1876] Kempe, A.B., "On a general method of describing plane curves of the nth degree by linkwork," *Proceedings of the London Mathematical Society*, 7, 1876, pp. 213–216.

[Kepler 1615] Kepler, Johannes, *Nova stereometria doliorum vinariorum . . . accessit stereometriae Archimedeae supplementum*, Linz, 1615.

[Kepler 1619] Kepler, Johannes, *Harmonices mundi libri V*, Linz, 1619 (standard edition: [Kepler 1937–1975] vol. 6; translations: [Kepler 1973]), [Kepler 1997].

[Kepler 1937–1975] Kepler, Johannes, *Gesammelte Werke* (ed. W. von Dyck, M. Caspar, F. Hammer), 19 vols., München, 1937–1975.

[Kepler 1973] Kepler, Johannes, *Weltharmonik* (tr. M. Caspar; repr. of edition 1939), München, Oldenbourg, 1973.

[Kepler 1997] Kepler, Johannes, *The harmony of the world* (tr. intr. annot. E.J. Aiton, A.M. Duncan, J.V. Field), Philadelphia, American Philosophical Society, 1997.

[Kinckhuysen 1660] Kinckhuysen, Gerard, *De grondt der meetkonst*, Haarlem, 1660.

[Kircher 1650] Kircher, Athanasius, *Musurgia universalis sive ars magna consoni et dissoni*, Rome, 1650.

[Klein 1968] Klein, Jakob, *Greek mathematical thought and the origin of algebra* (tr. Eva Brann; with appendix: Viète Introduction to the analytic art, tr. J.W. Smith), Cambridge, MA, MIT Press, 1968.

[Kline 1972] Kline, Morris, *Mathematical thought from ancient to modern times*, New York, Oxford University Press, 1972.

[Knorr 1986] Knorr, Wilbur R., *The ancient tradition of geometric problems*, Boston, Birkhäuser, 1986.

[Lachterman 1989] Lachterman, David Rapport, *The ethics of geometry; a genealogy of modernity*, New York, etc., Routledge, 1989.

[Lansbergen 1616] Lansbergen, Philippus van, *Cyclometriae novae libri duo . . . ⟨ho theos aei kuklometrei⟩*, Middelburg, 1616.

[Leibniz 1961–1962] Leibniz, Gottfried Wilhelm, *Mathematische Schriften* (C.I. Gerhardt ed., reprint of ed. 1849–1863), 7 vols., Hildesheim, 1961–1962.

[Loria 1902] Loria, Gino, *Spezielle algebraische und transcendente ebene Kurven : Theorie und Geschichte* (tr. (from Ital.) F. Schütte), Leipzig, Teubner, 1902.

[Mancosu 1992] Mancosu, Paolo, "Descartes' Geometrie and revolutions in mathematics," *Revolutions in mathematics* (Donald Gillies, ed, Oxford, Clarendon Press, 1992), pp. 83–116.

[Mancosu 1996] Mancosu, Paolo, *Philosophy of mathematics and mathematical practice in the seventeenth century*, Oxford, Oxford University Press, 1996.

[Mahoney 1968] Mahoney, Michael S., "Another look at Greek geometrical analysis," *Archive for history of exact sciences*, 5, 1968, pp. 318–348.

[Mahoney 1984] Mahoney, Michael S., "Changing canons of mathematical and physical intelligibility in the later 17th century," *Historia Mathematica*, 11, 1984, pp. 417–423.

[Mahoney 1994] Mahoney, Michael S., *The mathematical career of Pierre de Fermat*, Princeton, Princeton University Press, 1994 (first ed. Princeton, Pr.U.P. 1973).

[Manders 1995] Manders, Kenneth L., "Descartes et Faulhaber," *Archives de Philosophie (Bulletin Cartésien)* 58, 1995, pp. 1–12.

[Mersenne 1636] Mersenne, Marin, *Harmonie universelle contenant la théorie et la pratique de la musique*, Paris, 1636 (facsimile ed. Paris (CNRS) 1965).

[Mersenne 1636b] Mersenne, Marin, *Harmonicorum libri, in quibus agitur de sonorum natura, causis et effectibus*, Paris, 1636.

[Mersenne 1933–1986] Mersenne, Marin, *Correspondance*, Paris, 1933–1986.

[Milhaud 1921] Milhaud, Gaston, *Descartes savant*, Paris, Felix Alcan, 1921.

[Mittelstrass 1978] Mittelstrass, Jürgen, "Die Idee einer Mathesis Universalis bei Descartes," *Perspektiven der Philosophie*, 4, 1978, pp. 177–192.

[Mittelstrass 1979] Mittelstrass, Jürgen, "The philosopher's conception of Mathesis universalis from Descartes to Leibniz," *Annals of Science*, 36, 1979, pp. 593–610.

[Molland 1976] Molland, A.G., "Shifting the foundations: Descartes' transformation of ancient geometry," *Historia Mathematica*, 3, 1976, pp. 21–49.

[Molland 1991] Molland, George, "Implicit versus explicit geometrical methodologies : the case of construction," *Mathématiques et philosophie de l'antiquité à l'âge classique : hommage à Jules Vuillemin* (Roshdi Rashed, ed., Paris, Éditions du CNRS, 1991), pp. 181–196.

[Molther 1619] Molther, Johannes, *Problema Deliacum de cubi duplicatione hoc est de quorumlibet solidorum interventu Mesolabii secundi, quo duae capiantur mediae proportionales sub data ratione similium fabrica — nunc tandem post infinitos praestantissimorum mathematicorum conatus expedite et geometrice solutum*, Frankfurt, 1619.

[Mueller 1981] Mueller, I., *Philosophy of mathematics and deductive structure in Euclid's Elements*, Cambridge, MA, 1981.

[Murdoch 1963] Murdoch, John E., "The medieval language of proportions : elements of the interaction with Greek foundations and the development of new mathematical techniques," in: *Scientific change* (A. C. Crombie, ed, London, Heinemann, 1963), pp. 237–271.

[Mydorge 1631] Mydorge, C., *Prodromi Catopticorum et Dioptricorum : Sive conicorum Operis . . . Libri primus et sucundus . . .*, Paris, 1631.

[Nemore NumDat] Nemore, Jordanus de, *De numeris datis* (page references to translation in [Nemore 1970]).

[Nemore 1970] Nemore, Jordanus de, *The de numeris datis of Jordanus de Nemore; a critical edition, analysis, evaluation and translation* (ed. tr. B. Hughes), Ann Arbor, MI, 1970.

[Newton 1967–1981] Newton, Isaac, *The mathematical papers of Isaac Newton* (ed. D.T. Whiteside), 8 vols., Cambridge, 1967–1981.

[Nonius 1546] Nonius, Petrus, *De erratis Orontii Finaei*, Coimbra, 1546.

[Nonius 1567] Nonius, Petrus, *Libro di algebra*, Antwerp, 1567.

[Ozanam 1702] Ozanam, J., *Nouveaux elemens d'algèbre*, Amsterdam, 1702.

[Pappus Collection] Pappus, *Collection* (first printed edition: [Pappus 1588], standard edition: [Pappus 1965], references by Book-, Proposition- and §-number; page references to translation in [Pappus 1982], translation of Book 7: [Pappus 1986]).

[Pappus 1588] Pappus, *Pappi Alexandrini mathematicae collectiones a Federico Commandino Urbinate in latinum conversae at commentariis illustratae* (tr. ed. F. Commandino), Pesaro, 1588.

[Pappus 1589] Pappus, *Mathematicae collectiones : a Federico Commandino Urbinate in Latinum conversae et commentariis illustratae* (unchanged reprint of Pesaro 1588-ed. (tr. ed. F. Commandino), Venice, 1589.

[Pappus 1602] Pappus, *Federici Commandini Urbinatis . . . exactissima commentaria in libros octo mathematicarum Collectionum Pappi Alexandrini e Graeco in Latinum accuratissime conversos* (unchanged reprint of 1588-ed. of Commandino's translation of Pappus' Collectio), Pesaro, 1602.

[Pappus 1660] Pappus, *Pappi Alexandrini mathematicae collectiones a Federico Commandino Urbinate in latinum conversae at commentariis illustratae : in hac nostra editione ab innumeris quibus scatebant mendis et praecipuè in Graeco contextu diligenter vindicatae* (tr. ed. F. Commandino), Bologna, 1660.

[Pappus 1876–1878] Pappus, *Collectionis quae supersunt* (ed. F. Hultsch), 3 vols., Berlin, Weidman, 1876–1878.

[Pappus 1965] Pappus, *Collectionis quae supersunt* (ed. F. Hultsch, reprint of edition 1876–1878), 3 vols., Amsterdam, Hakkert, 1965.

[Pappus 1982] Pappus, *La collection mathématique* (tr. ed. Paul Ver Eecke, reissue of ed. 1933), 2 vols., Paris, Blanchard, 1982.

[Pappus 1986] Pappus, *Book 7 of the Collection* (tr., comm. A. Jones), 2 vols., New York, Springer, 1986.

[Pasini 1992] Pasini, Enrico, "Mathesis und Phantasie : die Rolle der Einbildungskraft im Umfeld der Descartesschen Regulae," *Studia Leibnitiana*, 24, 1992, pp. 160–176.

[Passalaqua 1994] Passalaqua, Lorena, "Le "Collezioni" di Pappo : polemiche editoriale e circolazione di manoscritti nella corrispondenza di Francesco Barozzi con il Duca di Urbino" (appendix pp. 129 – 156: Lettere e documenti), *Bollettino di storia delle scienze matematiche*, 14, 1994, pp. 91–156.

[Prado & Villalpando 1596–1605] Prado, H. & Villalpando, J. B., *In Ezechielem explanationes et apparatus urbis ac templi Hierosolymitani commentariis et imaginibus illustratus*, 3 vols., Rome, 1596–1605.

[Proclus 1560] Proclus, *In primum Euclidis Elementorum librum commentariorum ad universam mathematicam disciplinam principium eruditionis tradentium libri III* (ed. F. Barozzi), Padua, 1560.

[Proclus 1967] Proclus, *In primum Euclidis Elementorum librum commentarii* (ed. G. Friedlein; repr. of ed. 1873), Hildesheim, Olms, 1967.

[Proclus 1992] Proclus, *A commentary on the first book of Euclid's Elements* (tr. intr. Glenn R. Morrow, repr. of ed. 1970, pref. Ian Mueller), Princeton, Princeton University Press, 1992.

[Rabuel 1730] Rabuel, C., *Commentaires sur la Géométrie de M. Descartes* Lyon, 1730

[Ramus 1569] Ramus, Petrus, *Arithmeticae libri duo; geometriae septem et viginti*, Basel, 1569.

[Ramus 1569b] Ramus, Petrus, *Scholarum mathematicarum libri unus et triginta*, Basel, 1569.

[Ramus 1599] Ramus, Petrus, *Scholarum mathematicarum libri unus et triginta a Lazaro Schonero recogn. et emendati*, Frankfurt, 1599.

[Regiomontanus 1533] Regiomontanus, J., *De triangulis omnimodis libri quinque . . . Acesserunt . . . pleraque D. Nicolai Cusani de Quadratura circuli, deque recti et curvi commensuratione; itemque Io de Monteregio eadem de re ⟨elegktika⟩ hactenus a nemine publicata*, Nürnberg, 1533 (page references to edition in [Regiomontanus 1967]).

[Regiomontanus 1967] Regiomontanus, J., *Regiomontanus on triangles : De triangulis omnimodis by Johann Müller, otherwise known as Regiomontanus* (facs. of ed. 1533, with tr. intr. annot. by B. Hughes O.F.M.), Madison, University of Wisconsin Press, 1967.

[Renaldini 1670] Renaldini, Carolus, *Geometra promotus*, Padua, 1670.

[Roomen 1593] Roomen, Adriaan van, *Ideae mathematicae sive methodus polygonorum qua laterum, perimetrorum et arearum cujuscunque polygoni investigandorum ratio exactissima et certissima una cum circuli quadratura continetur*, Antwerp, 1593.

[Roomen 1596] Roomen, Adriaan van, *Problema Apolloniacum*, Würzburg, 1596.

[Roth 1608] Roth, Peter, *Arithmetica philosophica*, Nürnberg, 1608.

[Salignac 1574] Salignac, B., *Mesolabii expositio*, Geneva, 1574.

[Salinas 1577] Salinas, Fr., *De musica libri VII*, Salamanca, 1577.

[Sasaki 1985] Sasaki, Chikara, "The acceptance of the theory of proportion in the sixteenth and seventeenth centuries — Barrow's reaction to the analytic mathematics," *Historia scientiarum*, 29, 1985, pp. 83–116.

[Scaliger 1594] Scaliger, Joseph, *Cyclometrica elementa duo*, Leiden, 1594.

[Scaliger 1594b] Scaliger, Joseph, *Mesolabium ad nobiles academiae Lugdunensis Batavorum Curatores et magnifices eiusdem civitatis consules*, Leiden, 1594.

[Scaliger 1594c] Scaliger, Joseph, *Appendix ad cyclometrica sua : in qua asseritur quadratio circuli, contra oblationes quorundam, et castigantur quaedam errata in demonstrationibus cyclometricis*, Leiden, 1594.

[Schneider 1991] Schneider, Ivo, "Die Geschichte von der Begegnung zwischen dem Rechenmeister und dem Philosophen : besuchte Descartes den Rechenmeister Faulhaber im Winter 1619–1620?," *Kultur und Technik*, 1991, pp. (4) 46–53.

[Schneider 1993] Schneider, Ivo, *Johannes Faulhaber 1580–1635 : Rechenmeister in einer Welt des Umbruchs*, Basel etc., Birkhäuser, 1993.

[Schooten 1646] Schooten, Frans van, *De organica conicarum sectionum in plano descriptione tractatus*, Leiden, 1646.

[Schooten 1661] Schooten, Frans van, " Tractatus de concinnandis demonstrationibus geometricis ex calculo algebraico," in: [Descartes 1659–1661] vol. 2, pp. 342–420.

[Schuster 1980] Schuster, J.A., "Descartes' mathesis universalis: 1619–28," in *Descartes, philosophy, mathematics and physics* (ed. Stephen W. Gaukroger, Totowa, Barnes and Noble, Brighton, Harvester, 1980) pp. 41–96.

[Scriba 1961] Scriba, Christoph J., "Zur Lösung des 2. Debeauneschen Problems durch Descartes : ein Abschnitt aus der Frühgeschichte der inversen Tangentenaufgaben," *Archive for history of exact sciences*, 1, 1961, pp. 406–419.

[Serfati 1993] Serfati, Michel, "Les compas Cartésiens," *Archives de philosophie*, 56, 1993, pp. 197–230.

[Serfati 1994] Serfati, Michel, "Regulae et mathématiques," *Theoria, revista de teoria, historia y fundamentos de la ciencia*, 9, 1994, pp. 61–108.

[Shea 1991] Shea, William R., *The magic of numbers and motion; the scientific career of René Descartes*, Canton, MA, Watson Publishing, 1991.

[Sluse 1659] Sluse, René François de, *Mesolabum*, Liege, 1659.

[Sluse 1668] Sluse, René François de, *Mesolabum . . . Accessit pars altera de analysi et miscellanea . . .*, Liege, 1668.

[Snellius 1607] Snellius, Willebrord, *⟨Peri logou apotomès kai peri chooriou apotomès⟩ resuscita geometria*, Leiden, 1607.

[Snellius 1608] Snellius, Willebrord, *Apollonius Batavus seu exsuscitata Apollonii Pergaei ⟨Peri diorismenes tomes⟩ geometria*, Leiden, 1608.

[Snellius 1621] Snellius, Willebrord, *Cyclometricus. De circuli dimensione secundum logistarum abacos, et ad mechanicem accuratissima; atque omnium parabilissima. Eiusdemque usus in quarumlibet adscriptarum inventione longe elegantissimus, et quidem ex ratione diametri ad suam peripheriam data*, Leiden, 1621.

[Stefano 1992] Stefano, Maria Elena di, "Alcune considerazioni sull'In artem analyticem isagoge di François Viète," *La matematizzazione dell'universo : momenti della cultura matematica tra '500 e '600* (Lino Conti, ed., Assisi, Edizioni Porziuncola, 1992), pp. 153–164.

[Stevin 1583] Stevin, Simon, *Problematum geometricorum . . . libri V*, Antwerpen, 1583 (year of publication uncertain, page references to facsimile edition in [Stevin 1955–1966], vol. 2A, pp. 119–369, translation *ibid.*).

[Stevin 1585] Stevin, Simon, *L'Arithmetique . . . contenant les computations des nombres arithmetiques ou vulgaires : aussi l'algèbre avec les equations de cinc quantitez : Ensemble les quatre premiers livres d'algebre de Diophante d'Alexandrie maintenant premierement traducts en françois*, Leiden, 1585 (partial edition (facsimile with english translation) in [Stevin 1955–1966] vol. 2, pp. 477–708).

[Stevin 1955–1966] Stevin, Simon, *Principal works* (ed. E. Crone et al.), 5 vols., Amsterdam, Swets & Zeitlinger, 1955–1966.

[Sylla 1984] Sylla, E.D., "Compounding ratios: Bradwardine, Oresme and the first edition of Newton's Principia," in: *Transformation and tradition in the sciences* (E. Mendelsohn, ed., Cambridge, Cambridge University Press, 1984), pp. 11–43.

[Tamborini 1987] Tamborini, Massimo, "Tematiche algebraiche vietiane nelle Regulae e nel libro primo della Géométrie di Descartes," *Miscellanea secentesca: Saggi su Descartes, Fabri, White* (Milan, Cisalpino-Goliardica, 1987), pp. 51–84.

[Treweek 1957] Treweek, A.P., "Pappus of Alexandria : the manuscript tradition of the Collectio Mathematica," *Scriptorium*, 2, 1957, pp. 195–233.

[Tropfke 1980] Tropfke, J., *Geschichte der Elementarmathematik in systematischer Darstellung mit besonderer Berücksichtigung der Fachwörter* (ed. K. Vogel, K. Reich, H. Gericke), 4th ed., Berlin, de Gruyter, 1980.

[Ulivi 1985] Ulivi, Elisabetta, "Sulla "proporzionatrice seconda" di G. B. Villalpando," *Archimede*, 37, 1985, pp. 119–128.

[Valla 1501] Valla, Georgius, *De expetiendis et fugiendis rebus opus*, Venice, 1501.

[Verdonk 1969] Verdonk, J.J., "Über die geometrie des Petrus Ramus," *Sudhoffs Archiv*, 52, 1969, pp. 371–381.

[Viète 1591] Viète, François, *In artem analyticen isagoge : seorsim excussa ab opere restitutae mathematicae analyseos seu algebra nova*, Tours, 1591 (page references to edition [Viète 1970], pp. 1–12; translations: [Viète 1983], pp. 11–32, [Klein 1968] pp. 315–353).

[Viète 1592] Viète, François, *Effectionum geometricarum canonica recensio*, Tours, 1592 (page references to edition [Viète 1970], pp. 229–239; translation: [Viète 1983], pp. 371–388).

[Viète 1593] Viète, François, *Supplementum geometriae : ex opere restitutae mathematicae analyseos seu algebra nova*, Tours, 1593 (page references to edition [Viète 1970], pp. 240–257; translation: [Viète 1983], pp. 388–417).

[Viète 1593b] Viète, François, *Variorum de rebus mathematicis responsorum liber VIII*, Tours, 1593 (page references to edition [Viète 1970], pp. 347–435).

[Viète 1593–1600] Viète, François, *Zeteticorum libri quinque*, Tours/Paris, 1593–1600 (page references to edition [Viète 1970], pp. 42–81; translation: [Viète 1983] pp. 83–158).

[Viète 1595] Viète, François, *Ad problema quod omnibus mathematicis totius orbis construendum proposuit Adrianus Romanus Responsum*, Paris, 1595 (page references to edition [Viète 1970], pp. 305–324).

[Viète 1600] Viète, François, *Apollonius Gallus seu exsuscitata Apollonii Pergaei ⟨peri epafoon⟩ geometria. Ad V.C. Adrianum Romanum Belgam*, Paris, 1600 (page references to edition [Viète 1970], pp. 325–346).

[Viète 1600b] Viète, François, *De numerosa potestatum ad exegesim resolutione : ex opere restitutae mathematicae analyseos seu algebra nova*, Paris, 1600 (page references to edition [Viète 1970], pp. 163–228; translation: [Viète 1983] pp. 311–370).

[Viète 1615] Viète, François, *De aequationum recognitione et emendatione tractatus duo* (annot. Alexander Anderson), Paris, 1615 (page references to edition [Viète 1970], pp. 82–161; translation: [Viète 1983], pp. 159–310).

[Viète 1615b] Viète, François, *Ad angularium sectionum analyticen theoremata ⟨katholikootera⟩* (ed. annot. Alexander Anderson), Paris, 1615 (page references to edition [Viète 1970], pp. 287–304; translation: [Viète 1983], pp. 418–450).

[Viète 1631] Viète, François, *In artem analyticen isagoge : eiusdem ad logisticen speciosam notae priores nunc primum in lucem editae* (ed. annot. J. de Beaugrand), Paris, 1631.

[Viète 1631b] Viète, François, *Ad logisticen speciosam notae priores* (bound with [Viète 1631]), Paris, 1631 (page references to edition [Viète 1970], pp. 13–41; translation: [Viète 1983], pp. 33–82).

[Viète 1646] Viète, François, *Opera mathematica* (ed. F van Schooten), Leiden, 1646.

[Viète 1970] Viète, François, *Opera mathematica* (ed. F van Schooten, facsimile reprint of edition Leiden 1646, int. J.E. Hofmann), Hildesheim, Olms, 1970.

[Viète 1973] Viète, François, *Einführung in die neue Algebra* (tr. ed. Karin Reich and Helmuth Gericke), München, Werner Fritsch, 1973.

[Viète 1983] Viète, François, *The analytic art, nine studies in algebra, geometry and trigonometry* (tr. T. Richard Witmer), Kent (Ohio), 1983.

[Waerden 1961] Waerden, Bartel L. van der, *Science awakening*, New York, Oxford University Press, 1961.

[Werner 1522] Werner, Johannes, *In hoc opere haec continentur . . . Libellus super viginti duobus elementis conicis . . . commentarius . . . cubi duplicatio . . . commentatio in Dionysodori problema . . . De motu octavae sphaerae . . .*, Nürnberg, 1522.

[Whiteside 1960–1962] Whiteside, Derek T., "Patterns of mathematical thought in the later seventeenth century," *Archive for history of exact sciences*, 1, 1960–1962, pp. 179–388.

[Zarlino 1571] Zarlino, Gioseffo, *Dimostrationi armoniche*, Venice, 1571.

[Zarlino 1588] Zarlino, Gioseffo, *Sopplimenti musicali*, Venice, 1588.

[Zeuthen 1886] Zeuthen, H.G., *Die Lehre von den Kegelschnitten im Altertum*, Kopenhagen, Höst, 1886.

[Zeuthen 1896] Zeuthen, H.G., "Die geometrische Construktion als "Existenzbeweis" in der antiken Geometrie," *Mathematische Annalen*, 47, 1896, pp. 222–228.

[Zinner 1968] Zinner, Ernst, *Leben und Wirken des Johannes Müller von Königsberg, genannt Regiomontanus*, 2d ed., Osnabrück, Otto Zeller, 1968.

Name Index

Subject Index

Sources and Studies in the History of Mathematics and Physical Sciences

Continued from page ii

E. Kheirandish
The Arabic Version of Euclid's *Optics*, Volumes I and II

J. Lützen
Joseph Liouville 1809–1882: Master of Pure and Applied Mathematics

J. Lützen
The Prehistory of the Theory of Distributions

G.H. Moore
Zermelo's Axiom of Choice

O. Neugebauer
A History of Ancient Mathematical Astronomy

O. Neugebauer
Astronomical Cuneiform Texts

F.J. Ragep
Naṣīr al-Dīn al-Ṭūsī's Memoir on Astronomy
(al-Tadhkira fī ʿilm al-hay'a)

B.A. Rosenfeld
A History of Non-Euclidean Geometry

J. Sesiano
Books IV to VII of Diophantus' *Arithmetica*: In the Arabic Translation Attributed to Qusṭā, ibn Lūqā

B. Stephenson
Kepler's Physical Astronomy

N.M. Swerdlow/O. Neugebauer
Mathematical Astronomy in Copernicus's De Revolutionibus

G.J. Toomer (Ed.)
Appolonius *Conics* Books V to VII: The Arabic Translation of the Lost Greek Original in the Version of the Banū Mūsā, Edited, with English Translation and Commentary by G.J. Toomer

G.J. Toomer (Ed.)
Diocles on Burning Mirrors: The Arabic Translation of the Lost Greek Original, Edited, with English Translation and Commentary by G.J. Toomer

C. Truesdell
The Tragicomical History of Thermodynamics, 1822–1854

K. von Meyenn, A. Hermann, V.F. Weisskopf (Eds.)
Wolfgang Pauli: Scientific Correspondence II: 1930–1939

Sources and Studies in the History of Mathematics and Physical Sciences

Continued from the previous page

K. von Meyenn (Ed.)
Wolfgang Pauli: Scientific Correspondence III: 1940–1949

K. von Meyenn (Ed.)
Wolfgang Pauli: Scientific Correspondence IV, Part I: 1950–1952

K. von Meyenn (Ed.)
Wolfgang Pauli: Scientific Correspondence IV, Part II: 1953–1954

MIX
Papier aus verantwortungsvollen Quellen
Paper from responsible sources
FSC® C105338
FSC
www.fsc.org

If you have any concerns about our products,
you can contact us on
ProductSafety@springernature.com

In case Publisher is established outside the EU,
the EU authorized representative is:
Springer Nature Customer Service Center GmbH
Europaplatz 3, 69115 Heidelberg, Germany

Printed by Libri Plureos GmbH
in Hamburg, Germany